OIL PALM BIOMASS FOR COMPOSITE PANELS

OIL PALM BIOMASS FOR COMPOSITE PANELS

Fundamentals, Processing, and Applications

Edited by

S.M. SAPUAN

Professor of Composite Materials, Institute of Tropical Forestry and Forest Products (INTROP), Universiti Putra Malaysia, Serdang, Selangor, Malaysia

Department of Mechanical and Manufacturing Engineering, Faculty of Engineering, Universiti Putra Malaysia, Serdang, Selangor, Malaysia

M.T. PARIDAH

Professor of Wood Adhesives and Finishing, Institute of Tropical Forestry and Forest Products (INTROP), Universiti Putra Malaysia, Serdang, Selangor, Malaysia

S.O.A. SAIFULAZRY

Research Officer in the Laboratory of Biocomposite Technology (Biocomposite), Institute of Tropical Forestry and Forest Products (INTROP), Universiti Putra Malaysia, Serdang, Selangor, Malaysia

S.H. LEE

Research Fellow in the Laboratory of Biopolymer and Derivatives (BADs), Institute of Tropical Forestry and Forest Products, Universiti Putra Malaysia, Serdang, Selangor, Malaysia

ELSEVIER

Elsevier
Radarweg 29, PO Box 211, 1000 AE Amsterdam, Netherlands
The Boulevard, Langford Lane, Kidlington, Oxford OX5 1GB, United Kingdom
50 Hampshire Street, 5th Floor, Cambridge, MA 02139, United States

Copyright © 2022 Elsevier Inc. All rights reserved.

No part of this publication may be reproduced or transmitted in any form or by any means, electronic or mechanical, including photocopying, recording, or any information storage and retrieval system, without permission in writing from the publisher. Details on how to seek permission, further information about the Publisher's permissions policies and our arrangements with organizations such as the Copyright Clearance Center and the Copyright Licensing Agency, can be found at our website: www.elsevier.com/permissions.

This book and the individual contributions contained in it are protected under copyright by the Publisher (other than as may be noted herein).

Notices

Knowledge and best practice in this field are constantly changing. As new research and experience broaden our understanding, changes in research methods, professional practices, or medical treatment may become necessary.

Practitioners and researchers must always rely on their own experience and knowledge in evaluating and using any information, methods, compounds, or experiments described herein. In using such information or methods they should be mindful of their own safety and the safety of others, including parties for whom they have a professional responsibility.

To the fullest extent of the law, neither the Publisher nor the authors, contributors, or editors, assume any liability for any injury and/or damage to persons or property as a matter of products liability, negligence or otherwise, or from any use or operation of any methods, products, instructions, or ideas contained in the material herein.

ISBN: 978-0-12-823852-3

For information on all Elsevier publications
visit our website at https://www.elsevier.com/books-and-journals

Publisher: Matthew Deans
Acquisitions Editor: Edward Payne
Editorial Project Manager: Andrea R. Dulberger
Production Project Manager: Prem Kumar Kaliamoorthi
Cover Designer: Miles Hitchen

Typeset by STRAIVE, India

Contents

Contributors xiii

Part 1   Background to oil palm biomass and wood-based panel products

Chapter 1   Introduction to oil palm biomass **3**
R.A. Ilyas, S.M. Sapuan, M.S. Ibrahim, M.H. Wondi, M.N.F. Norrrahim,
M.M. Harussani, H.A. Aisyah, M.A. Jenol, Z. Nahrul Hayawin,
M.S.N. Atikah, R. Ibrahim, S.O.A. SaifulAzry,
C.S. Hassan, and N.I.N. Haris

1.1   Introduction 3
1.2   Is there a biomass industry? 4
1.3   Overview of oil palm plantation and its biomass 6
1.4   Types of oil palm biomass 10
1.5   By-products from oil palm plantations 12
1.6   By-products from oil palm mills 13
1.7   Properties of palm oil solid residues 24
1.8   Oil palm productions 27
1.9   Palm oil prices 28
1.10   Conclusions 30
Acknowledgments 30
References 30

Chapter 2   Basic properties of oil palm biomass (OPB) **39**
Syaiful Osman, Zawawi Ibrahim, Aisyah Humaira Alias,
Noorshamsiana Abdul Wahab, Ridzuan Ramli,
Fazliana Abdul Hamid, and Mansur Ahmad

2.1   Introduction 39
2.2   Physical and anatomical properties 41
2.3   Chemical properties 48
2.4   Mechanical properties 51

v

vi Contents

2.5   Conclusions 52

References 52

Chapter 3   Biological durability and deterioration of oil palm biomass **57**

Zaidon Ashaari, S.H. Lee, Wei Chen Lum, and S.O.A. SaifulAzry

3.1   Introduction 57

3.2   Physicochemical properties of oil palm biomass 59

3.3   Biological durability of oil palm biomasses 61

3.4   Improvement of biological durability of oil palm biomasses . . 64

3.5   Conclusions 65

Acknowledgments 66

References 66

Chapter 4   Wood-based panel industries **69**

B. Norhazaedawati, S.O.A. SaifulAzry, S.H. Lee, and R.A. Ilyas

4.1   Introduction 69

4.2   Types of wood-based panels 70

4.3   Wood-based panels industry market analysis 74

4.4   Wood-based industry in Malaysia 78

4.5   Way forward 85

References 86

Chapter 5   Formaldehyde-based adhesives for oil palm-based panels: Application and properties **87**

U.M.K. Anwar, M. Asniza, K. Tumirah, Y. Alia Syahirah, and A.H. Juliana

5.1   Introduction 87

5.2   Urea-formaldehyde 88

5.3   Melamine-urea-formaldehyde 89

5.4   Phenol-formaldehyde 90

5.5   Oil palm-based composites 91

Contents **vii**

| 5.6 | Conclusions | 94 |
| | References | 94 |

Chapter 6 Nonformaldehyde-based adhesives used for bonding oil palm biomass (OPB) 99

Nor Yuziah Mohd Yunus

6.1	Introduction	99
6.2	Resin for wood based industry	100
6.3	The adhesion theory	102
6.4	Bonding of oil palm biomass (OPB)	103
6.5	Conclusion	108
	References	108

Part 2 Processing and treatment based on oil palm biomass type

Chapter 7 Processing of oil palm trunk and lumber 113

Edi Suhaimi Bakar, S.H. Lee, Wei Chen Lum, S.O.A. SaifulAzry, and Ching Hao Lee

7.1	Introduction	113
7.2	Imperfections of oil palm trunks and lumber	115
7.3	Sawing of oil palm trunk and lumber	116
7.4	Drying of oil palm trunk and lumber	117
7.5	Quality enhancement of oil palm lumber and its products	121
7.6	Conclusion	127
	References	128

Chapter 8 Rotary veneer processing of oil palm trunk 131

M.T. Paridah

8.1	Introduction	131
8.2	Characteristics of oil palm trunk logs	132
8.3	Rotary OPT veneer processing	139
8.4	Challenges in rotary processing of OPT	150
	References	151

viii Contents

Chapter 9 Pretreatment of empty fruit bunch fiber: Its effect as a reinforcing material in composite panels **153**

Kit Ling Chin, Chuan Li Lee, Paik San H'ng, Pui San Khoo, and Zuriyati Mohamed Asa'ari Ainun

9.1 Introduction 153

9.2 The complex nature of EFB fibers 154

9.3 Constraints in producing composite panels from untreated EFB fiber 156

9.4 Types of lignocellulosic fiber pretreatment method........ 158

9.5 Pretreated EFB fibers for fiber-reinforced composites 164

References 169

Chapter 10 Microstructure, physical, and strength properties of compressed oil palm frond composite boards from *Elaeis guineensis*. **175**

Razak Wahab, Mohd Sukhairi Mat Rasat, Rashidah Kamarulzaman, Mohamad Saiful Sulaiman, Sofiyah Mohd Razali, Taharah Edin, and Nasihah Mokhtar

10.1 Introduction 175

10.2 Processing of oil palm composite boards 176

10.3 Physical properties of compressed oil palm fronds (COPF) composite.. 177

10.4 COPF composite board strength properties 180

10.5 Microstructure of COPF composite board................. 186

10.6 Conclusion 187

References 189

Chapter 11 The processing and treatment of other types of oil palm biomass. **191**

Norul Hisham Hamid, Mohd Supian Abu Bakar, Norasikin Ahmad Ludin, Ummi Hani Abdullah, and Asmaa Soheil Najm

11.1 Introduction 191

11.2 Source and characteristics of POME 193

11.3	POME treatment methods	195
11.4	Products from POME treatment	198
11.5	Waste aerobic granules (WAG)	199
11.6	Polyhydroxyalkanoate (PHA)	201
11.7	Generation of biohydrogen by using POME	204
11.8	PHA and biohydrogen processing issues	205
11.9	The by-product of oil palm kernel	206
11.10	Conclusions	208
	References	208

Part 3 Composite panel products from oil palm biomass and its applications

Chapter 12 Classification and application of composite panel products from oil palm biomass 217

R.A. Ilyas, S.M. Sapuan, M.S. Ibrahim, M.H. Wondi, M.N.F. Norrrahim, M.M. Harussani, H.A. Aisyah, M.A. Jenol, Z. Nahrul Hayawin, M.S.N. Atikah, R. Ibrahim, S.O.A. SaifulAzry, C.S. Hassan, and N.I.N. Haris

12.1	Introduction	218
12.2	Oil palm biomass for molded particleboard	222
12.3	Oil palm plywood	223
12.4	Oil palm biomass for medium-density fiberboard (MDF)	226
12.5	Oil palm lumber as a substitution for wood timber	226
12.6	Conclusions	228
	References	228

Chapter 13 Laminated veneer lumber from oil palm trunk 241

Aizat Ghani, S.H. Lee, Fatimah Atiyah Sabaruddin, S.O.A. SaifulAzry, and M.T. Paridah

13.1	Introduction	241
13.2	Laminated veneer lumber	242
13.3	Laminated veneer lumber from oil palm trunk	243

x Contents

13.4 Properties of laminated veneer lumber from oil palm trunk 245

13.5 Conclusions 248

References 249

Chapter 14 Enhancement of manufacturing process and quality for oil palm trunk plywood **253**

Y.F. Loh, Y.B. Hoong, A.B. Norjihan, B. Mohd Radzi, M. Mohd Fazli, and H. Mohd Azuar

14.1 Introduction 253

14.2 Plywood from oil palm trunk 255

14.3 Production process of oil palm trunk plywood 256

14.4 Enhancement of oil palm plywood manufacturing process . . 259

14.5 Conclusions 264

References 265

Chapter 15 Oriented strand board from oil palm biomass **267**

N.I. Ibrahim, S.O.A. SaifulAzry, M.T.H. Sultan, A.O. Fajobi, S.H. Lee, and R.A. Ilyas

15.1 Introduction 267

15.2 Preparation of OSB panels 270

15.3 Physical properties 272

15.4 Conclusions 278

Acknowledgments 279

References 279

Chapter 16 Particleboard from oil palm biomass **283**

S.O.A. SaifulAzry, S.H. Lee, and Wei Chen Lum

16.1 Introduction 283

16.2 Oil palm trunk 284

16.3	Oil palm frond 286
16.4	Empty fruit bunch 290
16.5	Challenges and future prospects 293
	Acknowledgments 294
	References 295

Chapter 17 Fiberboard from oil palm biomass 297

Mansur Ahmad, Zawawi Ibrahim, Aisyah Humaira Alias,
Noorshamsiana Abdul Wahab, Ridzuan Ramli,
Fazliana Abdul Hamid, and Syaiful Osman

17.1	Introduction 297
17.2	Fiberboard and its manufacturing process 298
17.3	Fiberboard from oil palm biomass 305
17.4	Fiberboard from oil palm frond (OPF) 307
17.5	Fiberboard from oil palm trunk (OPT) 311
17.6	Fiberboard from oil palm empty fruit bunches (EFB) 313
17.7	Conclusions 316
	References 317

Chapter 18 Other types of panels from oil palm biomass 321

A.H. Juliana, S.H. Lee, S.O.A. SaifulAzry, M.T. Paridah,
and N.M.A. Izani

18.1	Introduction 321
18.2	Laminated-based composites 322
18.3	Polymer-based composites 326
18.4	Cement composites 328
18.5	Application of other types of panels from oil palm biomass .. 332
18.6	Conclusions 333
	Acknowledgment 333
	References 333

xii Contents

Part 4 Current policy, environmental factors, and economic prospects for oil palm biomass and composite panels

Chapter 19 Policy and environmental aspects of oil palm biomass ... 339

R.A. Ilyas, S.M. Sapuan, M.S. Ibrahim, M.H. Wondi, M.N.F. Norrrahim, M.M. Harussani, H.A. Aisyah, M.A. Jenol, Z. Nahrul Hayawin, M.S.N. Atikah, R. Ibrahim, Syeed SaifulAzry Osman Al-Edrus, C.S. Hassan, and N.I.N. Haris

19.1 Policies regarding oil palm 339
19.2 Sustainable policy—NDPE policy......................... 340
19.3 Malaysia policy 341
19.4 A national strategy for the sustainable deployment of this untapped potential 343
19.5 Issues of oil palm/palm oil at national and international levels .. 344
19.6 Conclusions 349
Acknowledgments.. 349
References 349

Chapter 20 Corporate ownership structure of the major oil palm plantation companies in Malaysia and biomass agenda ... 353

K. Norfaryanti, A. Adhe Rizky, H.M. Omar Shaiffudin, A.M. Amira Nabilah, C.L. Ong, and J.M. Roda

20.1 Introduction 353
20.2 Ownership structure 354
20.3 Biomass utilization 355
20.4 Case study on the ownership structure of oil palm corporations.. 358
20.5 Findings 361
20.6 Biomass development progress and challenges 366
20.7 Conclusions 369
References 370

Index.. 375

Contributors

Fazliana Abdul Hamid Malaysian Palm Oil Board (MPOB), Persiaran Institusi, Kajang, Selangor, Malaysia

Noorshamsiana Abdul Wahab Malaysian Palm Oil Board (MPOB), Persiaran Institusi, Kajang, Selangor, Malaysia

Ummi Hani Abdullah Faculty of Forestry & Environment, Universiti Putra Malaysia, Serdang, Selangor, Malaysia

A. Adhe Rizky Institute of Tropical Forestry and Forest Products (INTROP), Universiti Putra Malaysia, Serdang, Selangor, Malaysia

Mansur Ahmad Faculty of Applied Sciences, Universiti Teknologi MARA, Shah Alam, Selangor, Malaysia

Norasikin Ahmad Ludin Solar Energy Research Institute (SERI), Universiti Kebangsaan Malaysia UKM, Bangi, Selangor, Malaysia

Zuriyati Mohamed Asa'ari Ainun Institute of Tropical Forestry and Forest Product, Universiti Putra Malaysia, Serdang, Selangor, Malaysia

H.A. Aisyah Institute of Tropical Forestry and Forest Products (INTROP), Universiti Putra Malaysia, Serdang, Selangor, Malaysia

Y. Alia Syahirah Forest Products Division, Forest Research Institute Malaysia, Kepong, Selangor, Malaysia

Aisyah Humaira Alias Institute of Tropical Forestry and Forest Products (INTROP), Universiti Putra Malaysia, Serdang, Selangor, Malaysia

A.M. Amira Nabilah Institute of Tropical Forestry and Forest Products (INTROP), Universiti Putra Malaysia, Serdang, Selangor, Malaysia

U.M.K. Anwar Forest Products Division, Forest Research Institute Malaysia, Kepong, Selangor, Malaysia

Zaidon Ashaari Institute of Tropical Forestry and Forest Products (INTROP); Faculty of Forestry and Environment, Universiti Putra Malaysia, Serdang, Selangor, Malaysia

M. Asniza Forest Products Division, Forest Research Institute Malaysia, Kepong, Selangor, Malaysia

M.S.N. Atikah Department of Chemical and Environmental Engineering, Universiti Putra Malaysia, Serdang, Selangor, Malaysia

Edi Suhaimi Bakar Institute of Tropical Forestry and Forest Products (INTROP); Faculty of Forestry and Environment, Universiti Putra Malaysia, UPM, Serdang, Selangor, Malaysia

Mohd Supian Abu Bakar Advance Engineering Materials and Composites Research Center, Department of Mechanical and Manufacturing Engineering, Faculty of Engineering, Universiti Putra Malaysia UPM, Serdang, Selangor, Malaysia

Kit Ling Chin Institute of Tropical Forestry and Forest Product, Universiti Putra Malaysia, Serdang, Selangor, Malaysia

Taharah Edin University of Technology Sarawak, Sibu, Sarawak, Malaysia

A.O. Fajobi Department of Aerospace Engineering, Faculty of Engineering, Universiti Putra Malaysia, Serdang, Selangor, Malaysia

Aizat Ghani Institute of Tropical Forestry and Forest Products (INTROP); Faculty of Forestry and Environment, Universiti Putra Malaysia, Serdang, Selangor, Malaysia

Paik San H'ng Institute of Tropical Forestry and Forest Product; Faculty of Forestry and Environment, Universiti Putra Malaysia, Serdang, Selangor, Malaysia

Norul Hisham Hamid Institute Of Tropical Forestry and Forest Products (INTROP); Faculty of Forestry & Environment, Universiti Putra Malaysia, Serdang, Selangor, Malaysia

N.I.N. Haris Institute of Sustainable and Renewable Energy, Universiti Malaysia Sarawak, Kota Samarahan, Sarawak, Malaysia

M.M. Harussani Advanced Engineering Materials and Composites Research Centre (AEMC), Department of Mechanical and Manufacturing Engineering, Faculty of Engineering, Universiti Putra Malaysia, Serdang, Selangor, Malaysia

C.S. Hassan Mechanical Engineering Department, UCSI University, Kuala Lumpur, Malaysia

Y.B. Hoong Fibre and Biocomposite Centre, Malaysian Timber Industry Board, Banting, Selangor, Malaysia

M.S. Ibrahim Integrated Ganoderma Management, Plant Pathology and Biosecurity Unit, Biology and Sustainability Research (BSR) Division, MPOB, Bandar Baru Bangi, Kajang, Malaysia

N.I. Ibrahim Institute of Tropical Forestry and Forest Products (INTROP), Universiti Putra Malaysia, Serdang, Selangor, Malaysia

R. Ibrahim Innovation & Commercialization Division, Forest Research Institute Malaysia, Kepong, Selangor, Malaysia

Zawawi Ibrahim Malaysian Palm Oil Board (MPOB), Persiaran Institusi, Kajang; Institute of Tropical Forestry and Forest Products (INTROP), Universiti Putra Malaysia, Serdang, Selangor, Malaysia

R.A. Ilyas School of Chemical and Energy Engineering, Faculty of Engineering; Center for Advanced Composite Materials (CACM), Universiti Teknologi Malaysia, Johor Bahru, Malaysia

N.M.A. Izani Faculty of Agriculture and Food Sciences, Universiti Putra Malaysia, Bintulu Sarawak Campus, Bintulu, Sarawak, Malaysia

M.A. Jenol Department of Bioprocess Technology, Faculty of Biotechnology and Biomolecular Sciences, Universiti Putra Malaysia, Serdang, Selangor, Malaysia

A.H. Juliana Faculty of Technology Management and Business, Universiti Tun Hussein Onn Malaysia, Parit Raja, Batu Pahat, Johor; Institute of Tropical Forestry and Forest Products (INTROP), Universiti Putra Malaysia, Serdang, Selangor, Malaysia

Rashidah Kamarulzaman University of Technology Sarawak, Sibu, Sarawak, Malaysia

Pui San Khoo Institute of Tropical Forestry and Forest Product, Universiti Putra Malaysia, Serdang, Selangor, Malaysia

Ching Hao Lee Institute of Tropical Forestry and Forest Products (INTROP), Universiti Putra Malaysia, Serdang, Selangor, Malaysia

Chuan Li Lee Institute of Tropical Forestry and Forest Product, Universiti Putra Malaysia, Serdang, Selangor, Malaysia

S.H. Lee Institute of Tropical Forestry and Forest Products (INTROP), Universiti Putra Malaysia, Serdang, Selangor, Malaysia

Y.F. Loh Fibre and Biocomposite Centre, Malaysian Timber Industry Board, Banting, Selangor, Malaysia

Wei Chen Lum Institute for Infrastructure Engineering and Sustainable Management (IIESM), Universiti Teknologi MARA, Shah Alam, Selangor, Malaysia

H. Mohd Azuar Fibre and Biocomposite Centre, Malaysian Timber Industry Board, Banting, Selangor, Malaysia

M. Mohd Fazli Fibre and Biocomposite Centre, Malaysian Timber Industry Board, Banting, Selangor, Malaysia

B. Mohd Radzi Fibre and Biocomposite Centre, Malaysian Timber Industry Board, Banting, Selangor, Malaysia

Nor Yuziah Mohd Yunus Faculty of Applied Sciences, Universiti Teknologi MARA Cawangan Pahang, Bandar Tun Abdul Razak, Jengka, Pahang, Malaysia

Nasihah Mokhtar University of Technology Sarawak; Centre of Excellence in Wood Engineered Products, UTS, Sibu, Sarawak, Malaysia

Z. Nahrul Hayawin Engineering and Processing Division, Malaysian Palm Oil Board (MPOB), Kajang, Selangor, Malaysia

Asmaa Soheil Najm Department of Electrical Electronic & Systems Engineering, Faculty of Engineering & Built Environment, Universiti Kebangsaan Malaysia, UKM, Bangi, Selangor, Malaysia

K. Norfaryanti Institute of Tropical Forestry and Forest Products (INTROP), Universiti Putra Malaysia, Serdang, Selangor, Malaysia

B. Norhazaedawati Fibre and Biocomposite Centre (FIDEC), Malaysian Timber Industry Board, Banting, Selangor, Malaysia

A.B. Norjihan Fibre and Biocomposite Centre, Malaysian Timber Industry Board, Banting, Selangor, Malaysia

M.N.F. Norrrahim Research Centre for Chemical Defence, National Defence University of Malaysia, Kuala Lumpur, Malaysia

H.M. Omar Shaiffudin Institute of Tropical Forestry and Forest Products (INTROP), Universiti Putra Malaysia, Serdang, Selangor, Malaysia

C.L. Ong Institute of Tropical Forestry and Forest Products (INTROP), Universiti Putra Malaysia, Serdang, Selangor, Malaysia

Syaiful Osman Faculty of Applied Sciences, Universiti Teknologi MARA, Shah Alam, Selangor, Malaysia

M.T. Paridah Institute of Tropical Forestry and Forest Products (INTROP); Faculty of Forestry and Environment, Universiti Putra Malaysia, Serdang, Selangor, Malaysia

Ridzuan Ramli Malaysian Palm Oil Board (MPOB), Persiaran Institusi, Kajang, Selangor, Malaysia

Mohd Sukhairi Mat Rasat University Malaysia Kelantan (UMK), Jeli, Kelantan, Malaysia

Sofiyah Mohd Razali Centre of Excellence in Wood Engineered Products, UTS, Sibu, Sarawak, Malaysia

J.M. Roda UR 105 Forests & Societies, The French Agricultural Research Centre (CIRAD), Montpellier, France

Fatimah Atiyah Sabaruddin Institute of Tropical Forestry and Forest Products (INTROP), Universiti Putra Malaysia, Serdang, Selangor, Malaysia

S.O.A. SaifulAzry Laboratory of Biocomposite Technology, Institute of Tropical Forestry and Forest Products (INTROP), Universiti Putra Malaysia, Serdang, Selangor, Malaysia

S.M. Sapuan Laboratory of Biocomposite Technology, Institute of Tropical Forestry and Forest Products (INTROP); Advanced Engineering Materials and Composites Research Centre (AEMC), Department of Mechanical and Manufacturing Engineering, Faculty of Engineering, Universiti Putra Malaysia, Serdang, Selangor, Malaysia

Mohamad Saiful Sulaiman University of Technology Sarawak; Centre of Excellence in Wood Engineered Products, UTS, Sibu, Sarawak, Malaysia

M.T.H. Sultan Institute of Tropical Forestry and Forest Products (INTROP), Universiti Putra Malaysia, Serdang; Department of Aerospace Engineering, Faculty of Engineering, Universiti Putra Malaysia, Serdang; Aerospace Malaysia Innovation Centre (944751-A), Prime Minister's Department, MIGHT Partnership Hub, Cyberjaya, Selangor, Malaysia

K. Tumirah Forest Products Division, Forest Research Institute Malaysia, Kepong, Selangor, Malaysia

Razak Wahab University of Technology Sarawak; Centre of Excellence in Wood Engineered Products, UTS, Sibu, Sarawak, Malaysia

M.H. Wondi Faculty of Plantation and Agrotechnology, Universiti Teknologi MARA, Mukah, Sarawak, Malaysia

PART

1

Background to oil palm biomass and wood-based panel products

1

Introduction to oil palm biomass

R.A. Ilyas[a,b], S.M. Sapuan[c,d], M.S. Ibrahim[e], M.H. Wondi[f], M.N.F. Norrrahim[g], M.M. Harussani[d], H.A. Aisyah[c], M.A. Jenol[h], Z. Nahrul Hayawin[i], M.S.N. Atikah[j], R. Ibrahim[k], S.O.A. SaifulAzry[c], C.S. Hassan[l], and N.I.N. Haris[m]

[a]School of Chemical and Energy Engineering, Faculty of Engineering, Universiti Teknologi Malaysia, Johor Bahru, Malaysia, [b]Center for Advanced Composite Materials (CACM), Universiti Teknologi Malaysia, Johor Bahru, Malaysia, [c]Laboratory of Biocomposite Technology, Institute of Tropical Forestry and Forest Products (INTROP), Universiti Putra Malaysia, Serdang, Selangor, Malaysia, [d]Advanced Engineering Materials and Composites Research Centre (AEMC), Department of Mechanical and Manufacturing Engineering, Faculty of Engineering, Universiti Putra Malaysia, Serdang, Selangor, Malaysia, [e]Integrated Ganoderma Management, Plant Pathology and Biosecurity Unit, Biology and Sustainability Research (BSR) Division, MPOB, Bandar Baru Bangi, Kajang, Malaysia, [f]Faculty of Plantation and Agrotechnology, Universiti Teknologi MARA, Mukah, Sarawak, Malaysia, [g]Research Centre for Chemical Defence, National Defence University of Malaysia, Kuala Lumpur, Malaysia, [h]Department of Bioprocess Technology, Faculty of Biotechnology and Biomolecular Sciences, Universiti Putra Malaysia, Serdang, Selangor, Malaysia, [i]Engineering and Processing Division, Malaysian Palm Oil Board (MPOB), Kajang, Selangor, Malaysia, [j]Department of Chemical and Environmental Engineering, Universiti Putra Malaysia, Serdang, Selangor, Malaysia, [k]Innovation & Commercialization Division, Forest Research Institute Malaysia, Kepong, Selangor, Malaysia, [l]Mechanical Engineering Department, UCSI University, Kuala Lumpur, Malaysia, [m]Institute of Sustainable and Renewable Energy, Universiti Malaysia Sarawak, Kota Samarahan, Sarawak, Malaysia

1.1 Introduction

Malaysia is a strategically situated country in Southeast Asia blessed with fertile agricultural land. With its rich agrobiomass resources as well as the booming agriculture industry, it is widely recognized that Malaysia has what it takes to develop its biomass industry. Fig. 1.1 shows the types of biomasses generated in Malaysia. In parallel

Oil Palm Biomass for Composite Panels. https://doi.org/10.1016/B978-0-12-823852-3.00015-5
Copyright © 2022 Elsevier Inc. All rights reserved.

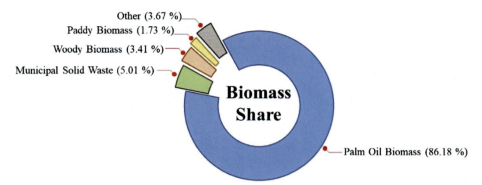

Fig. 1.1 Biomass share from various sources in Malaysia. From How, BS., Ngan, SL., Hong, BH., Lam, HL., Ng, WPQ., Yusup, S., Ghani, WAWAK., et al., 2019. An outlook of Malaysian biomass industry commercialisation: Perspectives and challenges. Renew. Sustain. Energy Rev. 113, 109277.

with the introduction of the National Green Technology Policy in 2009, the 10th Malaysia Plan (10MP) and the Economic Transformation Programme (ETP) in 2010, as well as the Renewable Energy Act 2011, Malaysia is gearing up to unlock the wealth creation potential of its biomass industry.

Areas of focus include:
- Biomass as energy by converting biomass into solid biofuel (pellets, briquettes), biomass power plant projects (co-firing, methane capture), and liquid biofuels, such as diesel and ethanol.
- Biomass as high-value chemicals, applying to the development of valuable chemical precursors such as sugar alcohols, ethanol, and lactic acid.
- Biomass as eco-products refers to the utilization of biomass feedstock into materials such as pulp and paper, biocomposite, and bioplastics as a sustainable substitution for fossil-based materials.

1.2 Is there a biomass industry?

The biomass industry represents several different industries brought together via utilizing renewable organic matters, including timber waste, oil palm waste, rice husk, coconut trunk fibers, municipal waste, sugar cane waste, etc. (refer to Fig. 1.2). These organic materials have the potential to be used in the manufacturing of value-added eco-products (e.g., bioplastics, biocomposites, biofertilizers, biopellets, etc.) and the generation of renewable energy (refer to Fig. 1.3).

While the biomass industry in Malaysia has enormous untapped potentials for commercialization given the minimum biomass

production of 168 million tons a year as well hundred types of biomass-related research and development (R&D) activities undertaken by local research institutions and universities, the full usage of biomass in the market is yet to be achieved. Among the barriers faced are:
- Many policies developed to facilitate the uptake of biomass and renewable energy among SMEs are still underway, limiting the coordination efficiencies among local agencies and the biomass industry in Malaysia.
- There is no reliable and clear data on the potential of biomass in the market.
- Limited incentives and funding support provided to bear the high cost of the initial investment.

In parallel to the need for increased utilization of renewable resources to combat climate change, biomass has a firm position in the

Fig. 1.2 Different types of biomasses available in Malaysia.

Fig. 1.3 Different types of products that can be derived from biomass.

national strategies to achieve sustainable consumption and production and fight climate change. The biomass commercialization issues are growing to be more complex and more diverse, opening opportunities for engagement between different industries, government agencies, and research institutions.

1.3 Overview of oil palm plantation and its biomass

Oil palm trees originally came from West and Southwest Africa. They were introduced to Indonesia and Malaysia in the late 19th and early 20th centuries. Oil palm trees became productive in Southeastern Asia since most of the diseases and insects that affected Africa's crops were absent in this region. Palm oil has been consumed as a staple food by human beings for more than 5000 years. It is also utilized for preventing vitamin A deficiency, cyanide poisoning, high cholesterol, high blood pressure, brain disease, cancer, and for treating malaria. Palm oil started to play a significant role during the colonial era, where European firms began to invest in the West African plantations and export the oil to Europe for edible cooking oil as well as raw materials for nonfood purposes, such as candles, soaps, and lubricants.

Oil palm flourishes in the humid tropics and produces high yield when grown 10 degrees north and south of the equator. Currently, oil palm plantations are developed in Indonesia, Malaysia, Thailand, Colombia, Nigeria, Guatemala, Ecuador, Honduras, Papua New Guinea, Brazil, Côte d'Ivoire, Cameroon, Ghana, Congo, Costa Rica, Peru, India, Mexico, Philippines, Benin, Dominican, Angola, Guinea, Liberia, Sierra Leone, Senegal, Togo, and Venezuela, as illustrated in Fig. 1.4. Among these, Indonesia (52.3%) and Malaysia (33.1%) are the world's largest palm oil producers, followed by Thailand (3.1%), Colombia (1.9%), and Nigeria (1.5%). In 2016, palm oil's global production was estimated at 62.6 million tons, 2.7 million tons more than in 2015. The palm oil production value was estimated at US $39.3 billion in 2016, an increase of US $2.4 billion (or +7%) against the production figure recorded in the previous year. In 2018, the total land area of oil palm plantations in Indonesia was approximately 12.8 million hectares, with a yield of 43.5 million tons, followed by Malaysia with about 3.31 million hectares with an output of 19.3 million tons. The third highest palm oil production by country is Thailand, followed

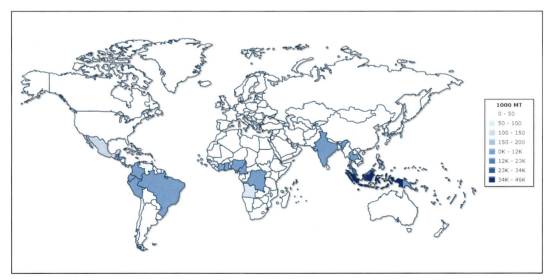

Fig. 1.4 Palm oil production by country in 1000 MT. https://www.usda.gov/.

by Colombia, Nigeria, Guatemala, Ecuador, Honduras, Papua New Guinea, and Brazil, with yields of 3.1, 1.67, 1.015, 0.852, 0.615, 0.580, 0.561, and 0.540 million tons, respectively.

As one of the commodity economies with the main agricultural activities, and one of the largest exporters and the second-largest producers of palm oil, Malaysia is blessed with the most abundant biomass available as bioresources that can be used for alternative renewable energy or useful biocomposites products. However, even though government policies and market incentives have been put in place to support green technology in the industry, the uptake of biomass commercialization needs further intervention.

Annually, about 168 million tons of biomass waste are generated in Malaysia. In general, oil palm (OP) waste accounts for 94% of biomass feedstock, while the remaining contributors are agricultural and forestry by-products, such as wood residues (4%), rice (1%), and sugarcane industry wastes (1%). By 2019, up to 5.9 million hectares of land were cultivated with oil palm, which is 18% of the country's total land area. The OP industry generates a substantial amount of by-products, especially through its processing. Fig. 1.5 shows the oil palm residues after oil extraction at the mill. With more than 443 mills in Malaysia, this industry generated around 80 million dry tons of biomass in 2010.

Out of the oil palm processing yield, only 10% are finished products, i.e., palm oil and palm kernel oil, and the remaining 90% are

Fig. 1.5 Oil palm residues after oil extraction at the mill. Source: http://www.besustainablemagazine.com/cms2/malaysias-biomass-potential/

harvestable biomass waste in the form of empty fruit bunches (EFB), palm kernel shell (PKS), palm oil mill effluent (POME), and palm kernel cake (PKC).

The abundant oil palm biomass is left on the plantations (i.e., palm fronds) or returned to the fields as a soil amendment or an organic fertilizer (i.e., EFBs) shown in Fig. 1.6. It is estimated that, by the year 2020, solid oil palm biomass will increase to 85–110 million tons. This biomass plays an essential part in ensuring the sustainability of plantations and preserving soil fertility. However, there is also the potential to utilize a share of this biomass for various additional end uses, including pellets, bioenergy, biofuels, and biobased chemicals, without depleting the soil.

In addition to oil palm biomass, it is estimated that 1.975 million m^3 of wood residues are generated every year in Malaysia. However, although the amount of residues obtained from the processing mills is considerably high, there are many competing uses for wood residues, i.e., particleboard and medium-density fiberboard production, furniture-making, and power generation. In response to the growing market for biomass as energy generation (heat or electricity), an increasing number of Malaysian companies are venturing into biomass pellets and briquettes production for the export market such as Europe, Japan, Korea, and China.

Apart from producing biomass as solid biofuel, many companies are also recognizing the cost- and energy-saving benefits of using biomass to generate heat and power for their production. To encourage the uptake of biomass as renewable energy resources, Malaysia provides many incentives to support the industry, such as Investment Tax

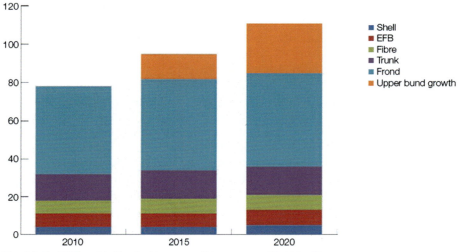

Fig. 1.6 By 2020, solid oil palm biomass will increase to 85–110 million tons.

Allowance or Pioneer Status, where companies are exempted from income tax for a certain period. Similarly, under the ETP, biogas trapping and utilization as energy is identified as one of the National Key Economic Areas (NKEAs) to encourage palm oil mill owners to take up methane captured from their POME as CDM projects (PEMANDU 2010). The Energy Commission of Malaysia had announced plans to boost renewable electricity share to 5.5% by 2015, compared with less than 1% in the year before.

Therefore, this chapter will present an overview of oil palm plantation and its biomass. Besides that, oil palm biomass types will also be discussed, such as by-products from oil palm plantations and oil palm mills. Properties of palm oil solid residues will also be reviewed. The oil palm productions, palm oil processing, and palm oil issues at national and international levels and policies regarding oil palm are also studied in this chapter. Finally, the potential advanced applications of palm oil biomass are also extensively elaborated in the last part of the chapter.

1.4 Types of oil palm biomass

In 2017, about 51.19 million tons (Mt) of oil palm waste was generated by Malaysia's palm oil industry (Hamzah et al., 2019). It is estimated that this volume will accelerate to 95–110 (Mt) by 2020 (Mokhtar et al., 2012). Yet, the oil production accounts for just about 10% of the industry's overall biomass. Thus, the other 90%, primarily in the form of lignocellulosic biomass, must be exploited and commercially used to support industry finances as well as the national economy. In addition to the extraction and processing of crude palm oil (CPO), the palm oil industry also generates by-products from oil palm plantations and oil palm mills. The palm oil industry produces large amounts of oil palm by-products biomass, such as the residues from oil palm mills that comprise POME, pressed fruit fibers, EFB, and shells. In contrast, the by-products from plantations consist of oil palm fronds (OPF) and oil palm trunk (OPT) (Mokhtar et al., 2012). Fig. 1.7 shows some of the oil palm lignocellulosic fiber biomass, such as oil palm EFB, OPF, and OPT. EFB is the by-product of the harvested oil palm from the refining process. Table 1.1 shows the availability of oil palm by-products in Malaysia in 2017. According to Hamzah et al. (2019), usually, the amount of EFB discharge from palm oil refineries is up to 7.78 Mt/year. The most abundant oil palm plantation residue is OPF with a value of 23.39 Mt. OPF is available daily when they are sheared during the fresh fruit bunches (FFB) harvesting for palm oil production. The abundance of OPF has led to this lignocellulosic fiber being used for ruminant livestock feed (Tomkins and Drackley, 2010; Wong and

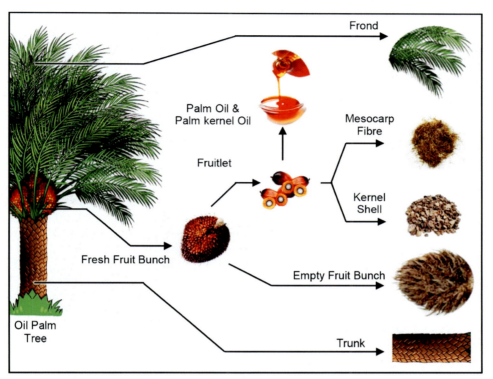

Fig. 1.7 Oil palm lignocellulosic fiber biomass.

Table 1.1 Availability of oil palm residues in Malaysia in 2017.

Biomass Type	OPT[a]	OPF[b]	POME[c]	MF[d]	PKS[e]	EFB[f]	Total
Availability (Mt)	3.74	23.39	3.38	8.18	4.72	7.78	51.19

[a] Oil palm trunks (OPT).
[b] Oil palm fronds (OPF).
[c] Palm oil mills effluent (POME).
[d] Mesocarp fibers (MF).
[e] Palm kernel shell (PKS).
[f] Empty fruit bunch (EFB).

Wan Zahari, 2011; Zamri-Saad and Azhar, 2015), pulp and paper production (Taiwo et al., 2017; Thongjun et al., 2012; Wang et al., 2012), biochar production (Lawal et al., 2020), bioethanol generation (Mad Saad et al., 2016), as well as for erosion control, soil conservation, and ultimately the long-term benefit of nutrient recycling in oil palm plantations (Kome and Tabi, 2020). Besides that, according to Hamzah et al. (2019), it is estimated around 75% of the oil palm residues consist of OPT and OPF, in which these biomasses can be directly yielded from plantation sites, whereas POME, PKS, MF, and EFB that account for the remaining 25% are obtained after palm oil extraction from FFB at the mill sites.

1.5 By-products from oil palm plantations

1.5.1 Oil palm frond

One of the by-products of oil palm plantation is OPF. According to Chan et al. (1981), recycling OPF in the plantation could benefit the palm oil industry due to its nutrient content. The fronds are organic matter that aid in soil moisture conservation and prevention of surface water evaporation. On the other hand, the vascular bundles originated from the OPF petiole were utilized to manufacture wood products, as represented in Fig. 1.8. From Chan et al., the dry weight of OPF available for biomass utilization from 5% of the mature palm during replanting is approximately 3 million tons. Plus, the overall OPFs dry weight harvested from annual pruning and normal routine harvesting from Peninsular Malaysia was 1.60 million tons.

1.5.2 Oil palm trunk

Kamarudin et al. (1997) stated that around 3.19 million palms generated more than 5.62 million m^3 of OPT. The trunks contributed to the plywood, fiber-based biocomposites, and sawn lumber productions. OPTs consist of three main parts: cortex, peripheral region, and central zone, as represented by Fig. 1.9. Generally, the cortex is just below the bark of the trunk, thus making up the outer layer of the stem. The peripheral region consisting of parenchymal layers and vascular bundles contributed the most to the mechanical support for the oil palm

(a) Oil palm fronds   (b) Petiole   Rachis

Fig. 1.8 Oil palm fronds.

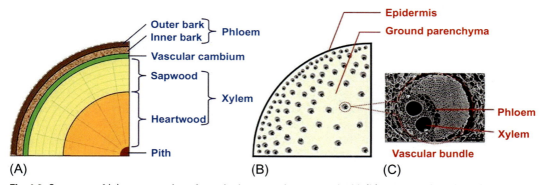

Fig. 1.9 Structure of (a) cross section of a typical tree trunk compared with (b) cross section of an oil palm trunk and (c) its vascular bundle. Source: Sap from various palms as a renewable energy source for bioethanol production. Chem. Ind. Chem. Eng. Q. 22 (4) 355–373 (2016). https://doi.org/10.2298/CICEQ160420024N.

tree. Killmann and Lim (1985) stated that the OPT density is about 220–550 kgm^{-3}, with a hardness of approximately 2450 N. The central zone is made up of larger amounts of vascular bundles embedded in the thin-walled parenchymatous ground tissue.

1.6 By-products from oil palm mills

1.6.1 Empty fruit bunch

Usually, the FFB will be unloaded on the loading ramp. The fruits will be assessed based on their ripeness and freshness to ensure only high-quality oil will be produced. Oil palm fruit processing is divided into several processing stages, which are fruit receiving, fruit sterilization, stripping, digestion, oil extraction, oil clarification, and nut cracking process. Fig. 1.10 shows the EFB and EFB stalk. These biomasses are collected after fresh fruit separation from FFB.

(a) Empty fruit bunch   (b) Empty fruit bunch stalk

Fig. 1.10 Empty fruit bunch.

1.6.1.1 Fruits sterilization process

The sterilization process is a pretreatment process of FFB using pressurized steam from the boiler at 40–45 psi for 90 min. Sterilization is a critical process for oil and kernel extraction. Sterilization of FFB should be done immediately upon EFB reaching the mill to deactivate the lipase enzyme activity and prevent free fatty acid (FFA) formation within the fruits, which can affect the CPO quality. The heat exerted by high-temperature steam during the sterilization process also assists in the detachment of the fruits from the bunch by softening the abscission layer (Tan et al., 2009a, b; Wondi et al., 2021). Besides, the sterilization is also aimed to soften the fruit and mesocarp, which will improve the efficiency of digestion prior to oil extraction process. Effective sterilization process also shrinks the kernel inside the nut and allows for efficient kernel separation from the shell, thus facilitating the nut cracking process in a later stage. There are several types of sterilizers used in the palm oil mills, which are (1) horizontal (Fig. 1.11), (2) vertical (Fig. 1.12), (3) oblique (Fig. 1.13), (4) spherical), and (5) continuous sterilizer (Fig. 1.14). Most sterilizers have a similar operational concept and use steam supplied from the boiler for heating purposes.

1.6.1.2 Fruit stripping

The fruit stripping process is a detachment or separation process of palm fruitlets from the whole bunch. The sterilized FFB is fed to a cylindrical perforated rotary drum, usually known as thresher (Fig. 1.15), which rotates at 23–25 rpm in speed. Through a centrifugal force, the fruitlets are stripped/detached from the bunch, fall through a gap, and

Fig. 1.11 Sterilization station. (A) Horizontal sterilizer, and (B) Sterilizer cage loaded with FFB.

Fig. 1.12 Horizontal sterilizer.

then are delivered to a digester by screw conveyor while the bunches remain inside the drum. The wholly stripped fruit bunches will be discharged at the end of the thresher as EFB. The EFB is commonly conveyed to the EFB hopper before being transported to the oil palm estate for mulching application.

There are two types of thresher drums used in the palm oil mill: shaft thresher drum and shaftless thresher drum. The shaftless thresher drum is an improved design, where the central shaft is eliminated. This design has reduced the thresher's breakdown and downtime because the central shaft tends to crack due to fatigue. The elimination of the central shaft also has increased the stripping efficiency (Corley and Tinker, 2015). The EFB discharged from the thresher drum is continuously inspected for hard or "unstrapped" bunches. These partially unstripped bunches are called an unstripped bunch (USB), as shown in Fig. 1.15. Stripping efficiency is measured by collecting 100 bunches from the thresher drum

Fig. 1.13 Oblique sterilizer.

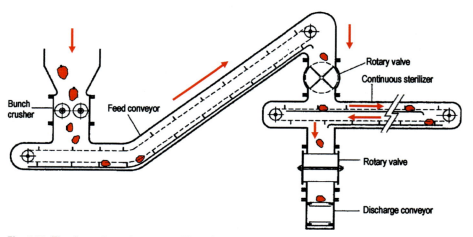

Fig. 1.14 The flow of continuous sterilizer. Source: Sivasothy, Kandiah, Basiron, Y., Suki, A., Mohd Taha, R., Tan, Y.H., Sulong, M., 2006. Continuous sterilization: the new paradigm for modernizing palm oil milling. J. Oil Palm Res. 144–152.

discharge. The USB counted from the collected bunches must be 3% and below the total bunches to indicate an efficient stripping process.

An inefficient stripping process can be observed when some fruitlets are still attached to the EFB spikelets after the stripping process. There are several factors that contribute to the increase in USB, which

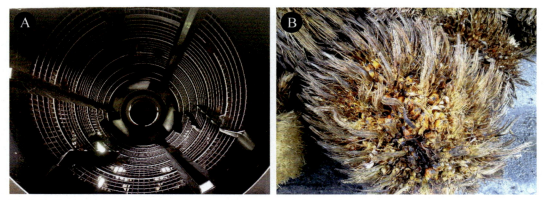

Fig. 1.15 Threshing station. (A) Thresher drum (shaftless type) and (B) the fruitlets remain attached on bunch after the stripping process.

are inefficient sterilization process, overfeeding of FFB into the thresher, unsuitable threshing speed, and a high percentage of unripe (hard-bunch) FFB being processed (Salleh et al., 2019; Sivasothy et al., 2005). To address these problems, the USB is recycled for resterilization. Most of the mills also implemented a double-threshing process to overcome the inefficient stripping process, provided that the recycling is efficient and adequate stripping of the USB is successful at the second attempt. Stripping efficiency of hard bunches can be improved by implementing the double or triple peak method during the sterilization process. Still, this method tends to increase oil losses on empty bunches and in sterilizer condensate. Moreover, splitting bunches in half before sterilization was proven to reduce the incidence of unstripped bunches (Corley and Tinker, 2015; Sivasothy et al., 2006).

1.6.1.3 Fruit digestion

The stripped fruitlets from the threshing machine are then conveyed into a digester for the fruit's digestion process prior to the oil extraction process. A digestion process is a crucial process for CPO extraction. A palm digester (Fig. 1.16) is a machine (Stork Amsterdam, 1960) used for macerating and rupturing the fruitlet mash into pulp prior to pressing. It is a cylindrical vessel equipped with a number of horizontal stirring arm blades inside rotating at 25–26 rpm. Usually, five sets of stirring arm blades are installed on the main rotating shaft comprised of long and short blades that rotate to stir the fruitlets inside the digester for fruit digestion purposes.

The stripped palm fruitlets enter from the digester's top through the hopper until around ¾ of the maximum digester volume. The stirring effect by the rotating blades operation will create rubbing and pressing action to the fruitlets, resulting in breaking down of the

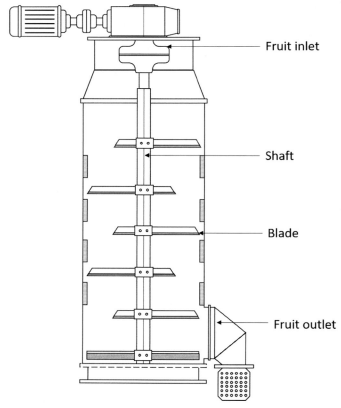

Fig. 1.16 Vertical digester illustration (Nurhidayat, 2014).

fruitlet's pericarp into mash and subsequent rupture of the oil-bearing cells (Wondi et al., 2020). The movement of fruitlets by the rotating blades creates a cutting and shearing effect that will loosen the mesocarp from its nuts (Tan et al., 2009a, b). With sufficient feeding volume of fruitlets inside the digester, an efficient cutting effect is created from the pressure exerted by the fruits themselves due to gravitational force.

Inside the digester, the heating process is provided by steam supplied by direct steam injection or in a steam jacket/coil to maintain the temperature at 90–100°C for 20–75 min (Baryeh, 2001). At this temperature, hot water or condensed steam acts as a solvent that would penetrate the cell wall, and dissolve and release the oil spontaneously from the MFs. The digester is also equipped with a perforated plate located at the bottom of the vessel to facilitate the drainage of water and oil mixture preventing it from entering the screw press. A mixture of oil and hot water inside the digester will lower the oil viscosity, thus reducing the flow resistance of oil. The continuous removal of oil during the digestion process through the perforated bottom plate will reduce the load of digested mash that enters the screw press, hence minimizing the oil losses in the pressed fiber. Effective digestion would take around 30 min to complete the process. A complete digested mesh (a mixture of mesocarp and nuts) are removed continuously to a screw press machine through the digester bottom discharge for the oil extraction process.

1.6.1.4 Oil extraction

The digested fruit mesh from the digester consisting of a mixture of mesocarp and nuts is then transferred into a screw press machine for the oil extraction process, as shown in Fig. 1.17. Screw press extraction is generally known as the dry extraction method due to the absence of water compared to the digester that requires water to enhance the oil extraction (Obibuzor et al., 2012). A screw press machine consists of a pair of the main worm screw shaft, perforated press cage, and a press cone.

As the digested mash enters the screw press, the worm screw is powered by an electric motor, transferring the digested mash to the press cone. Simultaneously, the press cone will move concurrently to the digested mesh movement, thus increasing the pressure resulting in the compression of digested mash. This compression enables the crude oil to be leached out from the MF and continuously flowed out through the perforated press cage, as shown in Fig. 1.18. The discharged digested mesh from the screw press machine is called a pressed cake. The cake, which consists of pressed fiber and nut, is then delivered to the depericarper system for the nut and fiber separation process.

Fig. 1.17 Pressing station. (A) The digester machine (at the top) attaches to a screw press (at the bottom). (B) Press cake, which consists of mesocarp fiber and nut.

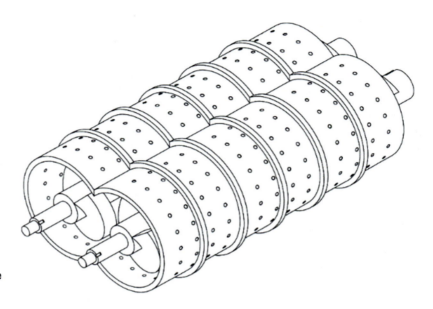

Fig. 1.18 Perforated press cage and worm screw illustration.

1.6.2 Mesocarp fiber

Oil palm mesocarp fiber (OPMF) biomass is a lignocellulosic material yielded after the oil extraction process. Fig. 1.19 shows the oil palm features comprised of exocarp, mesocarp, endocarp, and kernel. Thus, OPMF can be called the residue from the process and is left in the palm oil mill. The mesocarp is the edible part of the fruit with pulp rich in fatty acids, amino acids, and vitamins. According

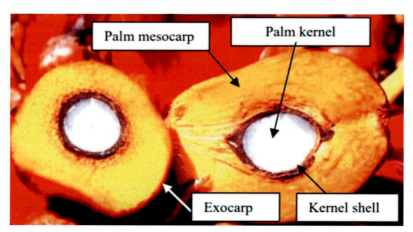

Fig. 1.19 Oil palm features comprised of exocarp, mesocarp, endocarp, and kernel.

to Sundram et al. (2003), the mesocarp pulp protects the palm oil in a fibrous matrix. In oil palm fruit, the palm oil can be extracted from two parts: mesocarp and kernel. The postoil extraction process of mesocarp pulp produces OPMF, which can be used for various applications.

Some approaches stated that the MFs are utilized as boiler fuel to generate steam and electricity (Subramaniam et al., 2004). The recent application of these materials has caused massive environmental contamination. Thus, a lot of works have been done to benefit the OPMF. For example, Then et al. (2013) studied the utilization of the fiber as a filler for biocomposites reinforcement. Poly (butylene succinate) (PBS) was melt-blended with OPMF biocomposites. The presence of the fiber enhanced the mechanical structure of the polymer/fiber biocomposites. Hence, the high flexural and tensile strength of the biocomposites suit for packaging applications. In placing more emphasis, Olusunmade et al. (2016) studied the hybridization of OPMF-reinforced thermoplastic (OPMFRT) composites for structural applications. The primary step of MF-reinforced biocomposites preparation includes fiber processing.

1.6.3 Palm kernel shell

According to Okoroigwe et al. (2014), PKS had been recognized as residue from processed palm oil. Fig. 1.20 shows the oil palm fruit and oil PKS. Approximately 15 tons of FFB are produced per hectare annually, and PKS encompasses about 64% of the bunch mass (Adewumi, 2009; Obisesan, 2004). Nowadays, PKS is applied as a reinforcing material in cement industries. In the Asia-Pacific region, kernel shell is utilized in biomass power plants for power generation. On the other hand, it is also used as an alternative fuel. However, these applications lead to massive emissions of smoke and air pollutants.

Fig. 1.20 (A) Oil palm fruit and (B) oil palm kernel shells (Yeboah et al., 2020).

1.6.3.1 Fiber processing

The fibers left inside the palm oil mill (Fig. 1.21) are washed with detergent and rinsed thoroughly with water to eliminate the remaining oil to ease the hybridizing process. Next, the OPMF is dried with direct sunlight and then pulverized into ground fibers using a grinding machine. Applying hot water and acetone after rinsing the fibers with distilled water is also preferred by some researchers. This step is paramount to remove impurities from the MFs. In addition, some works preferred to dry the OPMF using an oven with controlled moisture content (Iberahim et al., 2013; Then et al., 2013). This approach is suitable for laboratory-scale OPMF preparation.

1.6.4 Palm oil mill effluent (POME)

POME is the wastewater produced from the oil palm treatment, such as oil extraction, washing, and cleaning processes in the oil palm mill. Thus, it comprises various suspended materials. The characteristic of POME is tabulated in Table 1.2. Generally, the POME has high biochemical oxygen demand (BOD) and chemical oxygen demand (COD) due to a high concentration of organic nitrogen, phosphorus, and various suspended substances contained in the effluent (Kamyab et al., 2018).

The palm oil mill produces oily wastes that contain high toxic concentration toward oceanic environments and organisms. Improper POME treatment will create negative effects on the environment. Next, it will lead to a mass amount of nutrients in the water source, thus contributing to microalgae growth, also known as the algae bloom

Fig. 1.21 Oil palm mesocarp fiber collected from local palm oil mills (Olusunmade et al., 2016).

Table 1.2 Characteristics of POME.

Parameters	Concentration (mg/L)
pH	4.7
Oil and grease	4000
Total solids	40,500
Suspended solids	18,000
BOD	25,000
COD	50,000
Total nitrogen	750

phenomenon. Fig. 1.22 shows the treated POME that has the potential to be utilized as fertilizer. Ironically, the existence of microalgae was assumed to aid in removing the pollutants and reducing the presence of the organic compounds.

Ma et al. (1993) stated that palm oil mills in Malaysia have been practicing the ponding system, specifically for POME treatment, to avoid environmental pollution, as shown in Fig. 1.23. Kamyab et al. (2018) pointed out that there are four main types of POME treatment systems, which are waste stabilization ponds, activation sludge system, land application system, and closed anaerobic digester. Anaerobic and aerobic ponding system was widely used to treat POME waste. On the other hand, POME-derived biogas as biofuels could be earned from culturing the microalgae. This happened due to the role of microalgae that reacts with the nitrogen and phosphorus in wastewater to yield microalgae biomass that is rich in lipids.

Therefore, it can be concluded in this section that the remaining 90% of the total weight of the oil palm are in the form of OPF, OPT, EFB, PKS, POME, and PKC. Only 10% contribute to the main product, whereas the others are classified as biomass. The oil palm biomass can

Fig. 1.22 Treated POME contains a high level of nutrients for soil.

Fig. 1.23 POME treatment systems.

be categorized into two categories: plantation residual and palm oil mill residual, as shown in Fig. 1.24. Plantation residuals consist of OPF and OPT, whereas palm oil mill residuals comprise of EFB, PKS, MF, and POME.

1.7 Properties of palm oil solid residues

The main chemical compositions of oil palm biomass are cellulose, hemicellulose, and lignin. Other remaining constituents are ash and extractives, including minerals, proteins, sugar, starch, tanning agents, fats, and resins.

The highest cellulose content of oil palm biomass is EFB at 38%, followed by EPT, OPT, MF, OPF, and PKS at 35%, 35%, 34%, 30%, and 21%, respectively (Table 1.3). This high cellulose content can be converted into simple sugars and processed into biochemical or biofuels (Barbosa et al., 2020; Darwesh et al., 2020; Hamzah et al., 2019; Ko et al., 2020; Harussani et al., 2022; Paul et al., 2020). Besides that, cellulose can be processed into nanocellulose using chemical and mechanical treatments, described by Ilyas et al. Ilyas et al. (2017, 2018a, b, c, d, e, f, 2019a, b, c, d, 2020a, b, c). Nanocellulose from plant fiber can be used for wastewater treatment, including adsorptive removal of anionic dyes from aqueous solutions by Jin et al. (2015) and agent for wastewater treatment (Kardam & Rohit, 2014; Karim et al., 2016; Korhonen et al., 2011; Paltakari et al., 2011; Suopajärvi et al., 2015;

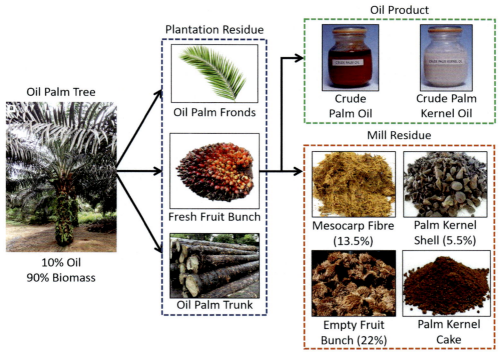

Fig. 1.24 Two categories of palm oil biomass are plantation residual and palm oil mill residual.

Table 1.3 Chemical composition of oil palm solid biomass from mills.

Biomass	Cellulose	Hemicellulose	Lignin	Ash	Extractives
Empty fruit bunches	38.3	35.3	22.1	1.6	2.7
Mesocarp fiber	33.9	26.1	27.7	3.5	6.9
Palm kernel shell	20.8	22.7	50.7	1.0	4.8
Oil palm frond	30.4	40.4	21.7	5.8	1.7
Oil palm trunk	34.5	31.8	25.7	4.3	3.7

Chemical Components (% Dry wt.)

Wang et al., 2013). Moreover, due to its outstanding mechanical properties, nanocellulose had been utilized by reinforcing it with other polymers to be adapted in various applications, including paper (Basta & El-Saied, 2009), treating tympanic membrane perforation (Biskin et al., 2016; Lou, 2016), shielding film (Lang et al., 2014), food packaging films (Azeredo et al., 2017; Ilyas et al., 2018a, c, f), audio speaker diaphragms (Gellner et al., 1993), and so on. Nanocellulose has gained interest in many fields, thereby opening the novel horizon for scientific research for scientists and engineers dealing with nanotechnology and nanoscience.

The highest hemicellulose content of oil palm biomass is OPF at 40%, followed by EFB, OPT, MF, and PKS at 35%, 32%, 26%, and 23%, respectively (Table 1.3). Hemicelluloses is a renewable plant polysaccharide, and it has been widely reported and extensively studied in the field of cosmetics (Carvalheiro et al., 2009), additives (Song et al., 2019), coatings in papermaking (Anthony et al., 2015; Farhat et al., 2017), food microencapsulation (Tatar et al., 2014a, b; Ubeyitogullari & Cekmecelioglu, 2016), active food packaging (Mugwagwa & Chimphango, 2020), thermoplastics (Farhat et al., 2018), and hydrogels (Arellano-Sandoval et al., 2020; Chen et al., 2020a, b; Hu et al., 2020; Tao & Sun, 2020). Due to its intrinsic properties such as innate immunological defense, promoting proliferation and cell adhesion, and inhibiting cell mutation and anticancer effect, hemicellulose is suitable for the hydrogel preparations to be used in biomedical engineering and pharmaceutical applications, such as drug carriers (Arellano-Sandoval et al., 2020; Liu et al., 2017; Sun et al., 2013; Wei et al., 2020). In medical science applications, hemicellulose had been used for histological scaffolds for tissue engineering, immunization barriers for the encapsulation of the cell, and controlled drug delivery systems owing to its unique properties (Darwesh et al., 2020; Liu et al., 2020; Yılmaz Celebioglu et al., 2012). Moreover, hemicellulose also has huge potential for applications in electronic devices and biosensors field due to its number of charged groups (Guo et al., 2020). Hemicellulose can be used as an alternative material for petroleum-based polyols as well as a nonfood-based substitute for starch-based polyols. Thus, hemicellulose would have great potential to supply raw materials for these industries.

The highest lignin content of oil palm biomass is PKS at 51%, followed by MF, OPT, EFB, and OPF at 28%, 26%, 22%, and 22%, respectively (Table 1.3). From Table 1.3, PKS is seen to have the highest lignin content, which has proven to be the most preferred fuel for thermal combustion compared to other parts of oil palm biomass (Hamzah et al., 2019; Harussani and Sapuan, 2021). Currently, more advanced materials from lignin have been developed and produced by using new techniques with a controlled structure down to the micro- and nanosize (Sameni et al., 2020). These micro- or nanoscale particles of lignin have been explored for their potential high-value applications in various areas, including in wastewater treatment (Ahmaruzzaman, 2011), nanosized coatings (Dastpak et al., 2020; Pandian et al., 2020; Shamaei et al., 2020), dye removal (de Araújo Padilha et al., 2020), filler in nanocomposites (Guo et al., 2020; Harussani et al., 2021; Younesi-Kordkheili and Pizzi, 2020; Zhang et al., 2020), purification (Rivière et al., 2020), supercapacitor electrodes (García-Mateos et al., 2020), biocarbon fiber (Navia, 2020; Shen et al., 2020), electrochemical energy storage (Espinoza-Acosta et al., 2018; Geng et al., 2017; Navarro-Suárez et al., 2018), food industry fat mimetics (Stewart et al., 2014; Tao et al., 2020), agricultural actives controlled release (Chen et al., 2020a, b), and biosensors (Lei et al., 2020). Lignin also has been utilized in pharmaceutical and medical science applications, such as delivery vehicles of DNA (Ten et al., 2014), encapsulation of hydrophobic/hydrophilic drugs (Kanomata et al., 2020), drug delivery (Alqahtani et al., 2019; Dai et al., 2017; Larrañeta et al., 2018), effective antioxidant and antimicrobial (do Nascimento Santos et al., 2020; Domínguez-Robles et al., 2020; Trevisan & Rezende, 2020). Moreover, lignin microcapsules have attracted huge attention from scientists, pharmacists, and biomedical engineers due to its noncytotoxic properties (Sameni et al., 2020).

1.8   Oil palm productions

Based on the data from the Malaysian Palm Oil Board (MPOB) until September 2020, Malaysia produced 14.59 million tons of CPO and 3.62, 1.68, and 1.89 million tons of palm kernel, crude palm kernel oil, and PKC, respectively, and were harvested throughout January until September of 2020 (MPOB, 2020a), as shown in Table 1.4. According to Mokhtar et al. (2012), the estimated yields of oil palm biomass from 2017 until September 2020 were 2.97, 7.14, and 2.86 million tons per

Table 1.4 Monthly production of oil palm products summary for the month of September 2020 (tons).

Months	Crude palm oil		Palm kernel		Crude palm kernel oil		Palm kernel cake	
	2019	2020	2019	2020	2019	2020	2019	2020
January	1,737,461	1,171,534	445,427	286,909	211,390	136,695	239,415	156,026
February	1,544,518	1,288,515	395,704	331,808	174,065	153,324	196,194	173,277
March	1,672,058	1,397,322	432,518	351,058	223,419	164,543	251,528	185,271
April	1,649,368	1,652,771	418,874	412,597	190,217	167,131	213,306	185,065
May	1,671,467	1,651,336	409,187	401,088	204,270	196,712	228,112	220,470
June	1,510,957	1,885,742	356,164	465,712	160,366	215,663	175,951	245,389
July	1,740,759	1,807,397	412,353	449,347	201,270	217,642	223,537	246,309
August	1,821,548	1,863,309	446,585	459,679	199,802	212,760	223,233	237,582
September	1,842,433	1,869,339	453,654	466,080	200,915	210,760	220,338	237,288
Total	15,190,569	14,587,265	3,770,466	3,624,278	1,765,714	1,675,230	1,971,614	1,886,677

year for OPT, OPF, and EFB, respectively. Besides that, according to Table 1.4, the estimated yields of oil palm FFB from 2010 until 2020 are increasing, from 19 tons/hectare until 26 tons/hectare. The increasing oil palm biomass is contributed by the increasing number of oil palm plantations. At 2020, there are 5.7 million hectares of estimated oil palm plantations. Thus, rapid increases in production of oil palm crude oil, palm kernel, and crude palm kernel oil, which is estimated for 2010 until 2020, 4.1%, 5.6% and 4.8%, respectively. The oil palm yields are also affected by the demand and competition existing in the market globally and locally.

1.9 Palm oil prices

From the recorded data collected from Malaysian Palm Oil Prices (MPOC, 2020), retrieved on October 26, 2020, the settlement price for CPO was increased almost RM 400 from the last 10 days, from October 19, 2020, until October 22, 2020, as shown in Fig. 1.25. The settlement price was unstable throughout the year 2020. Besides, MPOB (2020b) stated that the monthly average of CPO prices in Malaysia were RM 2924/ton for September 2020, RM 2815/tons for August 2020, RM 2519/ton for July 2020, and RM2411.50 for every ton for June, as shown in Table 1.5. This recorded an increasing pattern of palm oil prices throughout the year, which is supported by the data recorded by the MPOB shown in Table 1.6.

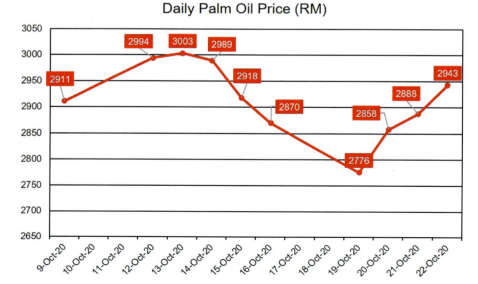

Fig. 1.25 Daily palm oil price.

Table 1.5 Monthly average crude palm oil (local delivered) throughout the year 2020.

Month	Average (RM/ton)
January	3013.50
February	2714.50
March	2382.00
April	2299.00
May	2074.00
June	2411.50
July	2519.00
August	2815.00
September	2924.00
October	–
November	–
December	–

Table 1.6 Summary of data on production of oil palm biomass and the growth in its export value as well as crude palm oil prices.

	Tahun (Juta Tan)			Pertumbuhan tahunan (%)		
Produk	2010	2015	2020	2011–15	2016–20	2011–20
Minyak sawit	17.2	20.4	23.7	3.5	3.0	3.3
Minyak Isirung Sawit	1	2.4	3.1	19.1	5.3	12.0
Dedak Isirung Sawit	2.2	3.1	3.5	7.1	2.5	4.8
Oleokimia	3.4	4	4.5	3.3	2.4	2.8
Produk Lain	0.5	2.5	4	38.0	3.7	19.6
JUMLAH	**24.3**	**32.4**	**37.8**	**5.9**	**3.1**	**4.5**
Hasil Eksport (RM bilion)	62.9	86	103.0	6.5	3.7	5.1
Harga CPO (RM/Tan)	2704	2600	2800	−0.8	1.5	0.3

The CPO prices kept fluctuating across several years owing to some paramount factors. This price trend is affected by the current demand and supply and other technical aspects, such as weather season, competition among producing countries, and with other edible oils. According to Fazlin (2019) in New Straits Times, production of

Malaysian palm oil reached almost 18% for local uses, while the rest are exported to Europe, China, India, and the Middle East. Thus, the global demands for edible oils affect the palm oil price in Malaysia. For example, Malaysian CPO prices have been dropping since June 2018. This was due to China's lower palm oil demands, which led to slower exportation of palm oil and increased the palm oil stocks in Malaysia.

On the other hand, the supply and production of other edible oils also influence the palm oil price, such that the vegetable oils like corn and sunflower oils lower prices and decrease global palm oil demands. This will consequently lead to lower palm oil prices and disturb the growth of Malaysia's economy.

1.10 Conclusions

Palm oil production accounts for just about 10% of the industry's overall biomass. Thus, the remaining 90%, primarily in lignocellulosic form, must be commercially exploited and used to support the industry finances and the national economy. In addition to the extraction and processing of CPO, the palm oil industry also generates by-products from oil palm plantations and oil palm mills. Oil palm industry produces large amounts of oil palm by-products biomass, such as the by-product from oil palm mill comprise of POME, pressed fruit fibers, EFB, and shells, whereas the by-products from plantations consist of OPF and OPT. The most abundant oil palm biomass is left on the plantations (i.e., palm fronds) or returned to the fields as soil amendment or organic fertilizer (i.e., EFB). This biomass plays a vital role in ensuring the sustainability of plantations and preserving soil fertility. However, there is also the potential to utilize a share of this biomass for various additional end uses, including pellets, bioenergy, biofuels, and biobased chemicals, without depleting the soil.

Acknowledgments

This project was funded by Universiti Putra Malaysia through Geran Putra Berimpak (GPB), UPM/800-3/3/1/GPB/2019/9679800.

References

Adewumi, I.K., 2009. Activated carbon for water treatment in nigeria: problems and prospects. In: Appropriate Technologies for Environmental Protection in the Developing World. Springer, pp. 115–122.

Ahmaruzzaman, M., 2011. Industrial wastes as low-cost potential adsorbents for the treatment of wastewater laden with heavy metals. Adv. Colloid Interf. Sci. https://doi.org/10.1016/j.cis.2011.04.005.

Alqahtani, M.S., Alqahtani, A., Al-Thabit, A., Roni, M., Syed, R., 2019. Novel lignin nanoparticles for oral drug delivery. J. Mater. Chem. B 7 (28), 4461–4473. https://doi.org/10.1039/C9TB00594C.

Anthony, R., Xiang, Z., Runge, T., 2015. Paper coating performance of hemicellulose-rich natural polymer from distiller's grains. Prog. Org. Coat. https://doi.org/10.1016/j.porgcoat.2015.09.013.

Arellano-Sandoval, L., Delgado, E., Camacho-Villegas, T.A., Bravo-Madrigal, J., Manríquez-González, R., Lugo-Fabres, P.H., Toriz, G., García-Uriostegui, L., 2020. Development of thermosensitive hybrid hydrogels based on xylan-type hemicellulose from agave bagasse: characterization and antibacterial activity. MRS Commun. 10 (1), 147–154. https://doi.org/10.1557/mrc.2019.165.

Azeredo, H.M.C., Rosa, M.F., Mattoso, L.H.C., 2017. Nanocellulose in bio-based food packaging applications. Ind. Crop. Prod. 97, 664–671. https://doi.org/10.1016/j.indcrop.2016.03.013.

Barbosa, F.C., Silvello, M.A., Goldbeck, R., 2020. Cellulase and oxidative enzymes: new approaches, challenges and perspectives on cellulose degradation for bioethanol production. Biotechnol. Lett. 42 (6), 875–884. https://doi.org/10.1007/s10529-020-02875-4.

Baryeh, E.A., 2001. Effects of palm oil processing parameters on yield. J. Food Eng. 48 (1), 1–6. https://doi.org/10.1016/S0260-8774(00)00137-0.

Basta, A.H., El-Saied, H., 2009. Performance of improved bacterial cellulose application in the production of functional paper. J. Appl. Microbiol. 107 (6), 2098–2107. https://doi.org/10.1111/j.1365-2672.2009.04467.x.

Biskin, S., Damar, M., Oktem, S.N., Sakalli, E., Erdem, D., Pakir, O., 2016. A new graft material for myringoplasty: bacterial cellulose. Eur. Arch. Otorhinolaryngol. 273 (11), 3561–3565. https://doi.org/10.1007/s00405-016-3959-8.

Carvalheiro, F., Silva-Fernandes, T., Duarte, L.C., Gírio, F.M., 2009. Wheat straw autohydrolysis: Process optimization and products characterization. Appl. Biochem. Biotechnol. https://doi.org/10.1007/s12010-008-8448-0.

Chan, K.W., Watson, I., Lim, K.C., 1981. Use of oil palm waste material for increased production. Planter 57 (658), 14–37.

Chen, J., Fan, X., Zhang, L., Chen, X., Sun, S., Sun, R., 2020a. Research progress in lignin-based slow/controlled release fertilizer. ChemSusChem 13 (17), 4356–4366. https://doi.org/10.1002/cssc.202000455.

Chen, T., Liu, H., Dong, C., An, Y., Liu, J., Li, J., Li, X., Si, C., Zhang, M., 2020b. Synthesis and characterization of temperature/pH dual sensitive hemicellulose-based hydrogels from eucalyptus APMP waste liquor. Carbohydr. Polym. 247, 116717. https://doi.org/10.1016/j.carbpol.2020.116717.

Corley, R.H.V., Tinker, P.B., 2015. The products of the oil palm and their extraction. In: The Oil Palm., https://doi.org/10.1002/9781118953297.ch15.

Dai, L., Liu, R., Hu, L.-Q., Zou, Z.-F., Si, C.-L., 2017. Lignin nanoparticle as a novel green carrier for the efficient delivery of resveratrol. ACS Sustain. Chem. Eng. 5 (9), 8241–8249. https://doi.org/10.1021/acssuschemeng.7b01903.

Darwesh, O.M., El-Maraghy, S.H., Abdel-Rahman, H.M., Zaghloul, R.A., 2020. Improvement of paper wastes conversion to bioethanol using novel cellulose degrading fungal isolate. Fuel 262, 116518. https://doi.org/10.1016/j.fuel.2019.116518.

Dastpak, A., Lourençon, T.V., Balakshin, M., Farhan Hashmi, S., Lundström, M., Wilson, B.P., 2020. Solubility study of lignin in industrial organic solvents and investigation of electrochemical properties of spray-coated solutions. Ind. Crop. Prod. 148, 112310. https://doi.org/10.1016/j.indcrop.2020.112310.

de Araújo Padilha, C.E., da Costa Nogueira, C., de Santana Souza, D.F., de Oliveira, J.A., dos Santos, E.S., 2020. Organosolv lignin/Fe3O4 nanoparticles applied as a β-glucosidase immobilization support and adsorbent for textile dye removal. Ind. Crop. Prod. https://doi.org/10.1016/j.indcrop.2020.112167.

do Nascimento Santos, D.K.D., da Barros, B.R.S., de Aguiar, L.M.S., da Cruz Filho, I.J., de Lorena, V.M.B., de Melo, C.M.L., Napoleão, T.H., 2020. Immunostimulatory and antioxidant activities of a lignin isolated from Conocarpus erectus leaves. Int. J. Biol. Macromol. 150, 169–177. https://doi.org/10.1016/j.ijbiomac.2020.02.052.

Domínguez-Robles, J., Larrañeta, E., Fong, M.L., Martin, N.K., Irwin, N.J., Mutjé, P., Tarrés, Q., Delgado-Aguilar, M., 2020. Lignin/poly(butylene succinate) composites with antioxidant and antibacterial properties for potential biomedical applications. Int. J. Biol. Macromol. https://doi.org/10.1016/j.ijbiomac.2019.12.146.

Dungani, R., Jawaid, M., Khalil, H.P.S.A., Jasni, J., Aprilia, S., Hakeem, K.R., Hartati, S., Islam, M.N., 2013. A review on quality enhancement of oil palm trunk waste by resin impregnation: Future materials. Bioresources 8 (2), 3136–3156.

Espinoza-Acosta, J.L., Torres-Chávez, P.I., Olmedo-Martínez, J.L., Vega-Rios, A., Flores-Gallardo, S., Zaragoza-Contreras, E.A., 2018. Lignin in storage and renewable energy applications: a review. J. Energy Chem. 27 (5), 1422–1438. https://doi.org/10.1016/j.jechem.2018.02.015.

Farhat, W., Venditti, R., Quick, A., Taha, M., Mignard, N., Becquart, F., Ayoub, A., 2017. Hemicellulose extraction and characterization for applications in paper coatings and adhesives. Ind. Crop. Prod. 107, 370–377. https://doi.org/10.1016/j.indcrop.2017.05.055.

Farhat, W., Venditti, R., Ayoub, A., Prochazka, F., Fernández-de-Alba, C., Mignard, N., Taha, M., Becquart, F., 2018. Towards thermoplastic hemicellulose: chemistry and characteristics of poly-(ε-caprolactone) grafting onto hemicellulose backbones. Mater. Des. 153, 298–307. https://doi.org/10.1016/j.matdes.2018.05.013.

Fazlin, A., 2019. Reasons for price fluctuations. New Straits Time.

García-Mateos, F.J., Ruiz-Rosas, R., María Rosas, J., Morallón, E., Cazorla-Amorós, D., Rodríguez-Mirasol, J., Cordero, T., 2020. Activation of electrospun lignin-based carbon fibers and their performance as self-standing supercapacitor electrodes. Sep. Purif. Technol. https://doi.org/10.1016/j.seppur.2020.116724.

Gellner, P.E.L., Dang, A.E., Jay, H., 1993. United States Patent (19).

Geng, X., Zhang, Y., Jiao, L., Yang, L., Hamel, J., Giummarella, N., Henriksson, G., Zhang, L., Zhu, H., 2017. Bioinspired ultrastable lignin cathode via graphene reconfiguration for energy storage. ACS Sustain. Chem. Eng. 5 (4), 3553–3561. https://doi.org/10.1021/acssuschemeng.7b00322.

Guo, J., Chen, X., Wang, J., He, Y., Xie, H., Zheng, Q., 2020. The influence of compatibility on the structure and properties of PLA/lignin biocomposites by chemical modification. Polymers. https://doi.org/10.3390/polym12010056.

Hamzah, N., Tokimatsu, K., Yoshikawa, K., 2019. Solid fuel from oil palm biomass residues and municipal solid waste by hydrothermal treatment for electrical power generation in Malaysia: a review. Sustainability (Switzerland) 11 (4), 1–23. https://doi.org/10.3390/su11041060.

Harussani, M.M., Sapuan, S.M., 2021. Development of Kenaf Biochar in Engineering and Agricultural Applications. Chemistry Africa, 1–17. https://doi.org/10.1007/s42250-021-00293-1.

Harussani, M.M., Sapuan, S.M., Umer, Rashid, Khalina, A., Ilyas, R.A., 2022. Pyrolysis of polypropylene plastic waste into carbonaceous char: Priority of plastic waste management amidst COVID-19 pandemic. Science of The Total Environment 803, 149911. https://doi.org/10.1016/j.scitotenv.2021.149911.

Harussani, M.M., Umer, Rashid, Sapuan, S.M., Khalina, A., 2021. Low-Temperature Thermal Degradation of Disinfected COVID-19 Non-Woven Polypropylene—Based Isolation Gown Wastes into Carbonaceous Char. Polymers 13 (22), 3980. https://doi.org/10.3390/polym13223980.

Hu, N., Chen, D., Guan, Q., Peng, L., Zhang, J., He, L., Shi, Y., 2020. Preparation of hemicellulose-based hydrogels from biomass refining industrial effluent for effective removal of methylene blue dye. Environ. Technol., 1–11. https://doi.org/10.1080/09593330.2020.1795930.

Iberahim, N.I., Jahim, J.M., Harun, S., Nor, M.T.M., Hassan, O., 2013. Sodium hydroxide pretreatment and enzymatic hydrolysis of oil palm mesocarp fiber. Int. J. Chem. Eng. Appl. 4 (3), 101.

Ilyas, R.A., Sapuan, S.M., Ishak, M.R., Zainudin, E.S., 2017. Effect of delignification on the physical, thermal, chemical, and structural properties of sugar palm fibre. Bioresources 12 (4), 8734–8754. https://doi.org/10.15376/biores.12.4.8734-8754.

Ilyas, R.A., Sapuan, S.M., Ishak, M.R., Zainudin, E.S., 2018a. Water transport properties of bio-nanocomposites reinforced by sugar palm (Arenga Pinnata) nanofibrillated cellulose. J. Adv. Res. Fluid Mech. Therm. Sci. 51 (2), 234–246.

Ilyas, R.A., Sapuan, S.M., Ishak, M.R., 2018b. Isolation and characterization of nanocrystalline cellulose from sugar palm fibres (Arenga Pinnata). Carbohydr. Polym. 181, 1038–1051. https://doi.org/10.1016/j.carbpol.2017.11.045.

Ilyas, R.A., Sapuan, S.M., Ishak, M.R., Zainudin, E.S., 2018c. Development and characterization of sugar palm nanocrystalline cellulose reinforced sugar palm starch bionanocomposites. Carbohydr. Polym. 202, 186–202. https://doi.org/10.1016/j.carbpol.2018.09.002.

Ilyas, R.A., Sapuan, S.M., Ishak, M.R., Zainudin, E.S., Atikah, M.S.N., 2018d. Characterization of sugar palm nanocellulose and its potential for reinforcement with a starch-based composite. In: Sugar Palm Biofibers, Biopolymers, and Biocomposites. CRC Press, pp. 189–220, https://doi.org/10.1201/9780429443923-10.

Ilyas, R.A., Sapuan, S.M., Sanyang, M.L., Ishak, M.R., Zainudin, E.S., 2018e. Nanocrystalline cellulose as reinforcement for polymeric matrix nanocomposites and its potential applications: a review. Curr. Anal. Chem. 14 (3), 203–225. https://doi.org/10.2174/1573411013666171003155624.

Ilyas, R.A., Sapuan, S.M., Ishak, M.R., Zainudin, E.S., 2018f. Sugar palm nanocrystalline cellulose reinforced sugar palm starch composite: degradation and water-barrier properties. IOP Conf. Ser.: Mater. Sci. Eng. 368. https://doi.org/10.1088/1757-899X/368/1/012006, 012006.

Ilyas, R.A., Sapuan, S.M., Ibrahim, R., Abral, H., Ishak, M.R., Zainudin, E.S., Atikah, M.S.N., Mohd Nurazzi, N., Atiqah, A., Ansari, M.N.M., Syafri, E., Asrofi, M., Sari, N.H., Jumaidin, R., 2019a. Effect of sugar palm nanofibrillated celluloseconcentrations on morphological, mechanical andphysical properties of biodegradable films basedon agro-waste sugar palm (Arenga pinnata(Wurmb.) Merr) starch. J. Mater. Res. Technol. 8 (5), 4819–4830. https://doi.org/10.1016/j.jmrt.2019.08.028.

Ilyas, R.A., Sapuan, S.M., Ibrahim, R., Abral, H., Ishak, M.R., Zainudin, E.S., Atikah, M.S.N., Mohd Nurazzi, N., Atiqah, A., Ansari, M.N.M., Syafri, E., Asrofi, M., Sari, N.H., Jumaidin, R., 2019b. Effect of sugar palm nanofibrillated cellulose concentrations on morphological, mechanical and physical properties of biodegradable films based on agro-waste sugar palm (Arenga pinnata (Wurmb.) Merr) starch. J. Mater. Res. Technol. 8 (5), 4819–4830. https://doi.org/10.1016/j.jmrt.2019.08.028.

Ilyas, R.A., Sapuan, S.M., Ishak, M.R., Zainudin, E.S., 2019c. Sugar palm nanofibrillated cellulose (Arenga pinnata (Wurmb.) Merr): effect of cycles on their yield, physic-chemical, morphological and thermal behavior. Int. J. Biol. Macromol. 123. https://doi.org/10.1016/j.ijbiomac.2018.11.124.

Ilyas, R.A., Sapuan, S.M., Ibrahim, R., Abral, H., Ishak, M.R., Zainudin, E.S., Asrofi, M., Atikah, M.S.N., Huzaifah, M.R.M., Radzi, A.M., Azammi, A.M.N., Shaharuzaman, M.A., Nurazzi, N.M., Syafri, E., Sari, N.H., Norrrahim, M.N.F., Jumaidin, R., 2019d. Sugar palm (Arenga pinnata (Wurmb.) Merr) cellulosic fibre hierarchy: a comprehensive approach from macro to nano scale. J. Mater. Res. Technol. 8 (3), 2753–2766. https://doi.org/10.1016/j.jmrt.2019.04.011.

Ilyas, R.A., Sapuan, S.M., Atiqah, A., Ibrahim, R., Abral, H., Ishak, M.R., Zainudin, E.S., Nurazzi, N.M., Atikah, M.S.N., Ansari, M.N.M., Asyraf, M.R.M., Supian, A.B.M., Ya, H., 2020a. Sugar palm (Arenga pinnata [Wurmb.] Merr) starch films containing sugar palm nanofibrillated cellulose as reinforcement: Water barrier properties. Polym. Compos. 41 (2), 459–467. https://doi.org/10.1002/pc.25379.

Ilyas, R.A., Sapuan, S.M., Atikah, M.S.N., Asyraf, M.R.M., Rafiqah, S.A., Aisyah, H.A., Nurazzi, N.M., Norrrahim, M.N.F., 2020b. Effect of hydrolysis time on the morphological, physical, chemical, and thermal behavior of sugar palm nanocrystalline cellulose (Arenga pinnata (Wurmb.) Merr). Text. Res. J. https://doi.org/10.1177/0040517520932393. 004051752093239.

Ilyas, R.A., Sapuan, S.M., Ibrahim, R., Abral, H., Ishak, M.R., Zainudin, E.S., Atiqah, A., Atikah, M.S.N., Syafri, E., Asrofi, M., Jumaidin, R., 2020c. Thermal, biodegradability and water barrier properties of bio-nanocomposites based on plasticised sugar palm starch and nanofibrillated celluloses from sugar palm fibres. J. Biobaased Mater. Bioenergy 14 (2), 234–248. https://doi.org/10.1166/jbmb.2020.1951.

Jin, L., Sun, Q., Xu, Q., Xu, Y., 2015. Adsorptive removal of anionic dyes from aqueous solutions using microgel based on nanocellulose and polyvinylamine. Bioresour. Technol. 197, 348–355. https://doi.org/10.1016/j.biortech.2015.08.093.

Kamarudin, H., Mohd, H., Ariffin, D., Jalani, S., 1997. An estimated availability of oil palm biomass in Malaysia. In: PORIM Occasional Paper 37. Palm Oil Research Institute of Malaysia, Kuala Lumpur.

Kamyab, H., Chelliapan, S., Din, M.F.M., Rezania, S., Khademi, T., Kumar, A., 2018. Palm oil mill effluent as an environmental pollutant. Palm Oil 13.

Kanomata, K., Fukuda, N., Miyata, T., Lam, P.Y., Takano, T., Tobimatsu, Y., Kitaoka, T., 2020. Lignin-inspired surface modification of nanocellulose by enzyme-catalyzed radical coupling of coniferyl alcohol in pickering emulsion. ACS Sustain. Chem. Eng. 8 (2), 1185–1194. https://doi.org/10.1021/acssuschemeng.9b06291.

Kardam, A., Rohit, K., 2014. Nanocellulose fibers for biosorption of cadmium, nickel, and lead ions from aqueous solution. Clean Technol. Environ. Policy, 385–393. https://doi.org/10.1007/s10098-013-0634-2.

Karim, Z., Claudpierre, S., Grahn, M., Oksman, K., Mathew, A.P., 2016. Nanocellulose based functional membranes for water cleaning: tailoring of mechanical properties, porosity and metal ion capture. J. Membr. Sci. 514, 418–428. https://doi.org/10.1016/j.memsci.2016.05.018.

Killmann, W., Lim, S.C., 1985. Anatomy and properties of oil palm stem. In: Proc. of the National Symposium on Oil Palm By-Products for Agro-Based Industries, Kuala Lumpur, pp. 18–42.

Killmann, W., Lim, S.C., 1985. Anatomy and properties of oil palm stem. In: Proceedings of the National Symposium on Oil Palm By-Products for Agro-Based Industries. Institiut Penyelidikan Minyak Kelapa Sawit Malaysia, Kuala Lumpur, pp. 18–42.

Ko, C.-H., Yang, B.-Y., Lin, L.-D., Chang, F.-C., Chen, W.-H., 2020. Impact of pretreatment methods on production of bioethanol and nanocrystalline cellulose. J. Clean. Prod. 254, 119914. https://doi.org/10.1016/j.jclepro.2019.119914.

Kome, G.K., Tabi, F.O., 2020. Towards sustainable oil palm plantation management: effects of plantation age and soil parent material. Agric. Sci. 11 (01), 54–70. https://doi.org/10.4236/as.2020.111004.

Korhonen, J.T., Kettunen, M., Ras, R.H.A., Ikkala, O., 2011. Hydrophobic nanocellulose aerogels as floating, sustainable, reusable, and recyclable oil absorbents. ACS Appl. Mater. Interfaces 3 (6), 1813–1816.

Lang, N., Merkel, E., Fuchs, F., Schumann, D., Klemm, D., Kramer, F., Mayer-Wagner, S., Schroeder, C., Freudenthal, F., Netz, H., Kozlik-Feldmann, R., Sigler, M., 2014. Bacterial nanocellulose as a new patch material for closure of ventricular septal defects in a pig model. Eur. J. Cardiothorac. Surg. 47 (6), 1013–1021. https://doi.org/10.1093/ejcts/ezu292.

Larrañeta, E., Imízcoz, M., Toh, J.X., Irwin, N.J., Ripolin, A., Perminova, A., Domínguez-Robles, J., Rodríguez, A., Donnelly, R.F., 2018. Synthesis and characterization of lignin hydrogels for potential applications as drug eluting antimicrobial coatings for medical materials. ACS Sustain. Chem. Eng. https://doi.org/10.1021/acssuschemeng.8b01371.

Lawal, A.A., Hassan, M.A., Farid, M.A.A., Yasim-Anuar, T.A.T., Yusoff, M.Z.M., Zakaria, M.R., Roslan, A.M., Mokhtar, M.N., Shirai, Y., 2020. Production of biochar from oil palm frond by steam pyrolysis for removal of residual contaminants in palm oil mill effluent final discharge. J. Clean. Prod. https://doi.org/10.1016/j.jclepro.2020.121643.

Lei, Y., Alshareef, A.H., Zhao, W., Inal, S., 2020. Laser-scribed graphene electrodes derived from lignin for biochemical sensing. ACS Appl. Nano Mater. https://doi.org/10.1021/acsanm.9b01795.

Liu, S., Chen, F., Song, X., Wu, H., 2017. Preparation and characterization of temperature- and pH-sensitive hemicellulose-containing hydrogels. Int. J. Polym. Anal. Charact. 22 (3), 187–201. https://doi.org/10.1080/1023666X.2016.1276257.

Liu, H., Chen, T., Dong, C., Pan, X., 2020. Biomedical applications of hemicellulose-based hydrogels. Curr. Med. Chem. 27 (28), 4647–4659. https://doi.org/10.2174/09298673 27666200408115817.

Lou, Z.C., 2016. A better design is needed for clinical studies of chronic tympanic membrane perforations using biological materials. Eur. Arch. Otorhinolaryngol. 273 (11), 4045–4046. https://doi.org/10.1007/s00405-016-4019-0.

Ma, A.N., Cheah, S.C., Chow, M.C., Yeoh, B., 1993. Current status of palm oil processing wastes management. In: Waste Management in Malaysia: Current Status and Prospects for Bioremediation, pp. 111–136.

Mad Saad, M.E., Farah Amani, A.H., Lee, C.K., 2016. Optimization of bioethanol production process using oil palm frond juice as substrate. Malays. J. Microbiol. https://doi.org/10.21161/mjm.84016.

Mokhtar, A., Hassan, K., Aziz, A.A., May, C.Y., 2012. Oil palm biomass for various wood-based products. In: Lai, O.-M., Tan, C.-P., Akoh, C.C. (Eds.), Palm Oil: Production, Processing, Characterization, and Uses, first ed. Elsevier, pp. 625–652, https://doi.org/10.1016/B978-0-9818936-9-3.50024-1.

MPOB, 2020a. Monthly Production of Oil Palm Products Summary for the Month of September 2020 2019 & 2020 (Tonnes).

MPOB, 2020b. Mpob Daily Malaysia Prices of Crude Palm Oil (RM/Tonne) 2020.

MPOC, 2020. Daily Palm Oil Price.

Mugwagwa, L.R., Chimphango, A.F.A., 2020. Enhancing the functional properties of acetylated hemicellulose films for active food packaging using acetylated nanocellulose reinforcement and polycaprolactone coating. Food Packag. Shelf Life. https://doi.org/10.1016/j.fpsl.2020.100481.

Navarro-Suárez, A.M., Saurel, D., Sánchez-Fontecoba, P., Castillo-Martínez, E., Carretero-González, J., Rojo, T., 2018. Temperature effect on the synthesis of lignin-derived carbons for electrochemical energy storage applications. J. Power Sources 397, 296–306. https://doi.org/10.1016/j.jpowsour.2018.07.023.

Navia, R., 2020. Lignin valorization into biocarbon materials. Waste Manag. Res. 38 (2), 109–110. https://doi.org/10.1177/0734242X20902892.

Nurhidayat, Y., 2014. Modeling and Simulation of Palm Oil Fruit Digestion Process Master Of Science. Universiti Putra Malaysia.

Obibuzor, J.U., Okogbenin, E.A., Abigor, R.D., 2012. Oil recovery from palm fruits and palm kernel. In: Palm Oil: Production, Processing, Characterization, and Uses. AOCS Press, https://doi.org/10.1016/B978-0-9818936-9-3.50014-9.

Obisesan, I.O., 2004. Yield, the Ultimate in Crop Improvement.

Okoroigwe, E.C., Saffron, C.M., Kamdem, P.D., 2014. Characterization of palm kernel shell for materials reinforcement and water treatment. J. Chem. Eng. Mater. Sci. 5 (1), 1–6.

Olusunmade, O.F., Adetan, D.A., Ogunnigbo, C.O., 2016. A study on the mechanical properties of oil palm mesocarp fibre-reinforced thermoplastic. J. Compos. 2016.

Paltakari, J., Jin, H., Kettunen, M., Laiho, A., Pynn, H., Marmur, A., Ikkala, O., Ras, R.H.A., 2011. Superhydrophobic and superoleophobic nanocellulose aerogel membranes as bioinspired cargo carriers on water and oil. Langmuir 27 (5), 1930–1934. https://doi.org/10.1021/la103877r.

Pandian, B., Arunachalam, R., Easwaramoorthi, S., Rao, J.R., 2020. Tuning of renewable biomass lignin into high value-added product: development of light resistant azo-lignin colorant for coating application. J. Clean. Prod. 256, 120455. https://doi.org/10.1016/j.jclepro.2020.120455.

Paul, M., Panda, G., Mohapatra, P., Das, K., Thatoi, H., 2020. Study of structural and molecular interaction for the catalytic activity of cellulases: an insight in cellulose hydrolysis for higher bioethanol yield. J. Mol. Struct. 1204, 127547. https://doi.org/10.1016/j.molstruc.2019.127547.

Rivière, G.N., Korpi, A., Sipponen, M.H., Zou, T., Kostiainen, M.A., Österberg, M., 2020. Agglomeration of viruses by cationic lignin particles for facilitated water purification. ACS Sustain. Chem. Eng. https://doi.org/10.1021/acssuschemeng.9b06915.

Salleh, S.F., Saree, S., Sanaullah, K., Khan, A., Yunus, R., Sawawi, M., Rajaee, N., Fadzli, N.A.M., Alhaji, M.H., 2019. Mathematical modelling and & simulation of thresher operation in palm oil mill. J. Sustain. Sci. Manag. 14 (5), 43–54.

Sameni, J., Jaffer, S.A., Tjong, J., Sain, M., 2020. Advanced applications for lignin micro- and nano-based materials. Curr. For. Rep. 6 (2), 159–171. https://doi.org/10.1007/s40725-020-00117-4.

Shamaei, L., Khorshidi, B., Islam, M.A., Sadrzadeh, M., 2020. Industrial waste lignin as an antifouling coating for the treatment of oily wastewater: creating wealth from waste. J. Clean. Prod. 256, 120304. https://doi.org/10.1016/j.jclepro.2020.120304.

Shen, Y., Li, Y., Yang, G., Zhang, Q., Liang, H., Peng, F., 2020. Lignin derived multi-doped (N, S, Cl) carbon materials as excellent electrocatalyst for oxygen reduction reaction in proton exchange membrane fuel cells. J. Energy Chem. 44, 106–114. https://doi.org/10.1016/j.jechem.2019.09.019.

Sivasothy, K., Halim, R.M., Basiron, Y., 2005. A new system for continuous sterilization of oil palm fresh fruit bunches. J. Oil Palm Res. 17 (December), 145–151.

Sivasothy, K., Basiron, Y., Suki, A., Mohd Taha, R., Tan, Y.H., Sulong, M., 2006. Continuous sterilization: the new paradigm for modernizing palm oil milling. J. Oil Palm Res., 144–152.

Song, F., Wei, Q., Ma, H., Liu, X., Wu, S., 2019. Preparation of carboxymethyl hemicellulose and its application with polyamide epichlorohydrin resin as wet strength additive. Chung-Kuo Tsao Chih/China Pulp and Paper. https://doi.org/10.11980/j.issn.0254-508X.2019.03.002.

Stewart, H., Golding, M., Matia-Merino, L., Archer, R., Davies, C., 2014. Manufacture of lignin microparticles by anti-solvent precipitation: Effect of preparation temperature and presence of sodium dodecyl sulfate. Food Res. Int. https://doi.org/10.1016/j.foodres.2014.08.046.

Stork Amsterdam, 1960. Stork Review., https://doi.org/10.1017/S0021911813002374.

Subramaniam, V., Chow, M.C., Ma, A.N., 2004. Energy database of the oil palm. In: Palm Oil Engineering Bulletin, Vol EB70. Malaysian Palm Oil Board (MPOB), Kajang.

Sun, X.-F., Wang, H., Jing, Z., Mohanathas, R., 2013. Hemicellulose-based pH-sensitive and biodegradable hydrogel for controlled drug delivery. Carbohydr. Polym. 92 (2), 1357–1366. https://doi.org/10.1016/j.carbpol.2012.10.032.

Sundram, K., Sambanthamurthi, R., Tan, Y.-A., 2003. Palm fruit chemistry and nutrition. Asia Pac. J. Clin. Nutr. 12 (3).

Suopajärvi, T., Liimatainen, H., Karjalainen, M., Upola, H., Niinimäki, J., 2015. Journal of water process engineering lead adsorption with sulfonated wheat pulp nanocelluloses. J. Water Process. Eng. 5, 136–142. https://doi.org/10.1016/j.jwpe.2014.06.003.

Taiwo, O.F.A.W., Alkarkhi, A.F.M., Ghazali, A., Wan Daud, W., 2017. Optimization of the strength properties of waste oil palm (Elaeis Guineensis) fronds fiber. J. Nat. Fibers, 1–13. https://doi.org/10.1080/15440478.2016.1215947.

Tan, C.H., Ghazali, H.M., Kuntom, A., Tan, C.P., Ariffin, A.A., 2009a. Extraction and physicochemical properties of low free fatty acid crude palm oil. Food Chem. 113 (2), 645–650. https://doi.org/10.1016/j.foodchem.2008.07.052.

Tan, K.T., Lee, K.T., Mohamed, A.R., Bhatia, S., 2009b. Palm oil: addressing issues and towards sustainable development. Renew. Sust. Energ. Rev. 13 (2), 420–427. https://doi.org/10.1016/j.rser.2007.10.001.

Tao, Z., Sun, X., 2020. Hemicellulose/poly(acrylic acid) semi-IPN magnetic nanocomposite hydrogel for lysozyme adsorption. Int. J. Nanomater. Nanotechnol. Nanomed. 6 (1), 001–005. https://doi.org/10.17352/2455-3492.000032.

Tao, J., Li, S., Ye, F., Zhou, Y., Lei, L., Zhao, G., 2020. Lignin—an underutilized, renewable and valuable material for food industry. Crit. Rev. Food Sci. Nutr. 60 (12), 2011–2033. https://doi.org/10.1080/10408398.2019.1625025.

Tatar, F., Cengiz, A., Kahyaoglu, T., 2014a. Effect of hemicellulose as a coating material on water sorption thermodynamics of the microencapsulated fish oil and artificial neural network (ANN) modeling of isotherms. Food Bioprocess Technol. 7 (10), 2793–2802. https://doi.org/10.1007/s11947-014-1291-0.

Tatar, F., Tunç, M.T., Dervisoglu, M., Cekmecelioglu, D., Kahyaoglu, T., 2014b. Evaluation of hemicellulose as a coating material with gum arabic for food microencapsulation. Food Res. Int. https://doi.org/10.1016/j.foodres.2014.01.022.

Ten, E., Ling, C., Wang, Y., Srivastava, A., Dempere, L.A., Vermerris, W., 2014. Lignin nanotubes as vehicles for gene delivery into human cells. Biomacromolecules. https://doi.org/10.1021/bm401555p.

Then, Y.Y., Ibrahim, N.A., Zainuddin, N., Ariffin, H., Wan Yunus, W.M.Z., 2013. Oil palm mesocarp fiber as new lignocellulosic material for fabrication of polymer/fiber biocomposites. Int. J. Polym. Sci. 2013.

Thongjun, N., Jarupan, L., Pechyen, C., 2012. Efficacy of activated carbon *in situ* to oil palm frond paper for active packaging on mechanical properties. Adv. Mater. Res. 506, 607–610. https://doi.org/10.4028/www.scientific.net/AMR.506.607.

Tomkins, T., Drackley, J.K., 2010. Applications of palm oil in animal nutrition. J. Oil Palm Res. 23, 1029–1035.

Trevisan, H., Rezende, C.A., 2020. Pure, stable and highly antioxidant lignin nanoparticles from elephant grass. Ind. Crop. Prod. https://doi.org/10.1016/j.indcrop.2020.112105.

Ubeyitogullari, A., Cekmecelioglu, D., 2016. Optimization of hemicellulose coating as applied to apricot drying and comparison with chitosan coating and sulfite treatment. J. Food Process Eng. https://doi.org/10.1111/jfpe.12247.

Wang, X., Hu, J., Liang, Y., Zeng, J., 2012. Tcf bleaching character of soda-anthraquinone pulp from oil palm frond. Bioresources. https://doi.org/10.15376/biores.7.1.0275-0282.

Wang, R., Guan, S., Sato, A., Wang, X., Wang, Z., Yang, R., Hsiao, B.S., Chu, B., 2013. Nanofibrous micro filtration membranes capable of removing bacteria, viruses and heavy metal ions. J. Membr. Sci. 446, 376–382. https://doi.org/10.1016/j.memsci.2013.06.020.

Wei, J., Wang, B., Li, Z., Wu, Z., Zhang, M., Sheng, N., Liang, Q., Wang, H., Chen, S., 2020. A 3D-printable TEMPO-oxidized bacterial cellulose/alginate hydrogel with enhanced stability via nanoclay incorporation. Carbohydr. Polym. https://doi.org/10.1016/j.carbpol.2020.116207.

Wondi, M.H., Shamsudin, R., Yunus, R., Alsultan, G.A., Iswardi, A.H., et al., 2020. Centrifugal separation-assisted and extraction of crude palm oil from separated mesocarp fiber: Central composite design optimization. J. Food Process Eng. 43 (7), 1–13. https://doi.org/10.1111/jfpe.13426. e13426 In press.

Wondi, M.H, Shamsudin, R., Yunus, R., Arnan, M.Z., Baharudin, M.S., Abdu Rahman, A.F., 2021. Physical and Mechanical Properties of Sterilized Oil Palm Fruits at Different Component. AMA Agric. Mech. Asia Afr. Lat. Am. 52 (3), 3985. 3977 In press.

Wong, H.K., Wan Zahari, M., 2011. Utilisation of oil palm by-products as ruminant feed in malaysia. J. Oil Palm Res. 23 (August 2011), 1029–1035.

Yeboah, M.L., Li, X., Zhou, S., 2020. Facile fabrication of biochar from palm kernel shell waste and its novel application to magnesium-based materials for hydrogen storage. Materials 13 (3), 625.

Yılmaz Celebioglu, H., Cekmecelioglu, D., Dervisoglu, M., Kahyaoglu, T., 2012. Effect of extraction conditions on hemicellulose yields and optimisation for industrial processes. Int. J. Food Sci. Technol. 47 (12), 2597–2605. https://doi.org/10.1111/j.1365-2621.2012.03139.x.

Younesi-Kordkheili, H., Pizzi, A., 2020. Ionic liquid- modified lignin as a bio- coupling agent for natural fiber- recycled polypropylene composites. Compos. Part B. https://doi.org/10.1016/j.compositesb.2019.107587.

Zamri-Saad, M., Azhar, K., 2015. Issues of ruminant integration with oil palm plantation. J. Oil Palm Res. 27 (4), 299–305.

Zhang, C., Nair, S.S., Chen, H., Yan, N., Farnood, R., Li, F., 2020. Thermally stable, enhanced water barrier, high strength starch bio-composite reinforced with lignin containing cellulose nanofibrils. Carbohydr. Polym. 230, 115626. https://doi.org/10.1016/j.carbpol.2019.115626.

2

Basic properties of oil palm biomass (OPB)

Syaiful Osman[a], Zawawi Ibrahim[b], Aisyah Humaira Alias[c], Noorshamsiana Abdul Wahab[b], Ridzuan Ramli[b], Fazliana Abdul Hamid[b], and Mansur Ahmad[a]

[a]*Faculty of Applied Sciences, Universiti Teknologi MARA, Shah Alam, Selangor, Malaysia,* [b]*Malaysian Palm Oil Board (MPOB), Persiaran Institusi, Kajang, Selangor, Malaysia,* [c]*Institute of Tropical Forestry and Forest Products (INTROP), Universiti Putra Malaysia, Serdang, Selangor, Malaysia*

2.1   Introduction

Oil palm is one of the most profitable perennial oil crops in tropical regions because of its important oil-producing fruits. The oil palm tree was first introduced to Southeast Asia in 1848 and to British Malaya (Malaysia) in early 1870. For many reasons, the planted oil palm area has increased in the last 30 years (Nambiappan et al., 2018). The key reason for the rapid growth of oil palm plantations is the rise in oil and fat consumption in developing economies such as China and India. Therefore, Malaysia has been identified as a suitable land for the establishment of oil palm plantations (Ratnasingam et al., 2008). The oil palm belongs to the *Elaeis guineensis* genus of the Palmaceae family and is monocotyledonous. Oil palm trees are planted for oil production, and the oil is used in food, fuel, cleansing agents, and cosmetic products.

In Malaysia, oil palm plantations cover approximately 5.9 million hectares (MPOB, 2020) and is the largest planting area. In 2020, Malaysia was the second largest producer of oil palm supply in the world, with about 19,300 million tons, after Indonesia that produces about 43,500 million tons. Both countries account for about 84% of the world's palm oil production (USDA, 2020). Malaysia generated more than 19 million tons of oil and 80 million tons of biomass, with just around 10% of the biomass extracted for the processing of oil and 90% used for waste (Abdullah and Sulaiman, 2013; Chin et al., 2013; Zwart, 2013; Umar et al., 2013). Abdul Khalil et al. (2011) stated that oil palm plantations produce more than 90 million tons of waste annually during the replanting phase. In addition, these lignocellulosic wastes contain about 50% cellulose, 25% hemicellulose, and 25% lignin in

Oil Palm Biomass for Composite Panels. https://doi.org/10.1016/B978-0-12-823852-3.00007-6
Copyright © 2022 Elsevier Inc. All rights reserved.

their cell walls and generate an abundant amount of biomass per year in the form of oil palm trunk (OPT), oil palm frond (OPF), pressed mesocarp fiber (PMF), palm oil mill effluent (POME), empty fruit bunch (EFB), and leaves as shown in Fig. 2.1.

From the total generated oil palm biomass (OPB) during harvesting and replantation, 70% is from OPF, 10% is from EFB, while OPT is only 5% . Based on research by Abdul Khalil et al. (2010a, b), the oil palm industries in Malaysia generated about 12.4 million tons/year of EFB and 24 million tons/year of pruned OPF. In addition, more than 30 million tons of total oil palms worldwide are often classified as agricultural sources, although approximately 8 million tons are EFB materials (Suradi et al., 2010; Ratnam et al., 2008). Thus, agricultural wastes are produced in huge quantities and are becoming an alternative raw material for fiber production. These raw materials play an important role in fiber production because of a lack of wood as a major raw material. The shortages are due to several reasons, including deforestation, forest degradation, and increasing demand for wood-based products. The advantages of using these materials include low cost, light in weight, low density, recyclability, and very high strength-to-weight ratio (Hill and Abdul Khalil, 2000).

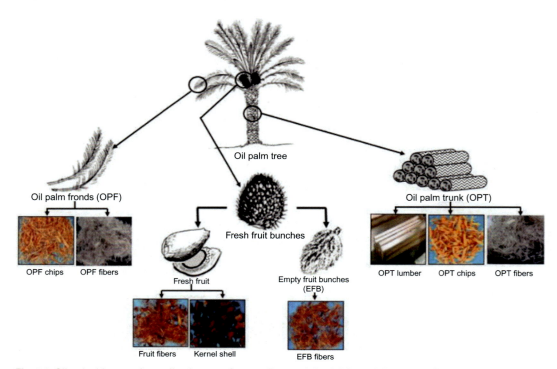

Fig. 2.1 Oil palm biomass from oil palm tree. Source: Dungani, R., Aditiawati, P., Aprilia, S., Yuniarti, K., Karliati, T., Suwandhi, I., Sumardi, I., 2018. Biomaterial from oil palm waste: properties, characterization and applications. Palm Oil 31. https://doi.org/10.5772/intechopen.76412.

2.2 Physical and anatomical properties

EFB is one of the main waste materials generated from the oil palm industry. The fresh fruit bunches (FFB) are transported to processing facilities after harvesting. Hundreds of fruits in the FFB produce a nut covered by a light orange pericarp of palm oil (Fig. 2.2A). The FFB contains around 21% palm oil, 6%–7% palm kernel, 14%–15% fiber, 6%–7% shell, and 23% EFB (Umikalsom et al., 1997). The FFB is sterilized by steaming at a high temperature, then stripped using a drum stripper to detach the fruit from the bunches. The oil is extracted from the mesocarp via a pressing process, which produces as crude palm oil (CPO).

The separation process of FFB produces EFB as a fibrous mass left behind from oil palm processing (Fig. 2.2B). The mass balance of various types of products produced from a palm oil mill is shown in Fig. 2.3.

According to Karina et al. (2008), for every ton of oil made, the palm oil industry discards around 1.1 tons of EFB. More than 70% of the processed FFB is residual oil palm waste produced during the palm oil mill processing. A significant environmental challenge, such as fouling and pest attraction caused by an excess of EFB, is known as waste and is usually used for mulching or burning, which causes environmental pollution.

Several researchers have reported that EFB fiber is hard, tough, and has a central part called lacuna (Figs. 2.4A and B). According to Sreekala et al. (1997), the presence of lacuna creates porous morphology in the EFB fibers and provides a better mechanical network with matrix resin for composite fabrication. However, the main disadvantage of EFB is that the shredded EFB still contains oil and hydrophilic properties of the cell (Abdul Khalil et al., 2007). In addition, significant amounts of bodies of silica, as shown in Fig. 2.5, are bound on the EFB fiber surface (Ibrahim et al., 2015; Sreekala et al., 1997; Law et al., 2007). This may cause difficulties in pulping and bleaching the EFB

(a)   (b)

Fig. 2.2 (A) Fresh fruit bunch (FFB) and (B) empty fruit bunch (EFB). Source: Mandang, T., Sinambela, R., Pandianuraga, N.R., 2018. Physical and mechanical characteristics of oil palm leaf and fruits bunch stalks for bio-mulching. IOP Conf. Ser. Earth Environ. Sci. 196 (1), 012015. https://10.1088/1755-1315/196/1/012015.

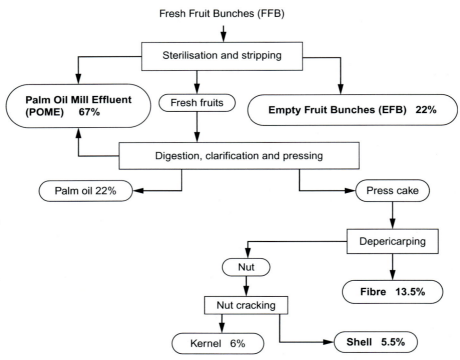

Fig. 2.3 Solid waste materials and by-products generated from fresh fruit bunches (FFB) in the palm oil extraction process. Source: Abdullah, N., Sulaiman, F. 2013. The oil palm wastes in Malaysia. Biomass Now Sustain. Growth and Use 1(3), 75–93. https://doi.org/10.5772/55302.

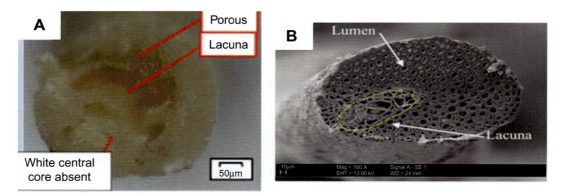

Fig. 2.4 (A) Image and (B) microscopic image of lacuna in empty fruit bunch (EFB) fiber. Source: Tan, Z.E., Liew, C.K., Yee, F.C., Talamona, D., Goh, K.L, 2017. Oil palm empty fruit bunch fibres and biopolymer composites: possible effects of moisture on the elasticity, fracture properties and reliability. Green Biocomposites 271–291. https://doi.org/10.1007/978-3-319-46610-1_12 and Anuar, N.I., Saffinaz, S.Z., Gan, S., Chia, C.H., Wang, C., Harun, J., 2019. Comparison of the morphological and mechanical properties of oil palm EFB fibres and kenaf fibres in nonwoven reinforced composites. Ind. Crop Prod. 127, 55–65. https://doi.org/10.1016/j.indcrop.2018.09.056.

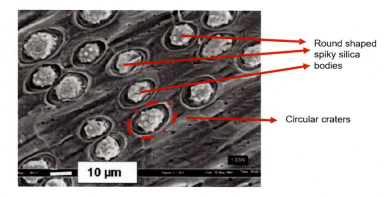

Fig. 2.5 Surface of EFB with silica bodies. Source: Law, K.-N., Daud, W.R.W., Ghazali, A., 2007. Morphological and chemical nature of fiber strands of oil palm empty-fruit-bunch (OPEFB). Bioresources 2 (3), 351–362.

fibers, as inorganic metals and substances may react with the chemicals used in the treatment process, which may reduce the quality of the fibers. Silica bodies are connected to spherical craters distributed thinly across the surface. The spiky silica bodies are round-shaped and are 10–15 μm in diameter. Because the silica bodies are tough, they can be mechanically removed by hammering the fibers.

The oil palm tree is a single-stemmed, non-wood tree with a trunk diameter of 45–65 cm, measuring 1.5 m above ground level and about 7–13 m in height (Abdul Khalil et al., 2010a, b). The oil palm tree has a life span of between 25 and 30 years and contains many residues. It is a monocotyledonous species and mainly consists of vascular bundles and parenchyma cells, which is different from wood's anatomical structure, as shown in Fig. 2.6.

The OPT is divided into two distinct areas: the main and the cortex area, as displayed in Fig. 2.7. A thin cortex protects the outer part of the OPT with a thickness of about 1.5–3.5 cm, which contains ground

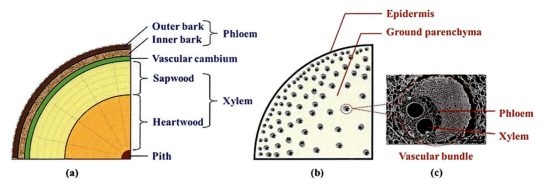

Fig. 2.6 Comparison of cross-section structure of (A) wood trunk, (B) oil palm trunk, and (C) oil palm trunk (OPT) vascular bundle. Source: Nguyen, V.D., Harifara, R., Shiro, S., 2016. Sap from various palms as a renewable energy source for bioethanol production. Chem. Ind. Chem. Eng. Q. 22 (4), 355–373. https://doi.org/10.2298/CICEQ160420024N.

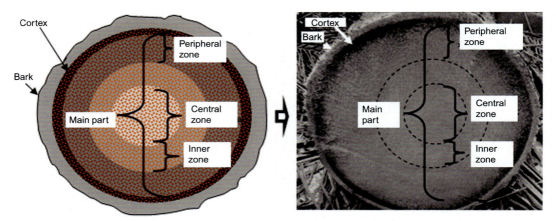

Fig. 2.7 Graphical illustration of oil palm trunk (OPT) actual cross-section. Source: Dungani, R., Mohammad Jawaid, H.P.S., Khalil, A., Jasni, J., Aprilia, S., Hakeem, K.R., Hartati, S., Islam, M.N., 2013. A review on quality enhancement of oil palm trunk waste by resin impregnation: future materials. Bioresources 8 (2), 3136–3156.

tissue parenchyma with multiple longitudinal fibrous strands with narrow and irregularly arranged fibrous strands and vascular bundles (Lim and Khoo, 1986). The main area can be separated into three distinct regions or zones: peripheral, central, and inner (Killmann and Lim, 1985). The periphery region comprises thin layers of parenchyma tissue and densely packed vascular bundles, which create the growth of the sclerotic zone and give the OPT its main mechanical support (Sulaiman et al., 2012).

The central zone is covered by 80% of the total area of the trunk, mainly composed of vascular bundles. In Fig. 2.8, it can be observed that there are broader and more broadly dispersed vascular bundles embedded in the thin-walled parenchymatous ground tissues. The number of vascular bundles is about $37/cm^2$, and as they grow, they become more uniformly dispersed toward the trunk's center (Killmann and Lim, 1985; Lim and Khoo, 1986; Lamaming et al., 2015).

Several researchers have stated that the physical properties of vascular bundles and parenchyma differ significantly, with vascular bundles being thick and fibrous and parenchyma being sparse and very spongy (Lim and Khoo, 1986; Abe et al., 2013). The vascular bundle numbers significantly decrease from the peripheral area to the middle of the cross-section and increase from the bottom to the top of the OPT, thus contributing to the variation in palm wood density. In addition, the number and thickening distribution of the vascular bundles vary radially and vertically from a radial to a vertical orientation of the trunk, resulting in a variation in physical and mechanical properties of the OPT (Sulaiman et al., 2012).

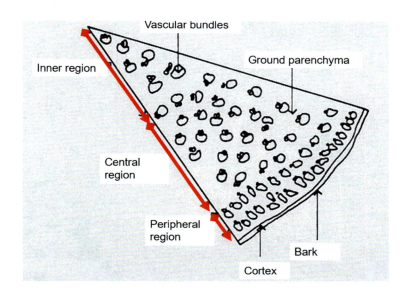

Fig. 2.8 Cross-section of oil palm trunk (OPT) that mainly consists of vascular bundles. Source: Killmann, W., Lim, S.C, 1985. Anatomy and properties of oil palm stems. In: National Symposium on Oil Palm by-Products for Agro-Based Industries, Kuala Lumpur, 5-6 Nov 1985. IPMKSM.

During oil production processing of EFB, the oil palm trees are pruned, which produces substantial quantities of OPF, about 24 million tons a year from oil palm plantations (Abdul Khalil et al., 2012). OPF consists of leaflets, rachis, and petioles, as shown in Fig. 2.9, which are mainly used for soil protection and nutrient recycling. The leaflets have a lamina, which is about 7.5–9.0 m, and each frond consists of around 250–400 strands. The rachis is where the young leaf is attached, and the petiole is part of the leaf and stems (Mandang et al., 2018).

The leaflet only accounts for about 30% of the frond, while the petiole accounts for about 70% and forms the solid woody part of the frond. Fig. 2.10 depicts a petiole cross-section, mostly made up of parenchymatous tissues with multiple fibrous strands and vascular bundles. The diameter of the petiole decreases from the base to the end of the petiole and from the bottom to the upper front (Kushairi et al., 2018; Hishamudin, 1987).

The OPT can retain high moisture compared to standard wood types due to the presence of a significant amount of sap in the OPT (Kosugi et al., 2010). The moisture content is usually between 40%–50%, and this remarkable distinctive property is found in the OPT. Furthermore, Killmann and Lim (1985) reported that the initial moisture content of the OPT ranges from 100% to 500%. It was observed that there is an increase in moisture content around the height of the trunk and toward the central zone because of the presence of parenchymatous cells, which hold more moisture than vascular bundles (Lim and Gan, 2005). These tissues are most abundant near the tip of

Fig. 2.9 Parts of oil palm frond (OPF). Source: Mandang, T., Sinambela, R., Pandianuraga, N.R., 2018. Physical and mechanical characteristics of oil palm leaf and fruits bunch stalks for bio-mulching. IOP Conf. Ser. Earth Environ. Sci. 196 (1): 012015. https://10.1088/1755-1315/196/1/012015.

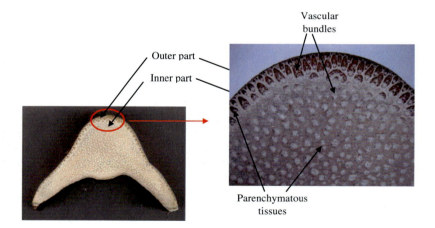

Fig. 2.10 The cross-section of the petiole of oil palm frond (OPF). Source: Rasat M.S.M., Wahab R., Izyan K., 2014. Bio-Composite Lumber from Compressed Oil Palm Fronds. Lambert Academic Publishing.

the palm trunk and radially from the perimeter to the inner and core regions. The moisture content of OPF is quite similar to OPT due to an increase in the number of vascular bundles, which have a high capacity for water absorption.

Fiber length is one of the important factors that determine most of the fiber composite performance. Fiber length-to-diameter (also referred to as the aspect ratio) significantly influences the properties of final composite materials. The high aspect ratio is very important in fiber composites, showing possible strength properties (Rowell Roger et al., 2000). Stark and Rowlands (2003) stated that the aspect ratio

Chapter 2 Basic properties of oil palm biomass (OPB) **47**

has the greatest effect on strength and stiffness rather than particle size. Additionally, Nourbakhsh and Ashori (2009) stated that the fiber length, specifically its aspect ratio, is one of the most important parameters that influences the composite fiber's mechanical properties.

Table 2.1 demonstrates the morphological properties of OPB fibers. The EFB fiber length shows an average of 0.89–1.42 mm, and this length is in the range of the length of hardwood and softwood fibers (Hassan et al., 2010). It was found that the OPT fibers displayed the highest fiber length and diameter, i.e., 2.04 and 26.1 mm, respectively. Researchers also reported that OPT has longer fibers than other OPB fibers (Abdul Khalil et al., 2008). The OPT fiber length decreases in the center's direction from the periphery and the lower to the top of the trunk. On the other hand, the OPF fibers are shorter and thicker than EFB and OPT fibers.

Abdul Khalil et al. (2008) conducted a comprehensive study on the cell structure of OPB fibers. They found that OPT fiber displayed the highest fiber lumen diameter (S_1), fiber lumen area (S_2), and wall thickness (S_3) as compared to EFB and OPF fibers. Fig. 2.11 shows the cell wall structure of OPB inside the vascular bundle under transmission electron micrographs (TEM). The cell wall structure of all OPB fibers is similar to the cell wall structure of wood in that it mainly consists of a primer layer (P) and secondary layers (S_1, S_2, and S_3). All fiber structures of OPB fibers are nearly round in shape, as shown in Figs. 2.11A, C, E; the S_1, S_2, and S_3 layers are observed bonded together, forming a sandwich-like structure. The sandwich structure is important because it gives additional strength to the fiber in the final composite properties. It was also observed that the S_2 layer is the main and thickest layer differentiated between other walls. Also,

Table 2.1 Fiber morphology properties of oil palm biomass (OPB) and wood fibers.

Properties	EFB	OPF	OPT	Hardwood	Softwood
Length (mm)	0.89–1.42	0.59–1.59	2.04	0.83	2.39
Diameter (µm)	19.1	15.1	26.1	14.7	26.8
Lumen width (µm)	8	8.2–11.6	17.6	10.7	19.8

Sources: Law, K.-N., Jiang, X., 2001. Comparative papermaking properties of oil-palm empty fruit bunch. TAPPI J. 84 (1), 95.; Erwinsyah, E., 2008. Improvement of Oil Palm Trunk Properties Using Bioresin. PhD diss., doctoral dissertation, Technische Universität Dresden, Germany; Abdul Khalil, H.P.S., Siti Alwani, M., Ridzuan, R., Kamarudin, H., Khairul, A., 2008. Chemical composition, morphological characteristics, and cell wall structure of Malaysian oil palm fibers. Polym.-Plast. Technol. Eng. 47 (3), 273–280. https://doi.org/10.1080/03602550701866840; Ahmad, Z., Saman, H.M., Tahir, P.M., 2010. Oil palm trunk Fiber as a bio-waste resource for concrete reinforcement. Int. J. Mech. Mater. Eng. 5 (2), 199–207; Hassan, A., Salema, A.A., Ani, F.N., Bakar, A.A., 2010. A review on oil palm empty fruit bunch fiber-reinforced polymer composite materials. Polym. Compos. 31 (12), 2079–2101. https://doi.org/10.1002/pc.21006; Mohanty, A.K., Misra, M., Drzal, L.T., 2005. Natural fibers, biopolymers, and biocomposites. In: Natural Fibers, Biopolymers, and Biocomposites. CRC Press, pp. 2–38; Law, K.-N., Daud, W.R.W., Ghazali, A., 2007. Morphological and chemical nature of fiber strands of oil palm empty-fruit-bunch (OPEFB). Bioresources 2 (3), 351–362.

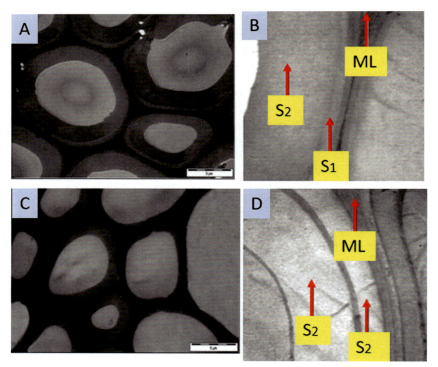

Fig. 2.11 Micrographs of OPB fibers at different magnifications: (A) EFB 3400 ×, (B) EFB 17000 ×, (C) OPF 3400 ×, (D) OPF 17000 ×. Note: ML, middle lamella; P, primer layer; S_1 and S_2, secondary layers. Source: Abdul Khalil, H.P.S., Siti Alwani, M., Ridzuan, R., Kamarudin, H., Khairul, A., 2008. Chemical composition, morphological characteristics, and cell wall structure of Malaysian oil palm fibers. Polym.-Plast. Technol. Eng. 47 (3), 273–280. https://doi.org/10.1080/03602550701866840.

OPT fiber is found to have the thickest S_2 layer, i.e., 3.43 mm as compared to OPF and EFB S_2 layers, as can be seen in Figs. 2.11E and 2.10F (Abdul Khalil et al., 2008).

2.3 Chemical properties

Lignocellulosic materials such as plant fiber, wood, crop, and agricultural wastes are a mixture of natural polymers composed of three basic elements: lignin, cellulose, and hemicellulose as shown in Fig. 2.12. Lignocellulosic fibers have recently gained attention due to their properties and being renewable in nature. However, these fibers' chemical composition significantly varies depending on their types, parts, and functions.

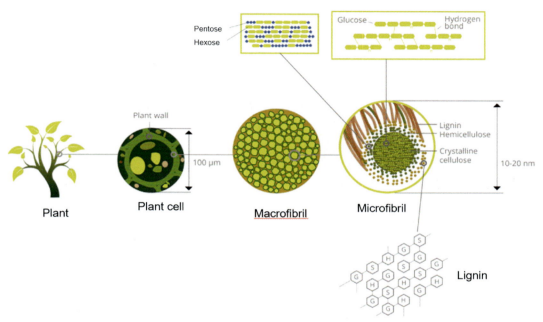

Fig. 2.12 Graphic illustration of cellulose, hemicellulose, and lignin in natural fiber structure. The hexagons represent the lignin subunits p-coumaryl alcohol (H), coniferyl alcohol (G), and sinapyl alcohol (S). Source: Streffer, F., 2014. Lignocellulose to biogas and other products. JSM Biotechnol. Biomed. Eng. 2 (1), 1–8.

The chemical composition of EFB, OPF, and OPT can be seen in Table 2.2. Compared to OPF and OPT, EFB has higher cellulose content, around 43% to 65%, reinforced by a lignin matrix similar to other lignocellulosic fibers. Cellulose is the main component that consists of a long linear chain of hydrophilic glucan polymer and hydroxyl groups, which is believed to provide high strength in fiber-based end products (Abdul Khalil et al., 2010a, b). The cellulose structure is composed of crystalline and amorphous regions, as shown in Fig. 2.13. Intermolecular hydrogen bonds in the macromoleculum are formed in groups of hydroxyl and cellulose macromolecules that are intermolecular in the hydroxyl groups, making all OPB hydrophilic in nature. Strong intermolecular hydrogen bonds with large molecules are formed by crystallite cellulose (Kabir et al., 2012).

The high cellulose content and good strength performance of EFB make it usable as a fiber source and ideal for polymer composite applications (Sreekala et al., 2004). Several studies have shown that EFB can be an efficient reinforcement for thermoplastics and thermosetting products (Karina et al., 2008; Izani et al., 2013a, b). Compared to mesocarp and OPF, EFB possesses better mechanical properties and less water absorption (Izani et al., 2013a, b; Ibrahim et al., 2015).

Table 2.2 Chemical composition of oil palm biomass (OPB) and woods.

Composition (%)	EFB	OPF	OPT	Softwood	Hardwood
α-cellulose	42.7	56.2	45.8	30–60	31–64
Cellulose	43–65	40–50	29–37	–	–
Hemicellulose	17–33	34–38	12–17	–	–
Holocellulose	70.0	75.5	71.8	60–80	71–89
Lignin	17.2	16.4	22.6	21–37	14–34
Ash	0.7	4.2	1.6	<1	<1

Sources: Peh, T.B., Khoo, K.C., Lee, T.W., 1976. Pulping studies on empty fruit bunches of oil palm (Elaeis guineensis Jacq.). Malays. For. 39 (1), 22–37; Tsoumis, George T., 1991. Science and Technology of Wood: Structure, Properties, utilization. New York: Van Nostrand Reinhold, p. 115; Jalaluddin H.B., 1993. Chemical Modification of Oil Palm Fibers and its Application in Composite Board. Doctoral dissertation, University of Wales, Bango; Abdul Khalil, Abdul, H.P.S., Jawaid, M., Hassan, A., Paridah, M.T., Zaidon, A., 2012. Oil palm biomass fibres and recent advancement in oil palm biomass fibres-based hybrid biocomposites. In: Composites and Their Applications: 187–220. https://doi.org/10.5772/48235.

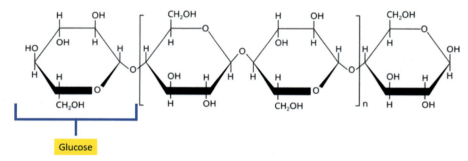

Fig. 2.13 Cellulose structure unit with hydroxyl groups. Source: Kabir, M.M., Hao Wang, Lau, K.T., Francisco Cardona, 2012. Chemical treatments on plant-based natural fibre reinforced polymer composites: an overview. Compos. Part B Eng. 43 (7), 2883–2892. https://doi.org/10.1016/j.compositesb.2012.04.053.

Compared to EFB and OPT, the hemicellulose content of OPF fiber is higher, ranging from 34% to 38%. According to Abdul Khalil et al. (2006), OPF fiber has the highest hemicellulose content compared to softwood and hardwood fibers, coir, kenaf, and banana fibers. The function of hemicellulose is to provide mechanical support and as a transport medium for water and food, and it is the most abundant polysaccharide polymer in natural fibers after cellulose. Hemicellulose is a complex, highly branched chain carbohydrate structure built by different sugars like pentoses, hexoses, and sugar acids used in the production of hydrogel, furfural, xylitol, and films (Sun et al., 2013; Egüés et al., 2013). Hemicelluloses are strongly linked to lignin and cellulose through hydrogen and covalent bonds in the network of plant cell walls.

As shown in Table 2.2, lignin content is highest in OPT fibers, i.e., 22.6% as compared to EFB and OPF fibers. Mansor and Ahmad (1990) also reported that OPT fibers contained a high amount of lignin, almost 19%, and it was similar to that found in hardwood trees (Law and Jiang, 2001). The values of lignin can vary depending on the height and zone area of the trunk. Lignin is a highly complex polymer composed of phenylpropanoid building units that act as a cementing material, determining the toughness and stiffness of the fiber and providing structural support. Lignin fills the spaces between the cellulose and hemicellulose (Mohanty et al., 2005).

2.4 Mechanical properties

The mechanical characteristics of fibers, including tensile, flexural, and rigidity strength, rely on cellulose and inner structure alignment that are usually organized along the length of fibers (John and Thomas, 2008). Dufresne (2008) also stated that the mechanical properties of fiber are strongly influenced by chemical composition, fiber structure, and microfibril angle. Therefore, it differs from different parts and functions of the plant as well as from different plants. Tensile strength and Young's fiber modulus increases with fiber cellulose content (Wang et al., 2011; Aji et al., 2009).

Table 2.3 shows the mechanical properties, namely tensile strength, Young's modulus, and elongation of OPB fibers. It can be seen that the OPT fiber has the highest mechanical properties as compared to EFB and OPF fibers, due to its fiber structure. These strength properties of OPT fibers increase due to the large lumen diameter and thickness of the secondary layers that provide stiffness, and the fiber is resistant to collapse (Ahmad et al., 2010). Fidelis et al. (2013) found a similar study where the curaua fiber with a thicker cell wall than jute and sisal cell wall displayed the highest fiber strength. Furthermore, the S_2 layer is reinforced by microfibril in 5–30 degrees to the axis and is calculated

Table 2.3 Fiber mechanical properties of oil palm biomass.

Properties	EFB	OPF	OPT
Tensile strength (MPa)	50–400	20–200	300–600
Young's modulus (GPa)	0.57–9	2–8	8–45
Elongation at break (%)	2.5–18	3–16	5–25

Sources: Ahmad, Z., H. M. Saman, and P. M. Tahir. 2010 "Oil palm trunk Fiber as a bio-waste resource for concrete reinforcement." Int. J. Mech. Mater. Eng. 5 (2): 199–207.; Bakar, A.A., Hassan, A., Mohd Yusof, A.F., 2006. The effect of oil extraction of the oil palm empty fruit bunch on the processability, impact, and flexural properties of PVC-U composites. Int. J. Polym. Mater. 55(9), 627–641. https://doi.org/10.1080/00914030500306446; Sreekala et al. (2004).

to be 40 times thicker than other layers (Harada and Côté, 1967; Abdul Khalil et al., 2008; Tabet and Aziz, 2013). The thick wall contributes a bulky structure layer, thus providing collapse resistance and bending stiffness toward bending forces (Daud and Law, 2011).

In addition, OPT fiber has a high tensile strength that is comparable to other natural fibers such as kenaf (426.4 MPa) and ramie (399.47 MPa) (Anuar et al., 2019; Wang et al., 2018). Several studies have stated that the tensile strength and Young's modulus of plant fibers improve with growing cellulose and crystalline organization (Lamaming et al., 2015; Aji et al., 2009; Bledzki and Gassan, 1999). It is well known that cellulose is the main structural fiber in the plant, and it has outstanding mechanical properties for a natural polymer attributed to its microfibrillar structure (Sosiati et al., 2013; Kabir et al., 2012). In the case of EFB fiber, the fiber cell structure of EFB has played a major role in mechanical properties (Nishino and Peijs, 2014) due to the crystal area, which provides the specimen with the highest modulus value.

2.5 Conclusions

OPB fibers consist mainly of cellulose, hemicellulose, and lignin, which can be potential fiber sources for reinforcement in composite manufacture for many applications. Among all oil palm fibers, EFB consists of higher cellulose content, but it still contains excess oil. The morphological structure of the oil palm differs from the various parts of the plant to the particular areas and functions of the plant. Chemical composition, fiber structure, microfibril angle, and cell dimensions all affect the mechanical properties of oil palm fibers. Biomass fibers from oil palms have special properties and the ability to be an outstanding matrix reinforcement material that can be used for composite, biocomposite, polymer composite, pulp, and paper industries. However, oil palm fibers have disadvantages such as hydrophilic, ununiform properties, high silica content, and high moisture content. Hence, to fully utilize the potential use of oil palm fibers as an alternative fiber, treatment and hybridization must be considered to improve its performance by enhancing and altering the fiber quality.

References

Abdul Khalil, H.P.S., Alwani, M.S., Omar, A.K.M., 2006. Chemical composition, anatomy, lignin distribution, and cell wall structure of Malaysian plant waste fibers. Bioresources 1 (2), 220–232.

Abdul Khalil, H.P.S., Issam, A.M., Shakri, M.T.A., Suriani, R., Awang, A.Y., 2007. Conventional agro-composites from chemically modified fibres. Ind. Crop Prod. 26 (3), 315–323. https://doi.org/10.1016/j.indcrop.2007.03.010.

Abdul Khalil, H.P.S., Siti Alwani, M., Ridzuan, R., Kamarudin, H., Khairul, A., 2008. Chemical composition, morphological characteristics, and cell wall structure of Malaysian oil palm fibers. Polym.-Plast. Technol. Eng. 47 (3), 273–280. https://doi.org/10.1080/03602550701866840.

Abdul Khalil, H.P.S., Nur Firdaus, M.Y., Jawaid, M., Anis, M., Ridzuan, R., Mohamed, A.R., 2010a. Development and material properties of new hybrid medium density fibreboard from empty fruit bunch and rubberwood. Mater. Des. 31 (9), 4229–4236. https://doi.org/10.1016/j.matdes.2010.04.014.

Abdul Khalil, H.P.S., Poh, B.T., Issam, A.M., Jawaid, M., Ridzuan, R., 2010b. Recycled polypropylene-oil palm biomass: the effect on mechanical and physical properties. J. Reinf. Plast. Compos. 29 (8), 1117–1130. https://doi.org/10.1177/0731684409103058.

Abdul Khalil, H.P.S., Nurul Fazita, M.R., Jawaid, M., Bhat, A.H., Abdullah, C.K., 2011. Empty fruit bunches as a reinforcement in laminated bio-composites. J. Compos. Mater. 45 (2), 219–236. https://doi.org/10.1177/0021998310373520.

Abdul Khalil, H.P.S., Abdul, M., Jawaid, A., Hassan, M.T.P., Zaidon, A., 2012. Oil palm biomass fibres and recent advancement in oil palm biomass fibres-based hybrid biocomposites. In: Composites and Their Applications, pp. 187–220, https://doi.org/10.5772/48235.

Abdullah, N., Sulaiman, F., 2013. The oil palm wastes in Malaysia. Biomass Now Sustain. Growth and Use 1 (3), 75–93. https://doi.org/10.5772/55302.

Abe, H., Murata, Y., Kubo, S., Watanabe, K., Tanaka, R., Sulaiman, O., Hashim, R., et al., 2013. Estimation of the ratio of vascular bundles to parenchyma tissue in oil palm trunks using NIR spectroscopy. Bioresources 8 (2), 1573–1581.

Ahmad, Z., Saman, H.M., Tahir, P.M., 2010. Oil palm trunk Fiber as a bio-waste resource for concrete reinforcement. Int. J. Mech. Mater. Eng. 5 (2), 199–207.

Aji, I.S., Sapuan, S.M., Zainudin, E.S., Abdan, K., 2009. Kenaf fibres as reinforcement for polymeric composites: a review. Int. J. Mech. Mater. Eng. 4 (3), 239–248.

Anuar, N.I., Saffinaz, S.Z., Gan, S., Chia, C.H., Wang, C., Harun, J., 2019. Comparison of the morphological and mechanical properties of oil palm EFB fibres and kenaf fibres in nonwoven reinforced composites. Ind. Crop Prod. 127, 55–65. https://doi.org/10.1016/j.indcrop.2018.09.056.

Bledzki, A.K., Gassan, J., 1999. Composites reinforced with cellulose based fibres. Prog. Polym. Sci. 24 (2), 221–274. https://doi.org/10.1016/S0079-6700(98)00018-5.

Chin, M.J., Poh, P.E., Tey, B.T., Chan, E.S., Chin, K.L., 2013. Biogas from palm oil mill effluent (POME): opportunities and challenges from Malaysia's perspective. Renew. Sustain. Energy Rev. 26, 717–726. https://doi.org/10.1016/j.rser.2013.06.008.

Daud, W.R.W., Law, K.N., 2011. Oil palm fibers as papermaking material: potentials and challenges. Bioresources 6 (1), 901–917.

Dufresne, A., 2008. Cellulose-based composites and nanocomposites. In: Monomers, Polymers and Composites from Renewable Resources, pp. 401–418, https://doi.org/10.1016/B978-0-08-045316-3.00019-3.

Egüés, I., Eceiza, A., Labidi, J., 2013. Effect of different hemicelluloses characteristics on film forming properties. Ind. Crop Prod. 47, 331–338. https://doi.org/10.1016/j.indcrop.2013.03.031.

Fidelis, M.E., Alves, T.V., Pereira, C., da Fonseca, O., Gomes, M., de Andrade Silva, F., Filho, R.D.T., 2013. The effect of fiber morphology on the tensile strength of natural fibers. J. Mater. Res. Technol. 2 (2), 149–157. https://doi.org/10.1016/j.jmrt.2013.02.003.

Harada, H., Côté, W.A., 1967. Cell Wall organisation in the pit border region of softwood Tracheids. Holzforschung- Int. J. Biol., Chem., Phys. Technol. Wood 21 (3), 81–85. https://doi.org/10.1515/hfsg.1967.21.3.81.

Hassan, A., Salema, A.A., Ani, F.N., Bakar, A.A., 2010. A review on oil palm empty fruit bunch fiber-reinforced polymer composite materials. Polym. Compos. 31 (12), 2079–2101. https://doi.org/10.1002/pc.21006.

Hill, C.A.S., Abdul Khalil, H.P.S., 2000. Effect of fiber treatments on mechanical properties of coir or oil palm fiber reinforced polyester composites. J. Appl. Polym. Sci. 78 (9), 1685–1697. https://doi.org/10.1002/1097-4628(20001128)78:9<1685::AID-APP150>3.0.CO;2-U.

Hishamudin, M.J., 1987. The Oil Palm Industry in Malaysia. A Guide Book. Palm Oil Research Institute of Malaysia, Bangi.

Ibrahim, Z., Aziz, A.A., Ramli, R., Jusoff, K., Ahmad, M., Jamaludin, M.A., 2015. Effect of treatment on the oil content and surface morphology of oil palm (Elaeis guineensis) empty fruit bunches (EFB) fibres. Wood Res. 60 (1), 157–166.

Izani, M.A., Norul, M.T., Paridah, U.M.K., Anwar, M.Y.M.N., H'ng, P.S., 2013a. Effects of fiber treatment on morphology, tensile and thermogravimetric analysis of oil palm empty fruit bunches fibers. Compos. Part B 45 (1), 1251–1257. https://doi.org/10.1016/j.compositesb.2012.07.027.

Izani, M.A.N., Paridah, M.T., Mohd Nor, M.Y., Anwar, U.M.K., 2013b. Properties of medium-density fibreboard (MDF) made from treated empty fruit bunch of oil palm. J. Trop. For. Sci. 25 (2), 175–183.

John, M.J., Thomas, S., 2008. Biofibres and biocomposites. Carbohydr. Polym. 71 (3), 343–364. https://doi.org/10.1016/j.carbpol.2007.05.040.

Kabir, M.M., Wang, H., Lau, K.T., Cardona, F., 2012. Chemical treatments on plant-based natural fibre reinforced polymer composites: an overview. Compos. Part B Eng. 43 (7), 2883–2892. https://doi.org/10.1016/j.compositesb.2012.04.053.

Karina, M., Holia Onggo, A.H., Abdullah, D., Syampurwadi, A., 2008. Effect of oil palm empty fruit bunch fiber on the physical and mechanical properties of fiber glass reinforced polyester resin. J. Biol. Sci. 8 (1), 101–106.

Killmann, W., Lim, S.C., 1985. Anatomy and properties of oil palm stems. In: National Symposium on Oil Palm By-products for Agro-based Industries, Kuala Lumpur, 5–6 Nov 1985. IPMKSM.

Kosugi, A., Tanaka, R., Magara, K., Murata, Y., Arai, T., Sulaiman, O., Hashim, R., et al., 2010. Ethanol and lactic acid production using sap squeezed from old oil palm trunks felled for replanting. J. Biosci. Bioeng. 110 (3), 322–325. https://doi.org/10.1016/j.jbiosc.2010.03.001.

Kushairi, A., Loh, S.K., Azman, I., Hishamuddin, E., Meilina Ong-Abdullah, Z.B.M.N., Izuddin, G.R., Sundram, S., Parveez, G.K.A., 2018. Oil palm economic performance in Malaysia and R & D progress in 2017. J. Oil Palm Res. 30 (2), 163–195. https://doi.org/10.21894/jopr.2018.0030.

Law, K.-N., Jiang, X., 2001. Comparative papermaking properties of oil-palm empty fruit bunch. TAPPI J. 84 (1), 95.

Lamaming, J., Hashim, R., Leh, C.P., Othman, S., Sugimoto, T., Nasir, M., 2015. Isolation and characterization of cellulose nanocrystals from parenchyma and vascular bundle of oil palm trunk (Elaeis guineensis). Carbohydr. Polym. 134, 534–540. https://doi.org/10.1016/j.carbpol.2015.08.017.

Law, K.-N., Daud, W.R.W., Ghazali, A., 2007. Morphological and chemical nature of fiber strands of oil palm empty-fruit-bunch (OPEFB). Bioresources 2 (3), 351–362.

Lim, S.C., Gan, K.S., 2005. Characteristics and utilization of oil palm stem. Timber Technol. Bull. 35, 1–7.

Lim, S.C., Khoo, K.C., 1986. Characteristics of oil palm trunk and its utilization. Malays. For. 49 (1–2), 3–22.

Mandang, T., Sinambela, R., Pandianuraga, N.R., 2018. Physical and mechanical characteristics of oil palm leaf and fruits bunch stalks for bio-mulching. IOP Conf. Ser. Earth Environ. Sci. 196 (1), 012015. https://10.1088/1755-1315/196/1/012015.

Mansor, H., Ahmad, A.R., 1990. Carbohydrates in the oil palm stem and their potential use. J. Trop. For. Sci. 2 (3), 220–226.

Mohanty, A.K., Misra, M., Drzal, L.T., 2005. Natural fibers, biopolymers, and biocomposites. In: Natural Fibers, Biopolymers, and Biocomposites. CRC Press, pp. 2–38.

MPOB (Ed.), 2020. Malaysian Oil Palm Statistics 2019, thirty-ninth ed. MPOB, Bangi.

Nambiappan, B., Ismail, A., Hashim, N., Nazlin Ismail, S., Nazrima, N.A., Idris, N., Noraida Omar, K.M., Saleh, N.A., Hassan, M., Kushairi, A., 2018. Malaysia: 100 years of resilient palm oil economic performance. J. Oil Palm Res. 30 (1), 13–25. https://doi.org/10.21894/jopr.2018.0014.

Nishino, T., Peijs, T., 2014. All-cellulose composites. In: Handbook of Green Materials, pp. 201–216, https://doi.org/10.1142/9789814566469_0028.

Nourbakhsh, A., Ashori, A., 2009. Preparation and properties of wood plastic composites made of recycled high-density polyethylene. J. Compos. Mater. 43 (8), 877–883. https://doi.org/10.1177/0021998309103089.

Ratnam, C.T., Raju, G., Ibrahim, N.A., Rahman, M.Z.A., Zin, W.M., Yunus, W., 2008. Characterization of oil palm empty fruit bunch (OPEFB) fiber reinforced PVC/ENR blend. J. Compos. Mater. 42 (20), 2195–2204. https://doi.org/10.1177/0021998308094555.

Ratnasingam, J.M.T.P., Ma, T.P., Manikam, M., Farrokhpayam, S.R., 2008. Evaluating the machining characteristics of oil palm lumber. Asian J. Appl. Sci. 1 (4), 334–340. https://doi.org/10.3923/ajaps.2008.334.340.

Rowell Roger, M., Han, J.S., Rowell, J.S., 2000. Characterization and factors effecting fiber properties. In: Natural Polymers And Agrofibers Based Composites. CRC Press, pp. 15–134.

Sosiati, H., Rohim, A., Maarif, K.T., Harsojo., 2013. Relationships between tensile strength, morphology and crystallinity of treated kenaf bast fibers. AIP Conf. Proc. 1554 (1), 42–46. https://doi.org/10.1063/1.4820279.

Sreekala, M.S, Kumaran, M.G, Geethakumariamma, M.L, Thomas, S, 2004. Environmental effects in oil palm fiber reinforced phenol formaldehyde composites: Studies on thermal, biological, moisture and high energy radiation effects. Advanced Composite Materials 13 (3–4), 171–197. https://doi.org/10.1163/1568551042580154.

Sreekala, M.S., Kumaran, M.G., Thomas, S., 1997. Oil palm fibers: morphology, chemical composition, surface modification, and mechanical properties. J. Appl. Polym. Sci. 66 (5), 821–835. https://doi.org/10.1002/(SICI)1097-4628(19971031)66:5%3C821::AID-APP2%3E3.0.CO;2-X.

Stark, N.M., Rowlands, R.E., 2003. Effects of wood fiber characteristics on mechanical properties of wood/polypropylene composites. Wood Fiber Sci. 35 (2), 167–174.

Sulaiman, O., Salim, N., Nordin, N.A., Hashim, R., Ibrahim, M., Sato, M., 2012. The potential of oil palm trunk biomass as an alternative source for compressed wood. Bioresources 7 (2), 2688–2706.

Sun, X.-F., Wang, H.-h., Jing, Z.-x., Mohanathas, R., 2013. Hemicellulose-based pH-sensitive and biodegradable hydrogel for controlled drug delivery. Carbohydr. Polym. 92 (2), 1357–1366. https://doi.org/10.1016/j.carbpol.2012.10.032.

Suradi, S.S., Yunus, R.M., Beg, M.D.H., Rivai, M., Yusof, Z.A.M., 2010. Oil palm bio-fiber reinforced thermoplastic composites-effects of matrix modification on mechanical and thermal properties. J. Appl. Sci. 10 (24), 3271–3276.

Tabet, T.A., Aziz, F.A., 2013. Cellulose microfibril angle in wood and its dynamic mechanical significance. In: Cellulose: Fundamental Aspects, pp. 113–142, https://doi.org/10.5772/51105.

Umar, M.S., Jennings, P., Urmee, T., 2013. Strengthening the palm oil biomass renewable energy industry in Malaysia. Renew. Energy 60, 107–115. https://doi.org/10.1016/j.renene.2013.04.010.

Umikalsom, M.S., Ariff, A.B., Zulkifli, H.S., Tong, C.C., Hassan, M.A., Karim, M.I.A., 1997. The treatment of oil palm empty fruit bunch fibre for subsequent use as substrate for cellulase production by Chaetomium globosum Kunze. Bioresour. Technol. 62 (1–2), 1–9. https://doi.org/10.1016/S0960-8524(97)00132-6.

USDA, 2020. Palm Oil Production by Country in 1000 MT - Country Rankings. https://www.indexmundi.com/agriculture/?commodity=palm-oil.

Wang, G., Shi, S.Q., Wang, J., Yan, Y., Cao, S., Cheng, H., 2011. Tensile properties of four types of individual cellulosic fibers. Wood Fiber Sci. 43 (4), 353–364.

Wang, C., Ren, Z., Li, S., Yi, X., 2018. Effect of ramie fabric chemical treatments on the physical properties of thermoset polylactic acid (PLA) composites. Aerospace 5 (3), 93. https://doi.org/10.3390/aerospace5030093.

Zwart, R., 2013. Opportunities and challenges in the development of a viable Malaysian palm oil biomass industry. J. Oil Palm, Environ. Health 4, 41–46. https://doi.org/10.5366/jope.2013.05.

3

Biological durability and deterioration of oil palm biomass

Zaidon Ashaari[a,b], S.H. Lee[a], Wei Chen Lum[c], and S.O.A. SaifulAzry[a]

[a]Laboratory of Biocomposite Technology, Institute of Tropical Forestry and Forest Products (INTROP), Universiti Putra Malaysia, Serdang, Selangor, Malaysia, [b]Faculty of Forestry and Environment, Universiti Putra Malaysia, Serdang, Selangor, Malaysia, [c]Institute for Infrastructure Engineering and Sustainable Management (IIESM), Universiti Teknologi MARA, Shah Alam, Selangor, Malaysia

3.1 Introduction

Angiosperms, or flowering plants, refer to plants that produce flowers and bear their seeds in fruits. Angiosperms could be further classified into two categories, namely monocotyledon and dicotyledon. The difference between monocotyledon and dicotyledon are summarized in Table 3.1. Monocotyledon can be easily distinguished from dicotyledon by comparing four distinct structure features: leaves, stems, roots, and flowers. Oil palm trees belong to monocotyledon, together with grasses, orchids, sugarcane, banana, and bamboo. Meanwhile, most of the woody plants belong to dicotyledon. Therefore, the physicochemical properties of the oil palm tree are essentially different from other woody plants.

According to Lamaming et al. (2013), a high amount of sugar and starch can be found in the parenchyma cells in the trunks of monocotyledons. Both sugar and starch are carbohydrates that can be easily consumed, and therefore parenchyma cells are a suitable substrate for many microorganisms (Schmidt, 2006). Oil palm trees, as a monocotyledon, are known to have experienced severe attacks by biodeterioration agents such as fungi and molds due to their high number of parenchyma cells compared to that of dicotyledons (Schmidt et al., 2016a). Oil palm wood is classified as Class V (5) in the timber durability classes, which is not durable (Bakar et al., 2013a). Because it falls into this class, oil palm wood is said to have a life expectancy of less than 5 years. Oil palm biomasses are one of the sources of lignocellulosic materials. It is well known that lignocellulosic materials can be

Oil Palm Biomass for Composite Panels. https://doi.org/10.1016/B978-0-12-823852-3.00002-7
Copyright © 2022 Elsevier Inc. All rights reserved.

Table 3.1 Difference between monocotyledon and dicotyledon.

Feature	Monocotyledon	Dicotyledon
Embryo	Single cotyledon	Two cotyledons
Pollen	Single furrow or pore	Three furrows or pores
Flower parts	Multiples of three	Multiples of four or five
Leaf veins	Parallel	Reticulated
Stem vascular bundles	Scattered	Ring-like
Root system	Fibrous	Taproot

degraded by a variety of microorganisms, especially white-rot fungi that exhibits a preference for lignin degradation (Sánchez, 2009).

The main wood-decaying agent is fungi, although bacteria is also known as a common wood-inhabiting microorganism (Greaves, 1971). Apart from fungi and bacteria, organisms such as insects also infest wood and other lignocellulosic materials. These wood-destroying insects include termites, carpenter ants, and powderpost beetles. Fungi requires specific conditions to grow. Those conditions include: (1) Nutrients in the form of cellulose, hemicellulose, or lignin; (2) temperature in the range of 10–30°C; (3) moisture content of the wood should be approximately 30%; and (4) a sufficient amount of oxygen (Townsend, 2016). Based on the damage caused, there are two major types of fungi, namely wood-destroying and wood-staining fungi. The former, also called decay fungi, are able to alter the physical and chemical properties of wood and subsequently reduce the strength of the decayed wood. The latter, consisting of sap stain fungi and mold fungi, can only cause discoloration of wood (Townsend, 2016).

Brown rot, white rot, and soft rot are the three main categories of wood-destroying fungi. Their names are inspired by the visual appearance of the decayed wood. Different wood fungi and their respective characteristics are presented in Table 3.2.

White-rot fungi (*Pycnoporous sanguineus*) and subterranean termites (*Coptotermes curvignathus*) are the two most important biodeterioration agents found in tropical countries such as Malaysia that cause severe damage to wood. White-rot fungi are the only microorganisms known to efficiently degrade all the components of wood, including lignin (Shary et al., 2007). *Pycnoporous sanguineus* is a common basidiomycete found in the tropics. It causes considerable deterioration even among supposedly durable timber. The Malaysian Timber Council uses white-rot fungi as the guide to measure the durability of Malaysian timbers. Meanwhile, degradation of wood by

Table 3.2 Characteristics of different fungi.

Rot type	Characteristics
White rot	— Degrades both lignin and cellulose — Infested wood shows whiter color — No cracking on wood
Brown rot	— Degrades cellulose and leaves lignin untouched — Infested wood shows dark brown color, excessive shrinkage, and can be easily crushed into brown powder — Cross-grain cracking
Soft rot	— Attacks green wood — Infested wood shows softening surface
Sap stain	— Penetrates and discolors sapwood — Unable to remove by brushing and planning — Blue stain is the most common of this group
Mold	— Can be easily removed from the wood surface — Does not affect wood strength — May increase the absorbed moisture and lead to attack by decay fungi

termites is a chronic problem in many tropical and even some temperate regions of the world, resulting in significant monetary and material losses with a far-reaching effect on the increasing demand for timber (Kirton et al., 2000). The subterranean termite *Coptotermes curvignathus* Holmgren is the most economically important termite species in Malaysia (Dhanarajan, 1969) and is a serious pest for structural timber, both inside and outside buildings (Sajap and Yaacob, 1997). Oil palm biomasses are known to have high lignocellulosic content as well as high sugar and starch. Therefore, they are very prone to attack by these fungi and termites. This chapter reviews the biological durability and deterioration of oil palm biomasses.

3.2 Physicochemical properties of oil palm biomass

Oil palm trunk (OPT), with high carbohydrates in its parenchyma cells, offers an easily consumable substate to many biodeteriorating agents. Schmidt (2006) found that the outer part of the OPT consists of 32 nmol glucose, 13 nmol fructose, 2 nmol sucrose, and 5 nmol starch per mg of dry wood. It was higher in the core part of oil palm wood where 162 nmol glucose, 33 nmol fructose, 5 nmol sucrose, and

34 nmol starch per mg of dry wood were recorded. Owing to this fact, the felled OPT, particularly with high moisture content, is very prone to colonization and damage by molds, blue-stain, and decay fungi.

OPT has low lignin content and high carbohydrate content (Tomimura, 1992). Table 3.3 lists the chemical compositions including cellulose, hemicellulose, lignin, and extractives content of OPT, oil palm frond (OPF), and empty fruit bunch (EFB). It can be seen that all of these biomasses have a high amount of cellulose and hemicellulose. Shrivastava et al. (2020) reported that the cellulose/lignin ratio for OPT, OPF, and EFB are approximately 6, 6, and 3, respectively, indicating a high amount of cellulose. The hemicellulose/lignin ratio for these biomasses are equally high. Both OPT and EFB have comparable cellulose content with softwood and hardwood, while OPF possesses higher cellulose content than both softwood and hardwood. However, all of the oil palm biomasses have lower lignin content compared to that of the softwood and hardwood species.

Glucose and xylose are the main sugar components found in OPTs; 34.8% xylose was found in both vascular bundles and parenchyma of OPTs. Meanwhile, 63.2% glucose was recorded in vascular bundles of oil pam trunks and 55.5% glucose in parenchyma (Tomimura, 1992). Omar et al. (2011) reported that the starch content increased from base upward and from the outer inward of the oil palm stem. Starch content is higher at the core part and increased from 5.64% to 10.37% along with increasing height. The middle and outer parts have lower starch content of 4.30% and 2.93% at the bottom part of the oil palm stem.

Sugar and starch found in oil palm parenchyma are an ideal nutrient for mold. According to Schmidt et al. (2016a), the content of sugar and starch increased from the outer to the inner part of the oil palm stem. The inner part of the oil palm stem contained glucose, fructose,

Table 3.3 Lignocellulose content of oil palm biomass samples.

Oil palm biomass	Cellulose (wt%)	Hemicellulose (wt%)	Lignin (wt%)	Extractives (wt%)
OPT[a]	39.4	25.97	6.64	27.99
OPF[a]	54.35	20.72	8.96	15.96
EFB[a]	37.82	21.85	12.16	28.17
Softwood[b]	40–44	30–32	25–32	5
Hardwood[b]	40–44	15–35	18–25	2

[a] Shrivastava et al. (2020).
[b] Doelle and Bajrami (2018).

sucrose, and starch of 196.5, 26.2, 18.5, and 31.1 nmol/g, respectively. The amount of sugar and starch decreased along with moving to the outer part. The middle part recorded 155.1, 22.0, 12.7, and 23.0 nmol/g, while the outer part gives 26.0, 8.1, 1.2, and 4.7 nmol/g of glucose, fructose, sucrose, and starch, respectively (Schmidt et al., 2016a). Consequently, the inner part, having the highest sugar and starch content, experienced the most severe weight loss against blue-stain fungi, *Alternaria alternata,* and *Aureobasidium pullulans.* Weight loss of more than 16% was recorded in inner part samples after 10 weeks' exposure to the blue-stain fungi. Meanwhile, the middle and outer part, respectively, experienced around 12% and 10% loss of weight after exposure (Schmidt et al., 2016a).

As for OPFs, it was stated in a study conducted by Suzuki et al. (1998) that polysaccharides and lignin are the main components in the walls of OPF. The core of OPF consists of 37.7% α-cellulose and 36.0% hemicellulose, while the peripheral part consists of 33.2% α-cellulose and 33.4% hemicellulose. The total lignin content was 14.3% for the core and 23.7% for the peripheral. One of the most prominent chemical characteristics of OPF is its significantly high level of acetyl groups, which is around 5.8 wt% of organic dry matter. The authors stated that it may be in compensation to the low arabinoxylan content in OPF. In comparison to OPT, OPF has lower xylose (26.0% in the core and 22.9% in the peripheral) and glucose content (31.6% in the core and 29.2% in the peripheral). EFBs have lower holocellulose content compared to OPT and OPFs. However, the lignin content of EFBs is higher.

3.3   Biological durability of oil palm biomasses

3.3.1   Oil palm trunk

Schmidt et al. (2016b) found that the fungal resistance differed by stem position. After exposed to brown rot (*Coniophora puteana*) and white rot (*Pleurotus ostreatus*) for 1 month, the basal portion of the oil palm stem experienced a weight loss of 9.3% and 7.8%, respectively. Fungal decay for the middle and top portion of oil palm stem is much higher than that of the basal portion. The middle and top portion experienced weight loss of 33.3% and 50.0%, respectively, when exposed to brown-rot fungus. On the other hand, weight loss of 17.3% and 21.3% were recorded in the middle and top portion, respectively, after 1 month exposure to white-rot fungus. It can be seen that brown-rot fungi are more destructive to the oil palm stem compared to white-rot fungus. Schmidt et al. (2016b) explained that oil palm stem has a high amount of parenchyma cells, which is favorable by the brown-rot fungus.

Apart from that, Schmidt et al. (2016b) also stated that the difference in decay resistance between stem positions is due to the density variance from basal to top of the oil palm stem. The basal portion of the oil palm stem is high in density and contains many vascular bundles with cells of highly lignified walls. On the contrary, the top portion of the stem normally has lower density and is full of easily accessible parenchyma cells, which makes it an easy target for fungi attack.

Bakar et al. (2013a) reported in their study that oil palm wood is heavily attacked by subterranean termites (*Coptotermes curvignathus* Holmgren), indicating low resistance of oil palm wood against termites. A block evaluation rating system was used to visually examine the oil palm wood after exposure to termites. A rating of 0–10 was assigned to the attacked sample in which 0 was the most severely attacked while 10 was sound and least attacked. In the study, oil palm wood was given a rating of 4, indicating heavily attacked by termites. Weight loss of 27.94% was recorded after 4 weeks' exposure to the termites. On the other hand, when exposed to white-rot fungus (*Pycnoporous sanguineus*), the oil palm wood recorded weight loss of 16.9%, and the entire surface of the oil palm wood samples was fully covered by the fungus (Bakar et al., 2013a). Oil palm wood, if not stored properly and chemically treated, will undergo tissue degradation. Only vascular bundles that have higher resistance remained (Schmidt et al., 2016b).

Naidu et al. (2017) employed various white-rot fungi of *Grammothele fuligo* and *Pycnoporus sanguineus* on a oil palm wood block and determined their decay pattern after an incubation period of 120 days. The results revealed that, at the end of the incubation period, oil palm wood blocks lost 58% of their initial weight when exposed to *G. fuligo* ST2 and 40% to *P. sanguineus* FBR. The purpose of the study was to identify a suitable biotechnological approach to degrade the oil palm-generated waste on site while minimizing the basal stem rot (BSR) disease of oil palm caused by *Ganoderma boninense*. As an obligatory practice of zero-burning, oil palm waste is commonly left on site to rot. This waste has, however, become an important source of inoculum for *G. boninense* and tends to spread to the oil palm trees through root contact. BSR is the most serious threat to the oil palm plantations in Southeast Asia where the infected trees showed progressive decay of the root system (Rees et al., 2009). Therefore, through the study, it was shown that both *G. fuligo* and *P. sanguineus* fungi could potentially be used to aid the degradation of oil palm debris left on plantation sites.

Tomimura (1992) reported that 25-year-old OPTs, at breast height, consisted of 71%–76% vascular bundles and 24%–29% parenchyma. High starch content of 55.5% was found in the parenchyma while 2.4% starch was recorded in the vascular bundles. This high starch content explained the reason of rapid fungal growth on the surface of the cross-section of OPTs.

3.3.2 Oil palm fronds

According to Loh (2017), as one of the largest groups of oil palm waste, approximately 21.03 million tons of OPFs were generated in 2014. In 2020, an estimated 56 million tons/year of OPFs would be generated during replanting activities (Lee and Ofori-Boateng, 2013). OPFs are often left rotting between the rows of oil palm trees for soil erosion control and nutrient recycling purposes (Abdul Khalid et al., 2015). In addition, the efficiency of OPFs as livestock feed is restricted by its high lignin content and high neutral detergent fiber, which resulted in low digestibility in cattle (Rahman et al., 2011). However, its high lignin content has become a favorable nutrient source for white-rot fungi as they degrade lignin preferentially and extensively (Rahman et al., 2011).

In a study by Rahman et al. (2011), the ruminal degradability of OPFs was improved by the introduction of white-rot fungi. In the study, the OPFs were colonized by 10 white-rot fungi, namely *Bjerkandera adusta, Ceriporiopsis subvermispora, Ganoderma lucidum, Lentinula edodes, Phanerochaete chrysosporium, Phlebia brevispora, Schizophyllum commune, Pleurotus eryngii, Pleurotus ostreatus,* and *Trametes versicolor.* When compared with the control OPFs without any treatment, those exposed to white-rot fungi showed a decrement in lignin content after 9 weeks' exposure. In conclusion, OPFs colonized by *C. subvermispora* for 3 weeks and *L. edodes* and *P. brevispora* for 9 weeks displayed the highest extent of improvement in terms of ruminal degradability.

3.3.3 Empty fruit bunches

Owing to its complex elemental form, the natural degradation process of EFBs normally takes a very long time, but the process could be accelerated by the introduction of ligninolytic fungi (Sapareng et al., 2017). Ligninolytic fungi are types of fungi characterized by their ability to depolymerize and mineralize lignin by secreting ligninolytic enzymes such as laccases and peroxidases. Sapareng et al. (2017) compared the composting process of EFBs using different rot fungi, namely *Trichoderma* sp., *Pleurotus ostreatus*, and *Tramella* sp. The authors found that the fungus inoculation treatment of *Trichoderma* sp. exhibited superior quality compared to that of other treatments, after taking several factors into consideration such as ratio of carbon and nitrogen (C/N), pH changes, and temperature.

Piñeros-Castro and Velásquez-Lozano (2014) exposed oil palm EFBs to two types of white-rot fungi, namely *Phanerochaete chrysosporium* and *Phanerochaete ostreatus,* and the weight loss of the EFBs were recorded every week. The results showed that EFBs lost 33.43% and 42.69% of weight after 4 weeks' exposure to *P. chrysosporium* and

P. ostreatus, respectively. Although the weight loss caused by *P. ostreatus* is higher at the end of the 4-week exposure period, it is interesting to note that, at the second week, EFBs exposed to *P. chrysosporium* had a considerably higher weight loss of 23.24% compared to 6.63% for the samples exposed to *P. ostreatus*. The author explained that this phenomenon was caused by the simultaneous attack of lignin and polysaccharides by *P. chrysosporium*. *P. ostreatus*, on the other hand, is considered as selective for delignification as cellulose degradation only started during the third week of exposure.

3.4 Improvement of biological durability of oil palm biomasses

Oil palm biomasses could be converted into a variety of panel products. However, their performance could be severely affected by the attacks of termites and fungi. Owing to that, many treatments could be applied to enhance the biological durability of oil palm biomasses and its resultant panel products. The treatments include treating with ammonia, organic acids, and amino resins such as phenol and urea formaldehyde. Phenolic resin treatment has been used to improve the durability of the low-density wood. Ang et al. (2014) treated mahang wood with 15% low molecular weight phenol formaldehyde (LMwPF) using a compregnation technique and found that there was no weight loss in the treated wood against white-rot fungi as opposed to 17.51% for untreated wood. The untreated wood was colonized by the fungus and the entire test block was covered by mycelium, while no mycelium was observed on the surface of treated test block. The results showed that the phenolic compregnation technique successfully increased the resistance of mahang wood to fungal decay by 100%. A similar finding was also reported by Adawiah et al. (2012) who revealed that the resistance of compregnated jelutong and sesenduk against white-rot fungi increased as much as 98%–99% in comparison to untreated wood.

Meanwhile, for the protection of oil palm wood, the application of boric acid and borax or the admixtures of both by a dipping method is a common practice. However, it only offers short-term protection to the oil palm wood. To achieve long-term protection, a certain chemical uptake rate has to be reached. Therefore, an impregnation technique is a better option in offering long-term protection to oil palm wood. For example, compregnated oil palm wood using LMwPF resin showed great improvement in resistance against both termites and fungal decay (Bakar et al., 2013a). Table 3.4 summarizes the enhancement of oil palm biomasses by various methods. In fact, Bakar et al. (2008) reported that the phenolic resin is nontoxic toward termites as most of the termites survived until the end of the test. The main reason

Table 3.4 Series of works on oil palm trunk enhancement against termites and fungal decay.

Material	Method—variables	Findings	References
Oil palm wood	Impregnation and compregnation—0%, 25%, 50% compression level using low molecular weight phenol formaldehyde resin	— 65%–88% improvement in resistance against termite (*Coptotermes curvignathus*) attack — 47%–93% improvement in resistance against white-rot fungi (*Pycnoporous sanguinues*)	Bakar et al. (2013a)
Oil palm wood	Impregnation and compregnation using low molecular weight phenol formaldehyde resin	— 66% improvement in resistance against termites (*Coptotermes curvignathus*)	Bakar et al. (2013b)
Oil palm stem plywood	Veneer pretreatment—soaking in 20% low molecular weight phenol formaldehyde resin for 20 s prior to curing at 60°C for 2 h	— 44% improvement in resistance against termite (*Coptotermes curvignathus*) attack — 69% improvement in resistance against white-rot fungi (*Pycnoporous sanguinues*)	Loh et al. (2011)
Oil palm wood	Dipped into a solution of 1%–5% acetic acid and propionic acid	— Free from molds even after 8 weeks — More resistance to white rot (*Pleurotus ostreatus*), brown rot (*Coniophora puteana*), and soft rot (*Chaetomium globosum*)	Bahmani and Schmidt (2017)

for the improvement in resistance against termites is the indigestibility of treated oil palm wood by the termites. Table 3.4 summarizes the enhancement of oil palm wood against termites and fungal decay via phenolic resin treatment. Apart from phenolic resin, organic acids such as acetic acid and propionic acid could also be used to protect oil palm wood from termites and fungal decay (Bahmani and Schmidt, 2017).

3.5 Conclusions

As a monocotyledon rich in sugar and starch, oil palm and its biomasses are inevitably prone to attack by biodeterioration agents such as fungi, termites, and molds. Brown-rot fungi is a more destructive fungus to oil palm wood compared to that of white-rot fungi. Meanwhile, OPF and EFB is a preferential choice by white-rot fungi due to its high lignin content. Various treatments can be employed to improve the performance of oil palm wood. One of the most widely applied techniques is an impregnation treatment using phenolic resin. Phenolic resin treatment has been known to greatly improve the biological durability of oil palm wood. This treatment opens up the possibility for oil palm wood to be utilized exteriorly.

Acknowledgments

This project was funded by Ministry of Higher Education, Malaysia, grant number 5540158, Reference code: FRGS/1/2018/WAB07/UPM//1.

References

Abdul Khalid, N.N., Ashaari, Z., Mohd Haniff, A.H., Mohamed, A., Lee, S.H., 2015. Treatability of oil palm frond and rubber wood chips with urea for the development of slow release fertiliser. J. Oil Palm Res. 27 (3), 220–228.

Adawiah, M.R., Zaidon, A., Izreen, F.N., Bakar, E.S., Hamami, S.M., Paridah, M.T., 2012. Addition of urea as formaldehyde scavenger for low molecular weight phenol formaldehyde-treated compreg wood. J. Trop. For. Sci. 24, 348–357.

Ang, A.F., Zaidon, A., Bakar, E.S., Hamami, S.M., Anwar, U.M., Jawaid, M., 2014. Possibility of improving the properties of mahang wood (Macaranga sp.) through phenolic compreg technique. Sains Malaysiana 43 (2), 219–225.

Bahmani, M., Schmidt, O., 2017. Prevention of fungal damage of oil and date palm wood by organic acids. In: The International Research Group on Wood Protection. IRG/WP, pp. 17–10877.

Bakar, E.S., Mohd Hamami, S., H'ng, P.S., 2008. A Challenge from the Perspective of Functional Wood Anatomy. Penerbit Universiti Putra Malaysia, Universiti Putra Malaysia, Serdang.

Bakar, E.S., Hao, J., Ashaari, Z., Yong, A.C.C., 2013a. Durability of phenolic-resin-treated oil palm wood against subterranean termites a white-rot fungus. Int. Biodeter. Biodegr. 85, 126–130.

Bakar, E.S., Paridah, M.T., Sahri, M.H., Mohd Noor, M.S., Zulkifli, F.F., 2013b. Properties of resin impregnated oil palm wood (Elaeis Guineensis Jack). Pertanika journal of Tropcial agricultural. Science 36, 93–100.

Dhanarajan, G., 1969. The termite fauna of Malaya and its economic significance. Malays. For. 32 (3), 274–278.

Doelle, K., Bajrami, B., 2018. Sodium hydroxide and calcium hydroxide hybrid oxygen bleaching with system. IOP Conf. Ser.:Mater. Sci. Eng. 301, 012136.

Greaves, H., 1971. The bacterial factor in wood decay. Wood Sci. Technol. 5 (1), 6–16.

Kirton, L.G., Wong, A.H., Cheok, K.S., 2000. An Overview of the Economic Importance and Control of Termites in Plantation Forestry and Wood Preservation in Peninsular Malaysia. The International Research Group on Wood Preservation Document No: IRG/WP, p. 10382. 00.

Lamaming, J., Sulaiman, O., Sugimoto, T., Hashim, R., Said, N., Sato, M., 2013. Influence of chemical components of oil palm on properties of binderless particleboard. Bioresources 8 (3), 3358–3371.

Lee, K.T., Ofori-Boateng, C., 2013. Sustainability of Biofuel Production from Oil Palm Biomass. Springer Science and Business Media LLC, Berlin, Germany.

Loh, S.K., 2017. The potential of the Malaysian oil palm biomass as a renewable energy source. Energ. Conver. Manage. 141, 285–298.

Loh, Y.F., Paridah, M.T., Yeoh, B.H., Bakar, E.S., Anis, M., Hamdan, H., 2011. Resistance of phenolic-treated oil palm stem plywood against subterranean termites and white rot decay. Int. Biodeter. Biodegr. 65, 14–17.

Naidu, Y., Siddiqui, Y., Rafii, M.Y., Saud, H.M., Idris, A.S., 2017. Investigating the effect of white-rot hymenomycetes biodegradation on basal stem rot infected oil palm wood blocks: biochemical and anatomical characterization. Ind. Crop Prod. 108, 872–882.

Omar, N.S., Bakar, E.S., Jalil, N.M., Tahir, P.M., Yunus, W.M.Z.W., 2011. Distribution of oil palm starch for different levels and portions of oil palm trunk. Wood Res. J. 2 (2), 73–77.

Piñeros-Castro, Y., Velásquez-Lozano, M., 2014. Biodegradation kinetics of oil palm empty fruit bunches by white rot fungi. Int. Biodeter. Biodegr. 91, 24–28.

Rahman, M.M., Lourenço, M., Hassim, H.A., Baars, J.J., Sonnenberg, A.S., Cone, J.W., De Boever, J., Fievez, V., 2011. Improving ruminal degradability of oil palm fronds using white rot fungi. Anim. Feed Sci. Technol. 169 (3–4), 157–166.

Rees, R.W., Flood, J., Hasan, Y., Potter, U., Cooper, R.M., 2009. Basal stem rot of oil palm (*Elaeis guineensis*); mode of root infection and lower stem invasion by *Ganoderma boninense*. Plant Pathol. 58 (5), 982–989.

Sajap, A.S., Yaacob, A.W., 1997. Termites in selected building premises in Selangor. Malays. For. 60, 1–4.

Sánchez, C., 2009. Lignocellulosic residues: biodegradation and bioconversion by fungi. Biotechnol. Adv. 27, 185–194.

Sapareng, S., Ala, A., Kuswinanti, T., Rasyid, B., 2017. The role of rot fungi in composting process of empty fruit bunches of oil palm. Int. J. Curr. Res. Biosci. Plant Biol. 4 (3), 17–22.

Schmidt, O., 2006. Wood and Tree Fungi. Biology, Damage, Protection, and Use. Springer, Berlin, p. 334.

Schmidt, O., Magel, E., Frühwald, A., Glukhykh, L., Erdt, K., Kaschuro, S., 2016a. Influence of sugar and starch content of palm wood on fungal development and prevention of fungal colonization by acid treatment. Holzforschung 70 (8), 783–791.

Schmidt, O., Bahmani, M., Koch, G., Potsch, T., Brandt, K., 2016b. Study of the fungal decay of oil palm wood using TEM and UV techniques. Int. Biodeter. Biodegr. 111, 37–44.

Shary, S., Ralph, S.A., Hammel, K.E., 2007. New insights into the ligninolytic capability of a wood decay ascomycete. Appl. Environ. Microbiol. 73 (20), 6691–6694.

Shrivastava, P., Khongphakdi, P., Palamanit, A., Kumar, A., Tekasakul, P., 2020. Investigation of physicochemical properties of oil palm biomass for evaluating potential of biofuels production via pyrolysis processes. Biomass Conserv. Biorefinery. https://doi.org/10.1007/s13399-019-00596-x.

Suzuki, S., Shintani, H., Park, S.Y., Saito, K., Laemsak, N., Okuma, M., Iiyama, K., 1998. Preparation of binderless boards from steam exploded pulps of oil palm (*Elaeis guneensis* Jaxq.) fronds and structural characteristics of lignin and wall polysaccharides in steam exploded pulps to be discussed for self-bindings. Holzforschung 52, 417–426.

Tomimura, Y., 1992. Chemical characteristics and utilization of oil palm trunks. JARQ 25, 283–288.

Townsend, L., 2016. Wood-Destroying Organisms (WDO). Available from https://www.uky.edu/Ag/Entomology/PSEP/cat17pests.html.

4

Wood-based panel industries

B. Norhazaedawati[a], S.O.A. SaifulAzry[b], S.H. Lee[b], and R.A. Ilyas[c,d]

[a]*Fibre and Biocomposite Centre (FIDEC), Malaysian Timber Industry Board, Banting, Selangor, Malaysia,* [b]*Institute of Tropical Forestry and Forest Products (INTROP), Universiti Putra Malaysia, Serdang, Selangor, Malaysia,* [c]*School of Chemical and Energy Engineering, Faculty of Engineering, Universiti Teknologi Malaysia, Johor Bahru, Malaysia,* [d]*Center for Advanced Composite Materials (CACM), Universiti Teknologi Malaysia, Johor Bahru, Malaysia*

4.1 Introduction

For decades, the timber industry has grown from a producer of planks and boards primarily for domestic consumption and an exporter of logs to be an exporter of primary products such as sawn log, sawn timber, plywood, and veneer (MTIB, 2009). The timber industry has since progressed to become a major exporter of value-added products especially in wood-based products such as plywood, particleboard, fiberboard, and others.

Wood-based panels are products made from fibers, particles, veneer, chips, strands, or any other wood derivate through a binding process with adhesive. These products are also known as engineered wood products, and they have a great scope of engineering properties with design specifications that meet national and international standards (Luchsinger et al., 2017). The manufacturing of wood-based products leverages the utilization of all wood parts including low-grade logs such as thin, bowed, and twisted logs. It also uses wood by-products and recycled materials as well as wood residues such as chips, sawdust, and slabs (Thoemen et al., 2010) and can manufactured into many kinds of particleboards and fiberboards. These panel products are commonly the choice for building materials because of wood's advantages such as its density, strength, and durability.

Generally, a wood-based panel can be categorized as either a structural or nonstructural panel. Structural panels are designed and manufactured as a structural component capable of withstanding applied loads in a specific application. This panel is usually used in houses, buildings, concrete formwork, and others. Nonstructural panels on

Oil Palm Biomass for Composite Panels. https://doi.org/10.1016/B978-0-12-823852-3.00018-0
Copyright © 2022 Elsevier Inc. All rights reserved.

the other hand are manufactured as free-standing panels and used for other applications such as furniture, flooring, doors, and others.

According to Barbu et al. (2013), compared with solid wood, wood-based panel products have been a good choice for building materials, especially for construction and furniture, for many reasons including being predesigned, having more uniform properties, and being more reliable in its mechanical properties in both a plane direction or cross-section. The most significant advantage for the use of these products is seen in load support, which can bolster the load with a smaller safety margin thus decreasing the clear contrast in quality between timber and composites.

Wood-based structural panels have changed user's perception, and many have been used in construction as they meet designer needs for great strength, lightweight, and renewable building material. Plywood and mat-formed panels such as oriented strand board (OSB) have now become the most in-demand product to be used as wood structural panels (APA, 2020). Plywood, as the oldest engineered wood, has been used in many structural applications. Meanwhile, OSB and waferboard began arising in the late 1970s. Surprisingly in 2001, more than 50% OSB has been used as structural panels in Northern America (O'Halloran and Abdullahi, 2017).

The wood-based industry was reported to have increased over the past years. In parallel, panel manufacturing capacities also have increased dramatically for more than 20 years in line with the demand and technology developed to produce panels that meet standard requirements. This chapter reviews several main wood-based panel products, global market trend analysis with projection, and finally highlights some issues and challenges as well as a way forward for wood-based panel industries.

4.2 Types of wood-based panels

In general, wood-based panel products can be made from four main raw materials: solid wood, veneer, particles, and fibers. These raw materials are processed to make various types of panel products as shown in Fig. 4.1. Most wood-based panel products are commonly used for structural applications and furniture production (TKH Technical Briefing Note, 2016) for the advantages of their engineered properties as well as other special properties that can be added such as low thermal conductivity, fire resistance, better bioresistance, and others.

4.2.1 Plywood

Plywood is by far the oldest established panel product. For years, plywood was the most influential export product. It is also one of the most in-demand products in any country. Plywood consists of a set

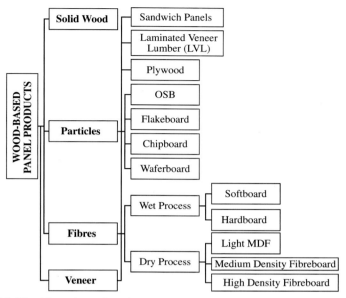

Fig. 4.1 Wood-based panel products.

of layers that are bonded together, with alternating grain directions. There are two types of plywood: veneer plywood and core plywood. Veneer plywood is made of veneer sheets oriented with their plane parallel to the surface of the panel, while core plywood consists of a central core made from wooden strips of solid wood or veneer laid on edge and then covered with two or more veneers on each side, such as blockboard and laminboards. Trada (2012) reported that veneer plywood is generally the quality used for interior, exterior, and marine applications. The endurance depends on the moisture resistance of the resin and the organic persistence of the wood.

The properties of plywood are generally influenced by the quality of veneer, the number of veneer plies, and the resin used for gluing (Stark et al., 2010). In terms of strength, plywood has good mechanical properties, especially in bending strength and shear both along and across the panel, making it advantageous for use due to its resistance toward splitting. Also, as the moisture content increases, plywood will show consistency on dimensional stability with the occurrence of small edge swelling (Curling and Kers, 2017).

Plywood naturally has great strength properties that are suitable to be processed into engineered products used in various applications such as flooring, ceiling, and sheathing for structural purposes, as well as cladding and doors for nonstructural purposes. Also, plywood also can be used in marine environments such as in boats and docks.

The largest volume of high strength and versatile plywood panels are produced in North America, Finland, Russia, South America, and

China. Germany, France, Italy, and Latvia also have significant capacity, but these countries only tend to produce more specialized panels.

Commonly, the main category of wood-based panel products consists of plywood, particleboard, OSB, and fiberboard. According to Booth (1980), wood-based panels are versatile, as they have innate properties and unique functions to perform in global trade. The uses continue to expand because of various advantages such as good strength, easy to work or finish, and available in various sizes and thicknesses.

4.2.2 Particleboard

Particleboard, also widely known as chipboard, is the second-oldest panel product in mass production. Particleboard has a great advantage as it can be made from recycled and lower-grade wood. Particleboard is manufactured by mixing wood particles or nonwood lignocellulose with an adhesive or mineral binder. It is then compressed at certain pressure under heat to form the board. Most of the particleboard is made using a layering system, whereby a larger fraction is arranged inside while a smaller fraction is used on the surface. Usually, the particle size used in particleboard is less than 2 mm in dimension. The moisture content in the particles is very crucial. To produce a panel with better quality, the particles usually need to be dried until the moisture content is in the range of 4%–8%, and then the moisture needs to be controlled, especially during the heating and pressure operation.

Particleboard has a consistent property throughout the board in all dimensions. Its mechanical properties are found to be lower with modulus of rapture (MOR) value of 16–22 MPa compared to 45 MPa for plywood (Stark et al., 2010; Cai and Robert, 2006). Particleboard offers a wide range of uses especially for furniture production, manufacturing, and structural applications. For structural applications, it can be made to meet specific requirements such as fire retardant, moisture resistance, and others. In terms of applications, particleboard is seen as the more universal wood-based product that can be applied in do-it-yourself (DIY) furniture products. It also offers a wide range of different uses such as nonstructural uses, load-bearing uses, and heavy-duty uses in line with the classification of particleboard according to EN 312 standard. Despite having a poor forested area, Italy has become the leading country in utilizing wood recycling technology for the particleboard-making industry. In 2010, Europe is way ahead in production with a capacity of more than 51 million m^3 in 2017.

4.2.3 Fiberboard

Fiberboard is a panel product constructed from wood fiber mixed with water or adhesive and compressed using heat and pressure. Commonly, the term of "fiberboard" refers to hardboard, medium-density fiberboard (MDF), and cellulosic fiberboard. Fiberboard can be manufactured using two different processes, namely the wet and dry process. Wet process is a process that is similar to paper manufacturing technology. In this process, the wet fibers are formed into a mat by either being rolled or rolled and pressed at high temperature. In this process, the main bonding agent is obtained from lignin that acts as an adhesive. The dry process involves a process similar to particleboard in which the fibers are bonded by a synthetic resin adhesive.

Fiberboard can be classified according to its density. The density of softboard is within the range of 230–400 kg/m^3, while medium board is 400–900 kg/m^3, and hardboard is more than 900 kg/m^3 (Wood Panel Industries Federation, 2014). In terms of properties, softboard shows great heat insulation properties with thermal conductivity of 0.045–0.05 W/mK and sound insulation of 22 dB. Hardboard is reported to have good strength properties and a high strength-to-weight ratio. However, this board is easily damaged upon exposure to water unless protected with laminate on both surfaces (Wood Panel Industries Federation, 2014). The application of softboard can be seen in nonstructural applications such as cladding and internal sound insulation while hardboard is commonly used in furniture as well as floor cover applications. Medium-density fiberboard has a wide range of applications including signage, moldings, and flooring.

4.2.4 Oriented strand board (OSB)

OSB is a panel made from wood strands, also known as flakes or wafers. Usually, the size of the strand used is approximately 15–25 mm wide, 50–75 mm long, and 0.3–0.7 mm thick. OSB is manufactured by mixing the strands with an adhesive to form a board of three layers on average with one on the outer oriented edge along the long axis of the board. The concept of orienting the flakes is because they impart greater rigidity and strength to the board. Principally, this method is similar to plywood in alternating the veneer. OSB panels vary in color depending on the selection of wood species, the resin used, and pressing condition (Stark et al., 2010).

It is reported that OSB has great strength properties resulting from the uninterrupted wood fibers, interweaving of the long strands, and degree of orientation of strands in the surface layers, thus making OSB comparable with plywood, even though it has higher shear strength. A combination of strands and additional properties such as waterproof and boilproof resin binders improves this panel in terms of internal strength, rigidity, and moisture resistance. Looking at the strength properties as mentioned, OSB can be used for structural purposes but also sees use for its aesthetic value for furniture, packaging, and others. From the market review, the total global capacity for OSB began in

North America. The industry then grew and was driven by Europe and joined the ranks later by China.

4.3 Wood-based panels industry market analysis

The wood-based industry has grown tremendously over the last few decades. The wood-based industry has encountered enormous demand that has influenced investments in the industry. The demand will have a long-term effect toward the industry due to several factors: the increasing of the world's population projected to be 8.2 billion in 2030; the increasing of economic growth, especially in Asia in the next 25 years; and an awareness and encouragement on energy policies, including the usage of wood biomass. The most influential factor is seen in the increasing of forest plantation due to the declining of raw material supply in natural forests and the development of wood-processing technology (FAO, 2009). The wood-based industry is comprised of four major sectors: sawn timber and logs, veneer, and panel products, while for panel products, it mainly includes four principal products: plywood, particleboard, fiberboard, and OSB.

Statistics by the Forest Agriculture Organization reported that the production of wood-based panels in 2018 was 408 million m^3 compared to 404 million m^3 in the previous year, which shows a 9% increase from the year of 2014–18 (Fig. 4.2). From this statistic, the Asia Pacific region accounted for 61% of the global production followed by

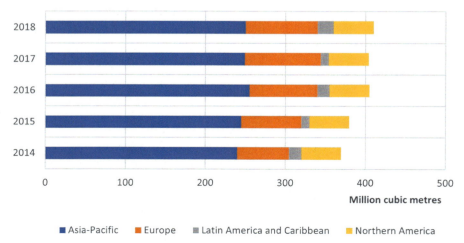

Fig. 4.2 Wood-based panel production. Source: FAO, 2019. Global Production and Trade in Forest Products in 2019. Available from http://www.fao.org/forestry/statistics/80938/en/.

Europe with 22%, Northern America with 12%, Latin America and the Caribbean with 4%, and Africa with 1%.

For global trade, this industry has increased gradually since 2014. It grew by 3% equivalent with 91 million m^3, equal to 22% of the total production in 2018. Europe and Asia Pacific dominated the largest international trade, which altogether accounted for 71% of all imports and 82% of exports in the same year. China, United States, Russian Federation, Germany, and Canada are the largest wood-based panels producers, which accounted for 69% or 282 million m^3 of the total global production for 2018, with China is the largest among them (Fig. 4.3). In terms of consumption, the four top consumers are the same as the four largest producer countries, and the product was consumed domestically.

Northern America was reported to be the main net importer while Asia Pacific was the main net exporter of wood-based panels in 2018. The combination of the five largest exporter countries, namely China, Canada, Germany, the Russian Federation, and Thailand, accounted for 41 million m^3, equal to 44% of global exports (Fig. 4.4). Meanwhile, the top five importer countries, namely United States, Germany, Japan, Poland, and United Kingdom, accounted for 33 million m^3 as the combined import value, equal to 33% of all global import in 2018 (Fig. 4.5).

As for the product analysis (Fig. 4.6), plywood has become the dominant product with a production volume of 163 million m^3 or 40% of all wood-based panel production in 2018. This value has increased tremendously by about 12% over the past 5 years, accounting for 72%

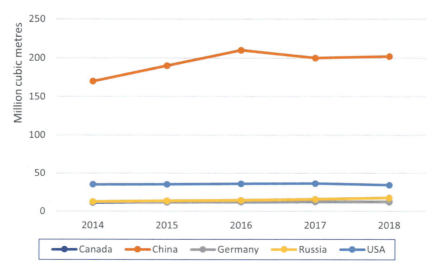

Fig. 4.3 Wood-based panel production by country. Source: FAO, 2019. Global Production and Trade in Forest Products in 2019. Available from http://www.fao.org/forestry/statistics/80938/en/.

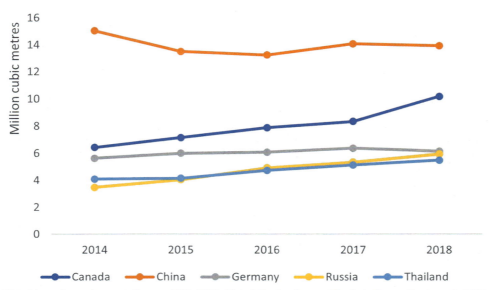

Fig. 4.4 Wood-based panel export. Source: FAO, 2019. Global Production and Trade in Forest Products in 2019. Available from http://www.fao.org/forestry/statistics/80938/en/.

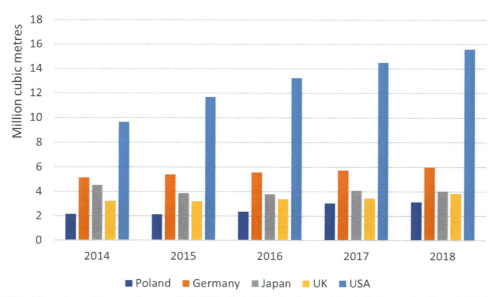

Fig. 4.5 Wood-based panel import. Source: FAO, 2019. Global Production and Trade in Forest Products in 2019. Available from http://www.fao.org/forestry/statistics/80938/en/.

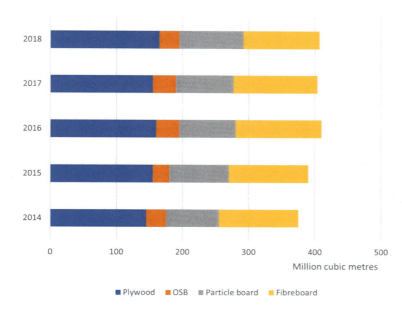

Fig. 4.6 Wood-based panel production by category. Source: FAO, 2019. Global Production and Trade in Forest Products in 2019. Available from http://www.fao.org/forestry/statistics/80938/en/

of global production in 2018. The main factor of the increased value is due to the rapid growth in plywood production in China.

There are regional differences in the composition of various wood-based panel products. Products such as OSB, particleboard, and fiberboard are leading other product categories in Northern America and Europe. Meanwhile, in the Asia Pacific region, especially in China, plywood has become the major wood-based panel product. A different scenario was seen in Latin America and the Caribbean, as every major product accounted for an equal share of total production.

Among panel production, OSB and particleboard recorded an increase of growth production with 25% and 13%, respectively, in 5 years from 2014 to 2018. As for fiberboard, global production reached its peak in 2016 and declined by 4% in 2018 with a total production of 16 million m^3. Meanwhile, for medium-density fiberboard/high-density fiberboard, the production increased by 5% over the last 5 years and accounted for 85% of all fiberboard production in 2018.

As for the current projection, some research such as Research and Market's "Wood Panels—Global Market Trajectory & Analytics" and Mordor Intelligence's "Wood-Based Panel Market" reported that the current market for wood-based panels is still promising with a production projection of 444.1 million m^3 for 2020 and is expected to reach at 658.1 million m^3 in 2027. This projection is said to grow at a Compound Annual Growth Rate (CAGR) of about 6% over the period 2020–27, most likely caused by an increasing demand from the infrastructure sector in developing countries, which will hamper the market growth. Among the factors that will contribute to the market growth are the strict regulations enforced by the government, the growth in residential construction in the Asia Pacific region, and the healthy growth of the wood-based panel market due to its wide usage in various application such as furniture, construction, and packaging.

Furthermore, with an increasing demand of wood-based panel for construction, especially in several emerging economies in the Asia Pacific and South American regions, this will globally drive the industry. Asia Pacific is estimated to continue dominating the global market share based on growing construction activities and the demand for furniture in countries like China, India, and Japan. With the program "Housing for All by 2022" launched by the India government, this will be a game-changer for the wood-based panel industry in that country. For product segmentation, plywood is projected to record 5.8% CAGR and will reach 259.6 million m^3 at the end of 2027, while particleboard is projected to reach 5.1% CAGR by the end of 2027. On the other hand, medium-density fiberboard is projected to have CAGR of 6% in the global market.

4.4 Wood-based industry in Malaysia

4.4.1 Wood-processing mills in Malaysia

According to The National Timber Industry Census Study conducted during 2017–19, there are 5195 wood-processing mills in Malaysia. Among these, 4638 mills are in Peninsular Malaysia, 271 mills are in Sabah, and 286 mills are in Sarawak. Table 4.1 lists a detail breakdown of the wood-based industry in Peninsular Malaysia where the main type of mills are furniture (677), followed by sawmill (458), wooden pallet (84), plywood and veneer (66), moldings (62), builder joinery and carpentry (BJC) (60), laminated board (37), kiln drying (33), particleboard/chipboard (4), wood preservation (4), and medium-density fiberboard (MDF) (4). The remainder (76) comprises other activities such as charcoal, woodchip utensils, frame/door, packaging and papers, etc.

Meanwhile, in Sabah, the main wood-processing activities are sawmilling (48), followed by plywood and veneer (20), moldings (13), furniture (8), BJC (4), kiln drying (2), wooden pallet (1), laminated board (1), and others (3) such as charcoal, as shown in Table 4.2. On the other hand, Table 4.3 displays the breakdown of wood-processing mills in Sarawak. About 28.3% (81 mills) of the wood-processing activities in Sarawak deal with sawmilling. The second highest number is in furniture manufacturing with 60 mills (21.0%) followed by plywood and veneer with 44 mills (15.4%), laminated board (8 mills), moldings (7 mills), MDF (3 mills), wooden pallet (2 mills), particleboard/

Table 4.1 Timber mills according to sectors in Peninsular Malaysia.

| No. | Timber sector | Northern | | | | Central | Southern | | | Eastern | | | Grand total |
		Pulau Pinang	Perak	Kedah	Perlis	Selangor and Wilayah Persekutuan	Negeri Sembilan	Melaka	Johor	Pahang	Terengganu	Kelantan	
1	Sawmill	26	53	16	1	26	25	6	55	105	72	73	**458**
2	Plywood and veneer	1	3	6	–	11	3	1	6	17	6	12	**66**
3	Medium-density Fiberboard (MDF)	–	–	–	–		2	–	1	1	–	–	**4**
4	Moldings	2	3	–	–	21	2	–	17	12	5	–	**62**
5	Laminated board	2	2	4	–	19	1	2	6	1	–	–	**37**
6	Particleboard/ chipboard	–	–	–	–	–	2	1		1	–	–	**4**
7	Builder joinery and carpentry (BJC)	8	14	3	–	12	1	2	8	3	6	3	**60**
8	Wooden pallet	7	8	–	–	15	4	–	28	10	9	3	**84**
9	Kiln drying	2	5	9	–	4	4	1	6	2	–	–	**33**
10	Wood preservation	1	1	–	–	1	1	–	–	–	–	–	**4**
11	Furniture	45	69	34	6	115	27	24	237	44	57	19	**677**
12	Others	7	17	8	–	3	9	1	13	11	2	5	**76**
	Grand total	**101**	**175**	**80**	**7**	**227**	**81**	**38**	**377**	**207**	**157**	**115**	**1565**
	Grand total by region	**363**				**227**	**496**			**479**			

Source: National Timber Industry Census Study, 2017. Malaysian Timber Industry Board. Kuala Lumpur, Malaysia.

80 Chapter 4 Wood-based panel industries

Table 4.2 Wood-processing mills according to sectors and subregions in Sabah.

No.	Timber sector	West coast division Kota Kinabalu	Interior division Keningau	Kudat division	Sandakan division	Tawau division	Total
1.	Sawmill	13	7	3	4	21	**48**
2.	Plywood and veneer	1	4	–	7	8	**20**
3.	Moldings	7	1	–	1	4	**13**
4.	Laminated board	–		–	1	–	**1**
5.	Builder joinery and carpentry (BJC)	1	1	–	–	2	**4**
6.	Wooden pallet	1		–	–	–	**1**
7.	Kiln drying	1		–	–	1	**2**
8.	Furniture	5	1	–	1	1	**8**
9.	Others	–	1	–	–	2	**3**
	Region total	**29**	**15**	**3**	**14**	**39**	**100**

Source: National Timber Industry Census Study, 2017. Malaysian Timber Industry Board. Kuala Lumpur, Malaysia.

chipboard (3 mills), BJC (2 mills), and others (23 mills) such as charcoal and woodchip operators.

4.4.2 Trade value of Malaysian major wood products

Malaysia is among the world's largest exporter of tropical logs, plywood, sawn timber, and furniture to international markets. The total export value for major timber products is RM 22,530.70 million in 2019. In 2020, the total export value decreased slightly to RM 22,022.45 million, a decrement of 2.14% compared to the previous year.

Fig. 4.7 compares the export value of five selected timber products for 2019 and 2020. From Fig. 4.7, one can see that the total export value of BJC, fiberboard, plywood, sawn timber, wooden furniture, particleboard, logs, moldings, and veneer decreased compared to the previous year. Only wooden furniture showed an increment in the export value in 2020 compared to the previous year.

Table 4.3 Wood-processing mills according to sectors and subregions in Sarawak.

No.	Timber sector	Kuching	Sibu	Bintulu	Miri	Limbang	Total
1	Sawmill	11	38	22	4	6	**81**
2	Furniture	34	22	2	2	–	**60**
3	Plywood and veneer	6	16	15	6	1	**44**
4	Moldings	–	4	1	2	–	**7**
5	Wooden pallet	1	–	1	–	–	**2**
6	Medium-density fiberboard (MDF)	–	–	2	1	–	**3**
7	Laminated board	5	1	–	2	–	**8**
8	Particleboard/chipboard	–	1	–	2	–	**3**
9	Builder joinery and carpentry (BJC)	–	–	1	1	–	**2**
10	Others	6	10	4	5	1	**26**
	Region total	**63**	**92**	**48**	**25**	**8**	**236**

Source: National Timber Industry Census Study, 2017. Malaysian Timber Industry Board. Kuala Lumpur, Malaysia.

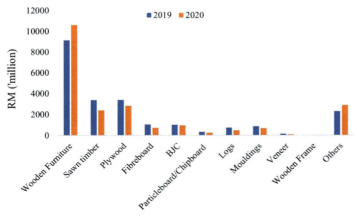

Fig. 4.7 Malaysia: Export value (RM) of major wood products in 2019 and 2020. Source: Malaysian Timber Industry Board (MTIB), 2021. Trade Performance Summary. Available from https://mytis.mtib.gov.my/csp/sys/bi/%25cspapp.bi.index.cls?scnH=0&scnW=1520.

The import value of major timber products has increased from RM 5964.47 million to RM 6810.61 million, an increment of 14.53%. Fig. 4.8 compares the import value of major timber products for year of 2019 and 2020. Wooden furniture, fiberboard, and BJC are among the timber products that experienced an increment in import value in 2020 compared to the previous year.

4.4.2.1 Wooden furniture

The total export value of wooden furniture in 2020 is RM 10,627.73 million. An increment of 16.2% in export value was reported compared to the export value recorded during the same period in 2019, which amounted to RM 9143.65 million. Fig. 4.9 displays the export of wooden furniture by region. Malaysian wooden furniture is particularly in America's favor, which contributed to 67.3% of the total export

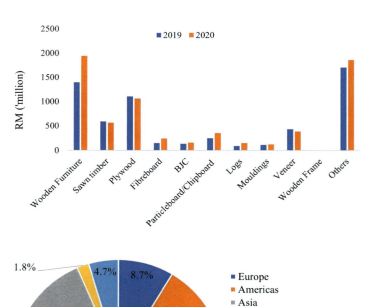

Fig. 4.8 Import value (RM) of major wood products in 2019 and 2020. Source: Malaysian Timber Industry Board (MTIB), 2021. Trade Performance Summary. Available from https://mytis.mtib.gov.my/csp/sys/bi/%25cspapp.bi.index.cls?scnH=0&scnW=1520.

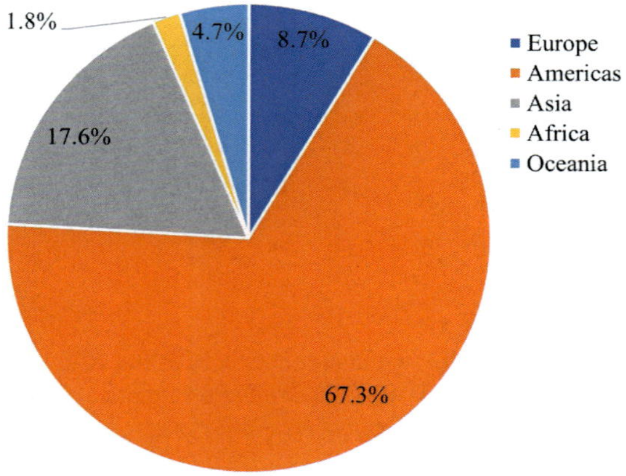

Fig. 4.9 Malaysia: Export value of wooden furniture by region in 2020. Source: Malaysian Timber Industry Board (MTIB), 2021. Trade Performance Summary. Available from https://mytis.mtib.gov.my/csp/sys/bi/%25cspapp.bi.index.cls?scnH=0&scnW=1520.

value of wooden furniture. Meanwhile, Asia (17.6%), Europe (8.7%), Oceanica (4.7%), and Africa (1.8%) made up for the remaining export value of wooden furniture.

4.4.2.2 Sawn timber

The total export value of sawn timber in 2020 is RM 2393.17 million. The sawn timber sector in Malaysia has experienced a substantial reduction of 29.2% in export value compared to the same period in 2019. During the same period in 2019, the export value of sawn timber amounted to RM 3378.84 million. Fig. 4.10 displays the export of sawn timber by region. Asia is the main export destination for Malaysian's sawn timber (82.7%), followed by Europe (10.6%), Africa (3.4%), Americas (2.1%), and Oceania (1.2%).

4.4.2.3 Plywood

The total export value of plywood in 2020 is RM 2840.56 million. Fig. 4.11 displays the export of plywood by region. Asia is the dominant export destination of plywood made by Malaysia, which accounted for 76.1%; 14.5% of the plywood was exported to Americas while Europe and Oceania shared 4.9% and 4.1%, respectively. A small amount of the plywood (0.4%) was exported to Africa. The export value of plywood during the same period in 2019 was recorded as RM 3402.10 million. That said, the export value has experienced a drastic drop of 16.5% in 2020 compared to that of the previous year.

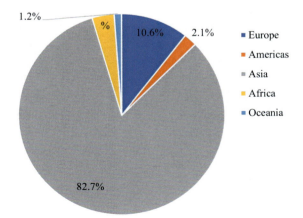

Fig. 4.10 Malaysia: Export value of sawn timber by region in 2020. Source: Malaysian Timber Industry Board (MTIB), 2021. Trade Performance Summary. Available from https://mytis.mtib.gov.my/csp/sys/bi/%25cspapp.bi.index.cls?scnH=0&scnW=1520

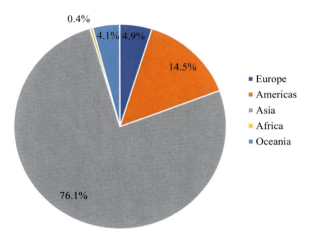

Fig. 4.11 Malaysia: Export value of plywood by region in 2020. Source: Malaysian Timber Industry Board (MTIB), 2021. Trade Performance Summary. Available from https://mytis.mtib.gov.my/csp/sys/bi/%25cspapp.bi.index.cls?scnH=0&scnW=1520.

4.4.2.4 Fiberboard

The total export value of fiberboard in 2020 is RM 726.93 million. Fig. 4.12 displays the export of fiberboard by region; 81.8% of the fiberboard was exported to Asia followed by Americas (8.7%). The remaining were exported to Africa (4.6%), Oceania (4.1%), and Europe (0.7%). The export value of fiberboard in 2020 has seen a drastic decrement of

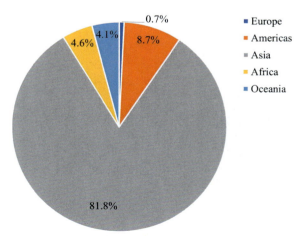

Fig. 4.12 Malaysia: Export value of fiberboard by region in 2020. Source: Malaysian Timber Industry Board (MTIB), 2021. Trade Performance Summary. Available from https://mytis.mtib.gov.my/csp/sys/bi/%25cspapp.bi.index.cls?scnH=0&scnW=1520.

31.5% compared to the export value of the same period of 2019, which was RM 1060.57 million.

4.5 Way forward

Indeed, the timber industry contributed significantly in 2018. To be sure, the world demand for wood products is growing, and this trend is expected to continue in the years to come. Seeing in this term, the industry prospect is undoubtedly bright. To ensure that this industry continues to grow, some direction should be adopted.

Exploring the use of panel products in numerous sectors is expected to drive the market over the forecast period. Demand for wood-based materials has accelerated interest for wood-based panel products over solid wood. This is mainly because of the flexibility, consistency in product quality, and diversification of the applications offered by the products that can be used in kitchen cabinets, flooring, furniture, and others. Wood-based products have huge potential across a range of sectors, from aerospace to furniture and construction. The wood-based panel industry has a low technology barrier, with new technologies emerging to help enhance many of the material's properties, and these potential application areas will continue to expand.

Wood-based panel products such as plywood, OSB, and other structural panels can be used as transportation materials due to its better properties such as being tough, durable, and resistant. This application is expected to gain market growth toward the increase of potential application of the products. A strengthening market has been seen due to demand for interior designing and furniture from developed countries such as the United States, United Kingdom, France, and Japan, as well as developing countries such as China, India, and Brazil. Furthermore, with an increase of renovation services toward residential properties such as in the United States, this panel industry will continue to undergo convincing market growth (Tathe, 2019).

Nonetheless, several key factors can hinder the growth of the industry. Among them are cost and compliance with policy. Logs transferred from the forest to the mills can incur high cost, which in return will increase raw material cost. A challenge is also found to occur in the manufacturing process due to the diversification of raw materials and repeatability of the manufacturing processing, which will also increase the cost of manufacturing. In addition to compliance with new regulations for formaldehyde emission, industries are expected to face challenges to meet these requirements with the standard established. Even though it seems to be tough for the industries, indeed it will offer huge potential opportunities for wood-based panel manufacturers to operate in the global panel market (Tsang, 2018).

References

APA-The Engineered Wood Association, 2020. Panel Design Specification. Available from http://www.ecodes.biz/ecodes_support/free_resources/Standards/APA/PDFs/D510A_2004.pdf.

Barbu, M.C., Reh, R., Irle, M., 2013. Wood-based composites. In: Aguilera, A., Davim, J.P. (Eds.), Research Developments in Wood Engineering and Technology. IGI Global, Pennsylvania, United States, pp. 1–45.

Booth, H., 1980. Asia's Wood-based Panels Industry and Trade. Available from http://www.fao.org/3/n7750e/n7750e01.htm.

Cai, Z., Robert, J.R., 2006. Mechanical Properties of Wood-based Composite Materials. General Technical Report FPL-GTR-190, U.S. Department of Agriculture, Forest Service, Forest Products Laboratory, Madison, Wisconsin.

Curling, S., Kers, J., 2017. Wood as a biobased building material—panels. In: Jones, D., Brischke, C. (Eds.), Performance of Biobased Building Materials, first ed. Woodhead Publishing Ltd., Duxford, UK, pp. 52–61.

FAO, 2009. Global Demand for Wood Products. Avaialble from ftp://ftp.fao.org/docrep/fao/011/i0350e/i0350e02a.pdf.ftp://ftp.fao.org/docrep/fao/011/i0350e/i0350e02a.pdf.

Luchsinger, T., Marcel, F., Miranda, B.A.E.M., 2017. Wood-Based Panels: Market Overview. Cargill Incorporated, Hochschule Rhine-Waal University, Minnesota, United States & Kleve, Germany.

Malaysian Timber Industry Board (MTIB), 2009. National Timber Industry Policy 2009–2020. Available from https://catalogue.nla.gov.au/Record/4726388.

O'Halloran, M.R., Abdullahi, A.A., 2017. Wood: structural panels. In: Reference Module in Materials Science and Materials Engineering, pp. 1–6, https://doi.org/10.1016/b978-0-12-803581-8.01997-4.

Stark, N.M., Cai, Z., Charles, C., 2010. Wood-based Composite Materials: Panel Products, Glued-Laminated Timber, Structural Composite Lumber, and Wood-Nonwood Composite Materials. General Technical report FPL-GTR-190, U.S. Department of Agriculture, Forest Service, Forest Products Laboratory, Madison, Wisconsin.

Tathe, S., 2019. Wood Based Panel Market By Type (Waferboard and OSB, Particleboard, Medium Density Fiberboard, Hardwood, Softwood, Plywood, Others) By End Use (Furniture, Construction, Doors, Others) And Region—Global Forecast To 2026. Available from https://marketresearch.biz/report/wood-based-panel-market/.

Thoemen, H., Irle, M., Sernek, M., 2010. Wood Based Panels: An Introduction for Specialists. Brunel University Press, London, England, ISBN: 978-1-902316-82-6.

TKH Technical Briefing Note 3, 2016. Dispersion Wood Glues. Available from https://webcache.googleusercontent.com/search?q=cache:rdwDLGVFiS-4J:https://www.klebstoffe.com/wp-content/uploads/2020/04/TKH_3_englisch.pdf+&cd=1&hl=en&ct=clnk&gl=my&client=firefox-b-d.

Trada, 2012. Introduction to Wood-based Panel Products. Available from https://www.trada.co.uk/publications/wood-information-sheets/introduction-to-wood-based-panel-products/.

Tsang, H.W., 2018. Understanding US EPA Regulations on Formaldehyde Emissions in Composite Wood Products. Available from https://www.sgs.com/en/news/2018/07/understanding-us-epa-regulations-on-formaldehyde-emissions-in-composite-wood-products.

Wood Panel Industries Federation, 2014. Panel Guide. Available from http://wpif.org.uk/uploads/PanelGuide/PanelGuide_2014_Preface.pdf.

5

Formaldehyde-based adhesives for oil palm-based panels: Application and properties

U.M.K. Anwar[a], M. Asniza[a], K. Tumirah[a], Y. Alia Syahirah[a], and A.H. Juliana[b]

[a]*Forest Products Division, Forest Research Institute Malaysia, Kepong, Selangor, Malaysia,* [b]*Faculty of Technology Management and Business, Universiti Tun Hussein Onn Malaysia, Parit Raja, Batu Pahat, Johor, Malaysia*

5.1 Introduction

Panel products such as medium-density fiberboard, particleboard, plywood, and laminated veneer lumber (LVL) have been successfully fabricated using oil palm biomass (Othman et al., 2009; Abdul Khalil et al., 2010a, b; Rokiah et al., 2011). An abundance of oil palm biomass, for example, 14.4 million dry tons of trunks annually (National Biomass Strategy, 2020), makes it one of the potential alternative raw materials to produce oil palm-based panels. The utilization of oil palm biomass can reduce dependency on wood, reduce environmental issues, and indirectly generate bioeconomy. Unfortunately, oil palm biomass has some limitations such as being low in strength, low dimensional stability, and less durable. Besides that, the natural anatomical structure of oil palm biomass, i.e., high portions of thin-walled parenchyma tissue, especially present in the trunk, cause water to be rapidly absorbed into the cell wall and lumen (Chai et al., 2011; H'ng et al., 2013). Therefore, the rate it shrinks and swells is much higher compared to wood (H'ng et al., 2013). Proper fabrication techniques and the right selection of adhesives are needed to produce high-quality oil palm-based panels.

Formaldehyde-based adhesive is a synthetic polymeric material, also known as resin or glue, chemically prepared using formaldehyde as one of its active ingredients. Basically, it softens when exposed to heat and solidifies when cooled to room temperature. Formaldehyde-based adhesive can be prepared through various ratios between formaldehyde and amino or phenolic compounds. According to

Oil Palm Biomass for Composite Panels. https://doi.org/10.1016/B978-0-12-823852-3.00017-9
Copyright © 2022 Elsevier Inc. All rights reserved.

Bono et al. (2006), variations in the formulation used during preparation of formaldehyde-based adhesives can cause different resin properties. Good formaldehyde-based adhesives should have good viscosity, maximum solubility in water, minimum curing period at room temperature, and maximum storage life. Basically, during the fabrication of an oil palm-based panel, formaldehyde-based adhesive should be capable of interacting physically, chemically, or both with the surface of the oil palm biomass in such a manner that stresses are transferred between bonded members, thus improving the dimensional stability and mechanical properties of oil palm biomass.

Common formaldehyde-based adhesives used in the panels industry are urea-formaldehyde (UF), melamine-urea-formaldehyde (MUF), and phenol-formaldehyde (PF). All these formaldehyde-based adhesives have different advantages and disadvantages. UF offers advantages like being low in cost, hardness, initial water solubility, low flammability, good thermal properties, ease of use under a wide variety of curing conditions, low cure temperatures, and no color changes in the cured end products (Ormondroyd, 2015; Conner and Bhuiyan, 2017). However, the major disadvantage of UF compared to MUF and PF is a lack of resistance to moist conditions (Conner, 2001; Conner and Bhuiyan, 2017). Under moist conditions with the presence of heat, a reversal of the bond-forming reaction can occur, which leads to the release of formaldehyde. Due to this reason, utilization of UF is limited to the manufacturing of panel products for interior purposes.

The previous findings show that the formaldehyde-based adhesives work well for keeping oil palm-based panels strong and durable. However, not all formaldehyde-based adhesives are suitable for the fabrication of oil palm-based panels. The different types of formaldehyde-based adhesives used for fabrication of oil palm-based panels have different advantages and disadvantages. In addition, the selection of formaldehyde-based adhesives depends on the end application of the panels.

This chapter attempts to discuss the comprehensive state-of-the-art formaldehyde-based adhesives used in fabrication of oil palm-based panels as well as to explore the properties of oil palm-based panels produced using either UF, MUF, or PF. This chapter may help bring more understanding on fabrication design of oil palm-based panels using formaldehyde-based adhesives.

5.2 Urea-formaldehyde

UF resins are opaque thermosetting synthetic polymers. They are formed by the condensation reaction of two monomers, urea and formaldehyde in the presence of an alkaline catalyst. UF resins are composed of linear or branched oligomeric and polymeric units. Even though UF resins consist of only two main components, they present a broad variety of possible reactions and structures. The basic characteristics of UF resins include high reactivity, water solubility, and the reversibility of the aminomethylene link (Dunky, 1998, 2003).

The current global production of UF resins exceeds 5 million metric tons annually. It is widely used for wood finishes, adhesives, and binding agents in particleboard, medium-density fiberboard, and interior plywood manufacturing (Papadopoulou et al., 2016). UF resins are the major adhesive in the wood industry due to low cost, ease of use under various curing conditions, low cure temperature, water solubility, resistance to microorganisms and abrasion, excellent thermal properties, and lack of color (Conner, 1996). When cured, it forms a very resilient finish. This resin can be distinguished from other formaldehyde resins, e.g., MF, MUF, and PF by its high reactivity and shorter press times. It also holds great tensile strength, elasticity, and heat distortion resistance (Dunky, 2003).

UF resins are mainly used for interior utilization due to low water or moisture resistance. Many household types of glue and finishing products are based on this resin because of the rise of usage of pressed wood products in cabinets, furniture, laminate flooring, and other indoor construction materials. However, the resin utilization may cause indoor air pollution and possible health problems due to the free formaldehyde emission from UF-bonded wood products (Hun et al., 2010). It is believed that the source of free formaldehyde from UF resin is primarily from unreacted formaldehyde and the reversibility of the aminomethylene link in the cured resin network (Tohmura et al., 2000).

Over the last 20 years, it has been challenging for UF chemists to reduce the content of formaldehyde in UF resins without introducing any other major changes in the performance of the resins. Many research efforts have been performed on reducing the unreacted formaldehyde content and improving the stability of the aminomethylene link. Various techniques have included adding a free formaldehyde scavenger in UF-bonded wood-based composites (Boran et al., 2012), lowering the formaldehyde-to-urea molar ratio in UF resin synthesis (Que et al., 2007), copolymerizing the UF resin with other stable chemicals (Yuan et al., 2015), and adding bioparticles (Petković et al., 2015). However, these modifications resulted in high cost products, deterioration, and complicated manufacturing operations, which indirectly affected the outstanding features of UF resins.

5.3 Melamine-urea-formaldehyde

MUF resin is a thermosetting polymer that is produced by the reaction between urea, melamine, and formaldehyde under an alkaline acid synthesis process. The properties of the MUF resin depend on the molar ratio of melamine, formaldehyde, and urea at each reaction stage in which the amino compounds are reacted (Pizzi and Mittal, 2003; No and Kim, 2007). The performance and durability of MUF resins closely rely on the resin formulation and synthesis parameters.

MUF resins have seen widespread use as an adhesive in wood production, coating technologies, paper industries, and as the main material in kitchenware production. In the wood industry, it has been successfully used for plywood, particleboard, medium-density fiberboard, and other wood products for decades due to its higher bond qualities (Tohmura et al., 2001). It is known that the melamine content has a very significant effect on both the resin and particleboard properties. High melamine content will increase gel time, solid content, and internal bond strength. In contrast, there is a decrease in thickness swelling, water absorption, and formaldehyde emission (Hse et al., 2008).

Similar to the UF resins, formaldehyde emission from the products bonded by MUF has become an indoor air pollution problem causing harm to human health. However, the containing of melamine in this resin not only increases the bonding strength but also reduces the formaldehyde emission compared to the other thermosetting resins such as MF and UF resins. Formaldehyde in the resin is mainly emitted from the following three sources: (1) residual formaldehyde in the resin; (2) formaldehyde released by hydrolytic degradation of cured resins, especially under high temperature and humidity circumstances; and (3) formaldehyde generated by a condensation reaction between hydroxymethyl groups and other aromatic carbons or two hydroxymethyl groups (Paiva et al., 2012a, b).

Recent efforts are engaged to reduce or prevent formaldehyde emissions from wood-based panels bonded with MUF resins. The most beneficial method to reducing formaldehyde emission involves lowering the formaldehyde/amino group molar ratio (F/A) during the synthesis of resin adhesives while maintaining the required reactivity and cross-linking. It has been projected that more substituted triazine and urea nuclei in the MUF resin produce lower amounts of formaldehyde as measured by '3C nuclear magnetic resonance (NMR) spectroscopy (Mercer and Pizzi, 1996).

5.4 Phenol-formaldehyde

PF resin is extensively used as adhesives and binders for lignocellulosic material (Pizzi, 1983). The resin is used by the timber industry, e.g., in the manufacturing of plywood. This resin is used as a binder for exterior application, which requires superior water resistance provided by all phenolic-type resins. Another advantage of PF resin is characterized by its high strength, resistance to moisture, good dimension stability, and low cost (Koch et al., 1987; Pizzi et al., 1994). The bond strength of PF resin is high and its deterioration at elevated temperature in the presence of moisture is slight compared to UF and MUF adhesives (Yazaki, 1996). Phenolic resin is the first true synthetic polymer to be developed commercially (Pizzi et al., 1994).

PF can be classified into two types: novolac and resol. Novolac is produced from the condensation process between phenol and formaldehyde in which the phenol molar is higher than formaldehyde and in the presence of acid catalyst. Novolac resin requires the addition of paraform or hexamethylenetetramine as hardening agent (Bain and Wagner, 1984). This resin, however, is relatively slow curing and is less tolerant to bonding at higher wood moisture contents and variations in tree species compared, for example, with MUF. Resol is also produced from a similar condensation process, but the formaldehyde molar is in excess of that of phenol and is carried out in the presence of an alkaline catalyst (Pizzi and Mittal, 2003).

Pizzi et al. (1994) explained that, in the resol PF reaction process, phenol will condense with formaldehyde initially in the presence of an alkaline to form a methylol phenol or phenolic alcohol, and further reacts to form dimethylol phenol. The initial attack may be either at ortho or para position of the phenol. The second stage of the reaction involves the reaction of the methylol groups with other available phenol or methyl phenol, leading to the formation of linear polymers and then to the formation of hard-cured, highly branched structures. A resol molecule contains reactive methylol groups, and heating will cause the groups to condense and form large molecules without the addition of a hardener. The function of phenols as nucleophiles is strengthened by ionization of the phenol without affecting the activity of the aldehyde (Pizzi and Mittal, 2003).

Generally, PF is applied in a broad spectrum of engineered wood products. It is very strong and highly durable in dry and wet conditions and exhibits a very high adhesion to wood. PF adhesives are mainly applied in moist and climate-resistant particle- or fiberboard, plywood, pressed laminated wood, and glued laminated timber.

5.5 Oil palm-based composites

Since the early 1980s, research on the utilization of oil palm as an alternative raw material for the wood industry has been conducted. It had been reported that a variety of products such as particleboard, medium-density fiberboard, plywood, nanocomposite, etc. could be produced using the discarded oil palm trunks (OPT), oil palm fronds (OPF) and empty fruit bunches (EFB) (Husin et al., 2003; Khalid et al., 2016). Apart from that, laminated-based composite made from OPT and frond was carried out by several studies (Srivaro et al., 2019;

Khalid et al., 2015; Aizat et al., 2014; Rasat Sukhairi et al., 2011). Most of the research on oil palm-based composites used UF, MUF, and PF as binders (Abdul Khalil et al., 2010a, b). Those composites for interior and nonstructural applications were made from UF, while those for exterior and structural applications were made from MUF and PF resins.

Recently, Onoja et al. (2019) compiled the previous studies on converting oil palm biomass into value-added products including composites. In their study, oil palm biomass, especially EFB, OPT, and OPF, are among the most common biomass used in the manufacture of particleboard, plywood, laminated veneer lumber (LVL) and medium-density fiberboard (MDF). The most available source of fiber comes from EFB, which has become a potential raw material for MDF mills in Malaysia. Many researchers studied the potential of EFB for MDF manufacturing using different types of adhesives. A study by Norul Izani et al. (2013) found that the properties of MDF increased from 8% to 12% when resin level of UF increased. The authors also noticed that the mechanical properties and internal bonding improved significantly when EFB was treated with alkali. While Ridzuan et al. (2002) used PF resin with a solid content of 48% to produce MDF from EFB. In addition, Lee et al. (2018a, b) used UF resin to fabricate particleboard from OPT. The bonding properties of particleboard decreased due to the heat applied to the substrate. These studies suggest that further investigation should be carried out using MUF and PF resins.

Apart from EFB, a composite made of OPT has gained much attention from researchers. A woody-like structure, high volume, and nearly cylindrical shape are the advantages of the OPT biomass. This makes OPT practical to be processed into any elements including lumber or scantling, strip, veneer, particle, and fiber. In a study conducted by Hoong et al. (2012), OPT can potentially be converted into high grade plywood with low molecular weight phenol-formaldehyde (LMwPF) resin treatment. In their study, the pretreatment of OPT plywood with LMwPF enhanced the mechanical and physical properties. Furthermore, Hafizah et al. (2014) studied the effect of different molecular weights of PF resin on the properties of prepreg palm veneers for plywood manufacturing. The research focused on using PF as an adhesive to glue between the layers. Apart from mechanical improvement, the surface quality of treated oil palm veneer was enhanced. It was found that the medium molecular weight phenol-formaldehyde (MMwPF) appears to be the most suitable resin for the prepreg process.

Meanwhile, the utilization of OPF in oil palm-based composites has also been reported in several works. In a study done by Khalid et al. (2015), the laminated OPF composite was manufactured using three different numbers of layers (6, 8, and 10 layers) and PF adhesive spread rates (200, 250, and $300\,g/m^2$). The findings stated that the 10-layer laminated board has better compaction, higher density, and greater mechanical properties compared to 8 and 6 layers.

Reported work on formaldehyde-based oil palm composites is illustrated in Table 5.1. The suitability of oil palm biomass in polymeric matrices (UF, MUF, and PF resin) in composite development as reported by various researchers in the recent past is compiled. The properties of these composites including physical, mechanical, thermal, electrical properties, water sorption, degradation, etc. are also available in the literature. Most of the studies proved that oil palm biomass loading significantly improved the properties and strength of the composites.

Apart from particleboard, plywood, and MDF, numerous studies on the development of hybrid composites are gaining attention to increase the mechanical and physical performance of oil palm by-product-based composites. The hybrid composites offer benefits to

Table 5.1 Reported work on formaldehyde-based oil palm composites.

Types of composite	Resin	References
Hybrid composite	UF	Suhaily et al. (2019)
Plywood	UF	Hermanto and Massijaya (2018)
Particleboard	UF	Lee et al. (2018a,b)
Particleboard	UF	Baskaran et al. (2017)
Medium density fibreboard (MDF)	UF	Norul Izani et al. (2013)
Hybrid plywood	UF	Abdul Khalil et al. (2010b)
Laminated veneer lumber (LVL)	UF	Othman et al. (2009)
Plywood	MUF	Ong et al. (2012, 2018)
Laminated veneer lumber (LVL)	MUF	Othman et al. (2009)
Particleboard	MUF	Hse et al. (2008)
Hybrid composite	PF	Ramlee et al. (2019)
Plywood	PF	Khalid et al. (2016)
Laminated board	PF	Khalid et al. (2015)
Plywood	PF	Hafizah et al. (2014)
Plywood	PF	Hoong et al. (2012)
Hybrid plywood	PF	Abdul Khalil et al. (2010b)
Laminated veneer lumber (LVL)	PF	Othman et al. (2009)
Hybrid composite	PF	Sreekala et al. (2004, 2005)
Medium density fibreboard (MDF)	PF	Ridzuan et al. (2002)
Hybrid composite	PF	Sreekala et al. (2002, 2000)

develop products with better strength properties for various industrial applications. Previous studies reported tensile and flexural characteristics of pure oil palm empty fruit bunch (OPEFB) composites can be improved by hybridization with natural/cellulosic fibers such as sugarcane bagasse (SCB) fiber, kenaf, pineapple leaf fiber, and various grasses with formaldehyde-based resin. For example, Ramlee et al. (2019) fabricated hybrid composites consisting of OPEFB and SCB fiber composites reinforced with PF resin. The authors investigated the potential effect of OPEFB and SCB fiber ratio on the tensile strength, density, water absorption, thickness swelling, void content, and morphological properties of pure and hybrid composite. It was found that hybridization of OPEFB/SCB fiber composites indicates high tensile strength and modulus, 5.56 and 661 MPa, respectively, with less porous and voids area compared to pure composites.

Furthermore, hybrid composites reinforced with oil palm fibers are very often combined with synthetic fibers such as glass fibers. The hybrid effect of glass fiber and OPEFB fiber on the tensile, flexural, and impact response of the PF resin-based composite was investigated by Sreekala et al. (2002), where it improved the overall performance of the composite. However, considering the environmental issues nowadays, utilization of natural/natural fiber in hybrid composites are more efficient and address safety concerns to the user compared to natural/synthetic fiber-based hybrid composites.

5.6 Conclusions

Formaldehyde-based adhesives are extensively used to fabricate oil palm-based panels with desired properties. The selection of formaldehyde-based adhesives is highly dependent on the specification of panel required. Although formaldehyde-based adhesives may seem an attractive adhesive for oil palm-based panel fabrication, there is still much to be explored and improved, especially the methodology for reducing formaldehyde emission.

References

Abdul Khalil, H.P.S., Aamir, H.B., Jawaid, M., Parisa, A., Ridzuan, R., Said, M.R., 2010a. Agro-wastes: mechanical and physical properties of resin impregnated oil palm trunk core lumber. Polym. Compos. 31 (4), 638–644.

Abdul Khalil, H.P.S., Nurul Fazita, M.R., Bhat, A.H., Jawaid, M., Nik Fuad, N.A., 2010b. Development and material properties of new hybrid plywood from oil palm biomass. Mater. Des. 31 (1), 417–424.

Aizat, A.G., Zaidon, A., Nabil, F.L., Bakar, E.S., Rasmina, H., 2014. Effects of diffusion process and compression on polymer loading of laminated compreg oil palm (Elaeis guineensis) wood and its relation to properties. J. Biobaased Mater. Bioenergy 8 (5), 519–525.

Bain, D.R., Wagner, J.D., 1984. Molecular weight distribution of phenol-formaldehyde resols by high performance gel permeation chromatography. Polymer 25 (3), 403–404.

Baskaran, M., Azmi, N.A.C.H., Hashim, R., Sulaiman, O., 2017. Properties of binderless particleboard and particleboard with addition of urea formaldehyde made from oil palm trunk waste. J. Phys. Sci. 28 (3), 151–159.

Bono, A., Duduku, K., Mariani, R., Nancy, J.S., 2006. Variation of reaction stages and mole composition effect on melamine-urea-formaldehyde (MUF) resin properties. In: 4th Asia-Pacific Chemical Reaction Engineering Symposium on New Opportunities of Chemical Reaction Engineering in Asia-Pacific Region, 12–15 Jun 2005, Gyeongju, South Korea.

Boran, S., Mustafa, U., Sedat, O., Esat, G., 2012. The efficiency of tannin as a formaldehyde scavenger chemical in medium density fibreboard. Compos. Part B Eng. 43 (5), 2487–2491.

Chai, L.Y., H'ng, P.S., Lim, C.G., Chin, K.L., Jusoh, M.Z., Bakar, E.S., 2011. Production of oil palm trunk core board with wood veneer lamination. J. Oil Palm Res. 23, 1166–1171.

Conner, A.H., 1996. Urea-formaldehyde adhesive resins. In: Polymeric Materials Encyclopedia. vol. 11, pp. 8496–8501.

Conner, A.H., 2001. Wood: adhesive. In: Encyclopedia of Materials: Science and Technology, second ed. Elsevier Science, Ltd., Amsterdam; New York, pp. 9583–9599.

Conner, A.H., Bhuiyan, M.S.H., 2017. Wood: adhesive. In: Reference Module in Materials Science and Materials Engineering. Elsevier, Amsterdam, Netherlands, https://doi.org/10.1016/B978-0-12-803581-8.01932-9.

Dunky, M., 1998. Urea–formaldehyde (UF) adhesive resins for wood. Int. J. Adhes. Adhes. 18 (2), 95–107.

Dunky, M., 2003. Adhesives in the wood industry. In: Handbook of Adhesive Technology. vol. 2, p. 50.

H'ng, P.S., Chai, L.Y., Chin, K.L., Tay, P.W., Eng, H.K., Wong, S.Y., Wong, W.Z., Chow, M.J., Chai, E.W., 2013. Urea formaldehyde impregnated oil palm trunk as the core layer for three-layered board. Mater. Des. 50, 457–462.

Hafizah, N., Wahab, A., Paridah, M.T., Nor Yuziah, M.Y., Zaidon, A., Adrian Choo, C.Y., Nor Azowa, I., 2014. Influence of resin molecular weight on curing and thermal degradation of plywood made from phenolic prepreg palm veneers. J. Adhes. 90 (3), 210–229.

Hermanto, I., Massijaya, M.Y., 2018. Performance of composite boards from long strand oil palm trunk bonded by isocyanate and urea formaldehyde adhesives. IOP Conf. Ser. Earth Environ. Sci. 141, 012012.

Hoong, Y.B., Yueh, F.L., Paridah, M.T., Harun, J., 2012. Development of a new pilot scale production of high-grade oil palm plywood: effect of pressing pressure. Mater. Des. 36, 215–219.

Hse, C.Y., Feng, F., Hui, P., 2008. Melamine-modified urea formaldehyde resin for bonding particleboards. For. Prod. J. 58 (4), 56–61.

Hun, D.E., Richard, L., Maria, T.M., Jeffrey, A.S., 2010. Formaldehyde in residences: long-term indoor concentrations and influencing factors. Indoor Air 20 (3), 196–203.

Husin, M., Anis, M., Wan Hasamudin, W.H., 2003. Oil Palm Plywood, MPOB Information Series. Malaysian Palm Oil Board, Bangi.

Khalid, I., Othman, S., Rokiah, H., Razak, W., Nadiah, J., Mohd Sukhairi, M.R., 2015. Evaluation on layering effects and adhesive rates of laminated compressed composite panels made from oil palm (*Elaeis guineensis*) fronds. Mater. Des. 68, 24–28.

Khalid, H., Zakiah, A., Paridah, M.T., Jamaludin, K., 2016. Investigation on the water absorption characteristics of plywood manufactured using veneers from oil palm stem. J. Teknol. 78 (5), 99–103.

Koch, G.S., Klareich, F., Exstrum, B., 1987. Adhesives for the Composite Wood Panel Industry. Noyes Publications, Park Ridge, New Jersey, United States.

Lee, S.H., Ashaari, Z., Lum, W.C., Ang, A.F., Halip, J.A., Halis, R., 2018a. Chemical, physico-mechanical properties and biological durability of rubberwood particleboards after post heat-treatment in palm oil. Holzforschung 72 (2), 159–167.

Lee, S.H., Zaidon, A., Aik, F.A., Juliana, A.H., Wei, C.L., Rasdianah, D., Rasmina, H., 2018b. Effects of two-step post heat-treatment in palm oil on the properties of oil palm trunk particleboard. Ind. Crop Prod. 116, 249–258.

Mercer, A.T., Pizzi, A., 1996. A 13C-NMR analysis method for MF and MUF resins strength and formaldehyde emission from wood particleboard. I. MUF resins. J. Appl. Polym. Sci. 61 (10), 1687–1695.

National Biomass Strategy, 2020. New Wealth Creation for Malaysia's Biomass Industry (Version 2.0, 2013). Available from https://www.nbs2020.gov.my/nbs2020-v20-2013.

No, B.Y., Kim, G.M., 2007. Evaluation of melamine-modified urea-formaldehyde resins as particleboard binders. J. Appl. Polym. Sci. 106 (6), 4148–4156.

Norul Izani, M.A., Paridah, M.T., Anwar, U.M.K., Mohd Nor, M.Y., H'ng, P.S., 2013. Effects of fibre treatment on morphology, tensile and thermogravimetric analysis of oil palm empty fruit bunches fibres. Compos. Part B Eng. 45 (1), 1251–1257.

Ong, H.R., Prasad, R., Rahman Khan, M., Chowdhury, M., Kabir, N., 2012. Effect of palm kernel meal as melamine urea formaldehyde adhesive extender for plywood application: using a Fourier transform infrared spectroscopy (FTIR) study. Appl. Mech. Mater. 121–126, 493–498.

Ong, H.R., Khan, M.M.R., Prasad, D.R., Yousuf, A., Chowdhury, M.N.K., 2018. Palm kernel meal as a melamine urea formaldehyde adhesive filler for plywood applications. Int. J. Adhes. Adhes. 85, 8–14.

Onoja, E., Sheela, C., Fazira Ilyana, A.R., Naji, A.M., Roswanira, A.W., 2019. Oil palm (Elaeis guineensis) biomass in Malaysia: the present and future prospects. Waste Biomass Valoriz. 10 (8), 2099–2117.

Ormondroyd, G.A., 2015. Adhesives for wood composites. In: Wood Composites. Woodhead Publishing, Sawston, United Kingdom, pp. 47–66.

Othman, S., Salim, N., Rokiah, H., Yusof, L.H.M., Razak, W., Yunus, N.Y.M., Hashim, W.S., Azmy, M.H., 2009. Evaluation on the suitability of some adhesives for laminated veneer lumber from oil palm trunks. Mater. Des. 30 (9), 3572–3580.

Paiva, N.T., Henriques, A., Cruz, P., Ferra, J.M., Carvalho, L.H., Magalhães, F.D., 2012a. Production of melamine fortified urea-formaldehyde resins with low formaldehyde emission. J. Appl. Polym. Sci. 124 (3), 2311–2317.

Paiva, N.T., Pereira, J., Ferra, J.M., Cruz, P., Luísa, C., Magalhães, F.D., 2012b. Study of influence of synthesis conditions on properties of melamine–urea formaldehyde resins. Int. Wood Prod. J. 3 (1), 51–57.

Papadopoulou, E., Sotiris, K., Zoe, N., Konstantinos, C., Chrysoula, M., Konstantinos, K., Angelos, A.L., 2016. Urea-formaldehyde (UF) resins prepared by means of the aqueous phase of the catalytic pyrolysis of European beech wood. COST action FP1105. Holzforschung 70 (12), 1139–1145.

Petković, B., Suzana, S.J., Vojislav, J., Biljana, D., Gordana, M., Milena, M.C., 2015. Effect of γ-irradiation on the hydrolytic stability and thermo-oxidative behavior of bio/inorganic modified urea–formaldehyde resins. Compos. Part B Eng. 69, 397–405.

Pizzi, A., 1983. Wood Adhesives Chemistry and Technology. Marcel Dekker, New York, pp. 59–104.

Pizzi, A., Mittal, K.L., 2003. Handbook of Adhesive Technology, Revised and Expanded. CRC Press, Boca Raton, Florida, United States.

Pizzi, A., Mtsweni, B., Parsons, W., 1994. Wood-induced catalytic activation of PF adhesives autopolymerization vs. PF/wood covalent bonding. J. Appl. Polym. Sci. 52 (13), 1847–1856.

Que, Z., Takeshi, F., Sadanobu, K., Yoshihiko, N., 2007. Effects of urea-formaldehyde resin mole ratio on the properties of particleboard. Build. Environ. 42 (3), 1257–1263.

Ramlee, N.A., Jawaid, M., Zainudin, E.S., Yamani, S.A.K., 2019. Tensile, physical and morphological properties of oil palm empty fruit bunch/sugarcane bagasse fibre reinforced phenolic hybrid composites. J. Mater. Res. Technol. 8 (4), 3466–3474.

Rasat Sukhairi, M.M., Razak, W., Othman, S., Janshah, M., Aminuddin, M., Tabet, A.T., Izyan, K., 2011. Properties of composite boards from oil palm frond agricultural waste. Bioresources 6 (4), 4389–4403.

Ridzuan, R., Shaler, S., Jamaludin, M.A., 2002. Properties of medium density ties of medium density fibreboard from oil palm empty fruit bunch fibre. J. Oil Palm Res. 14 (2), 34–40.

Rokiah, H., Wan Noor Aidawati, W.N., Othman, S., Fumio, K., Salim, H., Masatoshi, S., Tomoko, S., Tay, G.S., Ryohei, T., 2011. Characterization of raw materials and manufactured binderless particleboard from oil palm biomass. Mater. Des. 32 (1), 246–254.

Sreekala, M.S., Kumaran, M.G., Joseph, S., Jacob, M., Thomas, S., 2000. Oil palm fibre reinforced phenol formaldehyde composites: influence of fibre surface modifications on the mechanical performance. Appl. Compos. Mater. 7 (5), 295–329.

Sreekala, M.S., George, J., Kumaran, M.G., Thomas, S., 2002. The mechanical performance of hybrid phenol-formaldehyde-based composites reinforced with glass and oil palm fibres. Compos. Sci. Technol. 62 (3), 339–353.

Sreekala, M.S., Kumaran, M.G., Geethakumariamma, M.L., Thomas, S., 2004. Environmental effects in oil palm fibre reinforced phenol formaldehyde composites: studies on thermal, biological, moisture and high energy radiation effects. Adv. Compos. Mater. 13 (3–4), 171–197.

Sreekala, M.S., Thomas, S., Groeninckx, G., 2005. Dynamic mechanical properties of oil palm fiber/phenol formaldehyde and oil palm fiber/glass hybrid phenol formaldehyde composites. Polym. Compos. 26 (3), 388–400.

Srivaro, S., Nirundorn, M., Frank, L., 2019. Performance of cross laminated timber made of oil palm trunk waste for building construction: a pilot study. Eur. J. Wood Wood Prod. 77 (3), 353–365.

Suhaily, S.S., Gopakumar, D.A., Aprilia, N.S., Samsul, R., Paridah, M.T., Khalil, H.A., 2019. Evaluation of screw pulling and flexural strength of bamboo-based oil palm trunk veneer hybrid biocomposites intended for furniture applications. Bioresources 14 (4), 8376–8390.

Tohmura, S., Chung-Yun, H., Mitsuo, H., 2000. Formaldehyde emission and high-temperature stability of cured urea-formaldehyde resins. J. Wood Sci. 46 (4), 303–309.

Tohmura, S., Akio, I., Siti, H.S., 2001. Influence of the melamine content in melamine-urea-formaldehyde resins on formaldehyde emission and cured resin structure. J. Wood Sci. 47 (6), 451–457.

Yazaki, 1996. What comes after phenolic type adhesives for bonding wood to wood? In: 25th Forest Research Conference. vol. 1. CSIRO Divison of Forestry and Forest Products, Clayton, Victoria, Australia, pp. 4–7. 18–21 November 1991.

Yuan, J., Xiaowen, Z., Lin, Y., 2015. Structure and properties of urea-formaldehyde resin/polyurethane blend prepared via in-situ polymerization. RSC Adv. 5 (66), 53700–53707.

6

Nonformaldehyde-based adhesives used for bonding oil palm biomass (OPB)

Nor Yuziah Mohd Yunus

Faculty of Applied Sciences, Universiti Teknologi MARA Cawangan Pahang, Bandar Tun Abdul Razak, Jengka, Pahang, Malaysia

6.1 Introduction

Oil palm biomass (OPB) presents a myriad of materials with properties that considerably differ. These materials are derived from the oil palm trunk (OPT), oil palm fronds (OPF), kernel, empty fruit bunches (EFB), and other by-products of oil extraction (Anis et al., 2014). Out of these, the fibers attained from the trunks, fronds, and EFB provide an exciting resource that might be suitable for making composites or even a solid timber equivalent. The timber equivalent could come from processed trunks that need to undergo a specific cutting pattern then treated as the property gradient varies dramatically (Suthon et al., 2018). The OPB has been researched extensively and converted to products including plywood, particleboard, medium-density fiberboard (MDF), wood plastic composite, and solid wood equivalent. Some are still in the immature stage at a laboratory scale, while conversions to pilot scale or preproduction test have also been done. For plywood, with proper selection and treatment of the palm veneer, a high performance product is possible (Paridah et al., 2014). The properties of the high performance product rival the performance of normal tropical plywood.

In the history of wood-based composites preparation, the pre-1900 material relies on the naturally found gluing material (Frihart, 2015; Gadhave et al., 2017). Furniture-making deals with adhesives that utilized an animal base in the form of modified gelatin derived from bones and hides, blood glue, and glue resembling performance of phenolics, i.e., the casein. These gluing systems are found widely in plywood and furniture. Ever since the introduction of phenol formaldehyde (PF) at the onset of World War II, the development of synthetic materials escalated, causing more emphasis given to the likes

---

Oil Palm Biomass for Composite Panels. https://doi.org/10.1016/B978-0-12-823852-3.00009-X
Copyright © 2022 Elsevier Inc. All rights reserved.

of amino resins, phenolics, urethanes, and isocyanates. The urethanes and isocyanates are fast becoming the "go-to" adhesive systems, limited its cost and handling requirement (isocyanates). The enthusiasm toward recovery of research and development of natural- and renewable-based products is triggered by the awareness of the user toward the side issues of volatile emission and its detrimental effects (Adamová et al., 2020). Thus, the impetus for finding a more reliable, durable, sustainable, and comparable adhesive material has now become evident. The move toward nonformaldehyde or low emission formaldehyde-based resin becomes more attractive due to its major two prong benefit: i. reduction of volatile-emitting adhesive system and ii. the promise of renewable and potentially recyclable resources.

6.2 Resin for wood based industry

Global market watch by Grand View Research, Inc. (2019) generally forecasted a value of $6.34 billion by 2025 in resin usage. These values are linked to the demand for laminated flooring fixtures, plywood, and furniture applications. For the period of 2011–21 (estimate) (Fig. 6.1), an upward increment trend is seen. This increment is indirectly linked to the growth in population and expected wealth increase. It is expected that the younger generation is trendier but need materials that could be used in a lower footprint area such as condominiums and flats.

Urea formaldehyde (UF) value dominates at 37% followed by melamine urea-formaldehyde (MUF), isocyanates, and PF. The formaldehyde-based resin easily covers over 70% of resin usage in

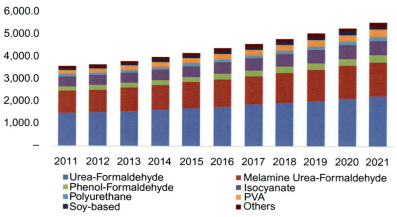

Fig. 6.1 Trend of market demand on wood-based adhesive (Grand View Research, 2019).

the wood-based industry. However, a move toward less utilization of formaldehyde-based and formaldehyde-free or even resin-free product has been pushed for. This is in the event of formaldehyde being classified as a carcinogen and the need of lower volatile content in composite product and those downstream, i.e., flooring and furniture.

For non-formaldehyde-based glue, isocyanate is highest (structural resin) followed by polyvinylacetate (nonstructural). The likelihood of isocyanate to emerge further is high as it is closely related to the polyurethanes (PUs). As seen later, PUs are formed in two parts in which the polyols could be derived from palm oil (Yeoh et al., 2020; Prociak et al., 2018). Soy-based adhesives are reported to have 7.5% growth revenue rate from 2019 to 2025 (estimate) and are largely replacing petrochemical-based adhesives in the woodworking industry. These adhesives are environmentally friendly and improved the overall performance of manufactured wood panels.

The formaldehyde-based resin, be it amino or phenolic, has dominated the wood-based industry, with amino comprising of UF, MUF, and melamine formaldehyde (MF) finding its way into the interior used composites. Being cheap, colorless, and flexible in its use during production, it's hard not to use the adhesive. The reliance of the furniture industry on lamination (heavily MF-based) to provide surfaces that are durable, heat resistant, coupled with design flexibility encourages more MF and MUF usage (Kandelbauer et al., 2010). A move toward the utilization of glulam and laminated veneer lumber product for structural purpose has popularized PF.

Keeping the formaldehyde-based resin advantage in mind, new or reinvention of earlier non-formaldehyde-based resins have raised the challenge of a well-established adhesive system. The closest competitor toward a nonformaldehyde-based adhesive is the isocyanate. Ranked third after UF and MF, isocyanate development need to address its monomer toxicity and processing handling. The positive part associated to the isocyanate is that it can bond on its own using the $=CNO$ functionality or be used with polyols to obtain PU (Li et al., 2015). The polyol could be contained from oil-based or cellulose-based, where an abundance of –hyxroxyl functionality can be made available.

The next potential are the carbohydrates available from the plant-based. The carbohydrates are found in abundance in roots, tubers, seeds (rice, wheat, corn), and parenchyma tissues as seen for OPT (Anis et al., 2014). Carbohydrates are rich in hydroxyl groups, which could be available for bonding. Starch, for example, has been used widely in the pulp and paper industry, in its native form or modified to be anionic or cationic. The modification is often done to cater to the bonding environment process as well as the final used destination of the resultant products. Often the starch is coupled with material such as nano silica to enhance its performance (Wang et al., 2011) or grafted to improve compatibility (Wang et al., 2012).

Next, research is interested in the soy-based adhesive system. The soy-based system is often quoted as soy hydrolysate, which has undergone modification with the likes of anhydrides (Qi et al., 2013) to improve its bonding with plywood. Recently, *Aspergillus niger* enzymic action was used to produce a workable hydrolyzed soy-based carbohydrate with 30% plywood strength enhancement by Zheng et al. (2019) to open more potential for this resource.

6.3 The adhesion theory

In adhesion, an angle of contact of less than 90 degrees is important to create good bonding. This angle is very much controlled by the resin droplet's behavior on the wood substrate surface. A good angle will ensure optimum wetting of the substrate creating good and even coverage, giving potential to intimate contact at the glueline. The resin types must consider wetting, which is controlled to a certain extent by viscosity. At the same time, the different interactions of the resin with the wood substrate will induce both a better environment toward wetting and strength development. The nature of the resin will control the behavior of the resin toward wood. Wood has an abundant hydroxyl group, derived from its major components, cellulose, hemicellulose, and maybe lignin. Fig. 6.2 depicts the wood component and their relationship.

With the availability of various components in wood, the resin used will also interact in a varying manner. The interaction could be chemical-based or mechanical-based. The best strength for chemical interaction would be from covalent bonding, followed by hydrogen

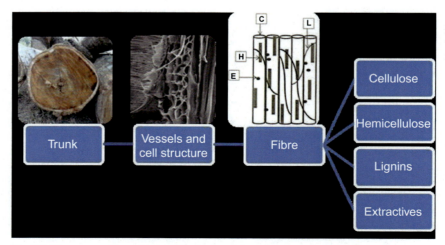

Fig. 6.2 The composition of wood, illustrating the positioning of each component (DeWild et al., 2011).

bonding, dipolar interaction, and finally Van der Waal forces. Mechanical interlocking tops the chart and is molecular weight, viscosity, and voids presence-dependent. Not to be excluded is diffusion of a similar material forming a welded interface at the glueline.

With bonding of oil palm, another challenge to be overcome is the variability of the components seen in trunk, fronds, and EFB. With fronds, it has a more fibrous form, which has a similar chemical characteristic to wood. The trunks harbor both a fiber bundle and parenchyma tissues, which need to accounted for depending on the types of products made. In term of parenchyma effect on interaction, it is expected for solid > veneers > strands > particles > fibers. Resin used to bond wood substrate needs to take advantage of the bonding strength. Thus, the popularity of formaldehyde-based resin occurs as the hydroxyl in methylol forms of the resin can easily form a covalent bond, hydrogen bonding, dipolar interaction, and, if the chain length is sufficient, Van de Waals forces. Amino has the added advantage of amine groups ready for covalent and hydrogen bonds. The next strongest glue for wood is isocyanate with −CNO group, where the double bond is ready to form a covalent bond with hydroxyl on wood. Subsequently PU, a polymer derived from polyol and diisocyanate, could also take advantage of covalent bonding (−NH, amine) and dipolar interaction (Fig. 6.3).

6.4 Bonding of oil palm biomass (OPB)

The bonding of OPB is widely reported with various types of resins and adhesive formulations. OPBs are varied in properties, which allow it to form innovative particleboards. Various innovative methods have been attempted by Balducci et al. (2008) and Haque et al. (2017) when dealing with annual and perennial farm plant waste. For OPB, the more common adhesive used would be formaldehyde-based

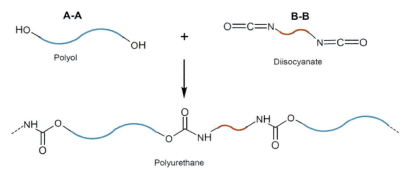

Fig. 6.3 Schematic of reaction of polyol and diisocyanate (anon.polymerdatabase.com 2020).

adhesives, both amino and phenolic. This is expected as ~ 70% of the adhesive used by the composite producers are formaldehyde-based, namely UF, MUF, MF, PF resorcinol-PF, and resorcinol formaldehyde (Hartono and Sucip, 2018).

With the exception of a binderless board, other potential adhesives could be either natural or synthetic. These include the PUs (both synthetic and natural), isocyanate, carbohydrate, epoxy, proteins, tannins, and lignin. The olefins used in fiber plastic composite are considered as a matrix for comparison purposes. All these materials have been studied at one point or another to access their suitability. It is worth noting that the utilization of resin in oil palm comes in the form of binding and/or stabilization. Binding is when the substrate surface interfaces with the resin, while stabilization normally involves impregnation of the resin to fill lumen and void in the substrate.

6.4.1 Polyurethane

PU, as shown in Fig. 6.4, is made from two components. An interesting point on PU, oil palm can form the polyol base for its production. In an interesting study by Badri et al. (2005), initially a PU based on palm kernel oil was made. The PU was foamed, and the researchers tried to enhance the stability of the foam by incorporation of fillers. Their finding showed that EFB fibers of different sizes at levels namely 45–56 μm (5.5%), 100–160 μm (4.5%), and 200–315 μm (2.5%) could enhance the foam produced. A composite with 5.5% fiber loading at size 45–56 μm exhibited best compression and modulus. Larger particles cause tearing of foam, while smaller sizes filled the voids in the cell structure. The small size fiber filled the void in the PU, and its compactness was shown by SEM micrograph. The smaller the particle size, the higher the strength observed. Here, factors other than fiber length such as mode of failure, plasticization, and better tendency for consolidation in molding play a significant role. Similar failure caused by fiber pull-out is more common in larger or longer size fiber, and the manner of disruption was also captured by SEM. Fiber pull-out is common as the fibers are embedded and not covalently bonded to the PU matrix.

6.4.2 Epoxy

A study on randomly reinforced epoxy with short oil palm fiber by Mohd Zuhri et al. (2009, 2010) showed decreasing tensile and flexural as loading of fiber increased owing to the fiber pull-out. Fiber orientation and aspect ratio is said to affect both properties. The presence of hydroxyl groups on epoxies is supposed to bond the material owing to

The covalent and dipolar interaction. The remarks were linked to findings by Myrtha et al. (2008), where the flexural strength of EFB/polyester composites for longer fiber is 8% higher than short fiber at 18 vol%.

6.4.3 Isocyanate

Indra Hermanto and Massijaya (2018) compared performance of UF against isocyanate in bonding long strands of 150 cm from the trunk. The strand was compressed by 40%, followed by glue spread. The resinated strands were then arranged to three layers board following a plywood technique. The isocyanate performed better than UF, needing only 225 g/m^2 spread value and UF needing a 33% higher dose to obtain JIS A 5908 (2003) requirement. The performance is expected as isocyanate is a structural resin. Another binding of oil palm xylem using isocyanate at three glue spread of 200, 250, and 300 g/m^2 was performed by Atmawi et al. (2014). The strength comparison of 250 and 300 g/m^2 did not show a significant shear value increase as the dosage increased. However, the higher glue spread was observed to have 100% wood failure as opposed to 78% in 250 g/m^2. This indicated cohesive failure in the substrate showing the limitation of the material. The damage can easily be seen in Fig. 6.4 where the breakage occurs at the soft parenchyma tissue. Thus, there is some study on the potential of oil palm wood densification, which could increase the strength of bonded product (Hartono and Sucip, 2018; Choowang and Hiziroglu, 2015).

6.4.4 Carbohydrate

Carbohydrate is a renewable resource that includes glucose, sucrose, fructose, and xylose. Its polymer could be formed at high temperature aiming for acid-catalyzed dehydration with reaction extent measured by moles of water loss. The degraded carbohydrate can be reacted with urea, phenol, and formaldehyde. The formulation (35% replacement of phenol) was then alkaline cured and showed promising plywood bonding. The sensitivity of the formulation is the molar ratio of reacting ingredient (Phenol: Carbohydrate-1 $\geq$ 1 and Phenol: Formaldehyde-1 $\geq$ 2) (Christiansen and Gillespie, 1986). Conner et al. (1986) incorporated reducing carbohydrates partially into phenol-formaldehyde resol-resins. The formulation did not adversely affect the dry- or wet-shear strength of bonded 2-ply Douglas fir panels. For OPB, direct utilization of carbohydrate as a binder has yet to be reported. Recent work by the author on the utilization of carbohydrate derived from OPT showed promise in a working formulation. Again, partial replacement of formaldehyde-based resin is necessary. The resulting resin is still to be tested in plywood made from oil palm veneers.

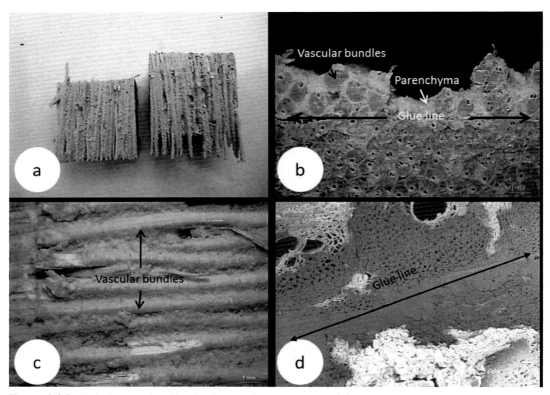

Fig. 6.4 (A) Optical micrographs of longitudinal-section wood failure; (B) cross-section view of the parenchymal tissue damage; (C) less damage of vascular bundles in longitudinal section; (D) and a SEM micrograph showing a well-bonded cross-section of oil palm xylem by isocyanate adhesive (Atmawi et al., 2014).

6.4.5 Binderless

Studies by Rokiah et al. (2012) utilized the natural bonds such as proteins, hemicellulose, lignin, and cellulose content of the trunk to bond particles of OPT and leaves. It is cited as binderless, as the particles were subjected to a temperature of 180°C under pressure for 20 min to induce bonding. Acceptable board was obtained from the frond and trunk mid/core but not from the leaves and bark. The nature of biomass in frond and trunk has been shown to be suitable for bonding. Temperature of 180°C was proven to be the optimal value for activation of various components of the substrate with a higher temperature causing degradation in protein (Milawarni et al., 2019). Hemicelluloses quickly decomposed in a temperature range from 200°C to 300°C, and cellulose is the least stable polymer, decomposing from 300°C to 400°C. Lignin exhibits intermediate thermal degradation behavior from 250°C to 500°C (Theapparat et al., 2018). Further comparative testing showed using 10% UF versus binderless particleboard using palm oil waste confirm 15%–25% stronger modulus of rupture and 70%–85% stronger IB for UF (Mohana et al., 2017).

6.4.6 Protein

Utilization of protein as a wood-based adhesive has been seen in the likes of casein, animal glue, and blood glue. Soy protein experimentation had shown formulation that needs fortification with UF and PF. Recent work by Wang et al. (2019) showed potential with soy protein adhesive containing 4% epoxy resin and 0.05% polyacrylamide (PAM) dosage. Reduction of viscosity from 77% to 58% for the formulation allows the modified soy protein to spread and wet the veneers used in the bonding test. With the addition of PAM, which lowers the cure temperature, the resultant plywood showed good water resistance and wet shear strength (interior use plywood of the Chinese Industrial Standard). Further study on viscosity reduction may render this combination suitable for other wood-based composites. This type of resin is yet to be tested on OPB.

6.4.7 Olefins

Olefins are thermoplastic in nature and are basically hydrophobic. Utilization of the material as a matrix for OPB usually showed the need of the fiber to be treated or modified to change its surface tension. Addition of compatibilizer that deals with both hydrophobic and hydrophilic surfaces is possible. Otherwise, other cross-linking techniques such as irradiation remains an option. Ratnam et al. (2007) worked with irradiated polyvinylchloride, epoxidized natural rubber, and EFBs composite where the radiation triggers a cross-link between fiber and matrix. In fact, oil palm flour was tested in the 1990s by Zaini et al. (1996), which showed that larger size of particles does show better tensile and impact strength, which is a similar trend seen with particles with a higher aspect ratio. Their additive-free composite exhibited a reduction of mechanical properties with addition of more filler. Russita and Bahruddin (2018) utilized polypropylene with and without maleated polypropylene (MAPP) with a touch palm frond paraffin plasticized with OPFs. Their attempt at large production with a twin extruder shows the difficulty of controlling two materials with different bulk densities. The result of the trial showed the polypropylene:palm frond powder and 40:60 ratio combination could meet the strength requirement of a commercial standard.

6.4.8 Others

There are other potential bonding materials that can be examined. Tannin or tannin incorporated into PF is possible (Jahanshaei et al., 2012), and lignin extracted from EFB was done by Risanto et al. (2014) and applied to plywood. Options of resin made from casein, starch, and other natural resources are still open for exploration.

6.5 Conclusion

OPB has diversified properties for its biomass, starting from the root to shoots and including the EFBs. Each location has its own peculiarity and needs to be accounted for when dealing with bonding. With treatment, the material can be modified to be more approachable by both hydrophobic and hydrophilic binders and matrix types. The utilization of nonformaldehyde-based adhesive systems has shown that the compatibility of the biomass is critical. The option of using synthetics includes urethanes, isocyanate, epoxy, and olefins. In the natural line, more emphasis can be given to soy-based, carbohydrate, tannins, lignin, and better still a binderless system that takes advantage of the binding properties of the biomass components.

References

Adamová, T., Hradecký, J., Pánek, M., 2020. Volatile organic compounds (VOCs) from wood and wood-based panels: methods for evaluation, potential health risks, and mitigation. Polymers 12, 2289. https://doi.org/10.3390/polym12102289. 2020.

Anis, M., Kamaruddin, H., Loh, Y.F., 2014. Availability of oil palm trunk. In: Handbook of Oil Palm Trunk Plywood Manufacturing. Malaysian Timber Industrial Board, pp. 1–15. Chapter 1.

Anon, 2020. Polymer Properties Database. https://polymerdatabase.com/polymer%20 chemistry/Urethanes.html. Accessed on14 Aug 2020, 2.27pm.

Atmawi, D., Muhammad, Y.M., Naresworo, N., Eka, M.A., Dodik, R.N., 2014. Bond ability of oil palm xylem with isocyanate adhesive. J. Ilmu dan Teknologi Kayu Tropis 12 (1), 39–47.

Badri, K., Zulkefly, O., Ilyati, M.R., 2005. Mechanical properties of polyurethane composites from oil palm resources. Iran. Polym. J. 14 (5), 441–448.

Balducci, F., Harper, C., Meinlschmidt, P., Dix, B., Sanasi, A., 2008. Development of innovative particleboard panels. Drv. Ind. 59 (3), 131–136.

Choowang, R., Hiziroglu, S., 2015. Properties of thermally-compressed oil palm trunks (*Elaeis Guineensis*). J. Trop. For. Sci. 27 (1), 39–46.

Christiansen, A.W., Gillespie, R.H., 1986. Potential of carbohydrates for exterior-type adhesives. For. Prod. J. 36 (7/8), 20–28.

Conner, A.H., River, B.H., Lorenz, L.F., 1986. Carbohydrate modified phenol-formaldehyde resins. J. Wood Chem. Technol. 6 (4), 591–613.

DeWild, P.J., Reith, H., Heeres, H.J., 2011. Biomass pyrolysis for chemicals. Biofuels 2, 185–208.

Frihart, C.R., 2015. Wood adhesives: past, present, and future. For. Prod. J. 65 (1/2), 4–8. https://doi.org/10.13073/65.1-2.4.

Gadhave, R.V., Mahanwar, P.A., Gadekar, P.T., 2017. Starch-based adhesives for wood/wood composite bonding: review. Open J. Polym. Chem. 7, 19–32. https://doi.org/10.4236/ojpchem.2017.72002.

Grand View Research, 2019. https://www.grandviewresearch.com/press-release/global-wood-adhesives-market. Accessed on Sep 2019.

Haque, A., Mondal, D., Khan, I., Usmani, M., Bhat, A.H., Gazal, U., 2017. Fabrication of composites reinforced with lignocellulosic materials from agricultural biomass. In: Lignocellulosic Fibre and Biomass-Based Composite Materials: Processing, Properties and Applications., https://doi.org/10.1016/B978-0-08-100959-8.00010-X.

Hartono, R., Sucipto, T., 2018. Quality improvement of laminated board made from oil palm trunk at various outer layer using phenol formaldehyde adhesive. IOP Conf. Ser.: Mater. Sci. Eng. 309, 012049. https://doi.org/10.1088/1757-899X/309/1/012049. 2018.

Hermanto, I., Massijaya, M.Y., 2018. Performance of composite boards from long strand oil palm trunk bonded by isocyanate and urea formaldehyde adhesives. IOP Conf. Ser.: Earth Environ. Sci. 141, 1–10.

Jahanshaei, S., Tabarsa, T., Asghari, J., 2012. Eco-friendly tannin-phenol formaldehyde resin for producing wood composites. Pigm. Resin Technol. 41 (5), 296–301.

JIS A 5908, 2003. Particleboards. Japanese Standards Association.

Kandelbauer, A., Petek, P., Medved, S., Pizzi, A., Teischinger, A., 2010. On the performance of a melamine–urea–formaldehyde resin for decorative paper coatings. Eur. J. Wood Wood Prod., 63–75. Springer-Verlag https://doi.org/10.1007/s00107-009-0352-y.

Li, Y., Luo, X., Hu, S., 2015. Bio-based Polyols and Polyurethanes, Springer Briefs in Green Chemistry for Sustainability. pp. 15–43, https://doi.org/10.1007/978-3-319-21539-6.

Milawarni, Nurlaili, Sariadi, Amra, S., Yassir, 2019. Influence of press temperature on the properties of binderless particleboard. IOP Conf. Ser.: Mater. Sci. Eng. 536, 012066.

Mohana, B., Nur Adilah, C.H.A., Rokiah, H., Othman, S., 2017. Properties of binderless particleboard and particleboard with addition of urea formaldehyde made from oil palm trunk waste. J. Phys. Sci. 28 (3), 151–159.

Mohd Zuhri, M.Y., Sapuan, S., Napisah, l., 2009. Oil palm fibre reinforced polymer composites: a review. Prog. Rubber Plast. Recycl. Technol. 25, 233–246. https://doi.org/10.1177/147776060902500403.

Mohd Zuhri, M.Y., Sapuan, S., Napsiah, I., Riza, W., 2010. Mechanical properties of short random oil palm fibre reinforced epoxy composites. Sains Malays. 39 (1), 87–92. 2010.

Myrtha, K., Holia, O., Dawam, A.A.H., Anung, S., 2008. Effect of oil palm empty fruit bunch fibre on the physical and mechanical properties of fibre glass reinforced polyester resin. J. Biol. Sci. 8 (1), 101–106.

Paridah, M.T., Hoong, Y.B., Nor Yuziah, M.Y., Loh, Y.F., Hashim, W.S., 2014. Properties enhancement of oil palm plywood. In: Handbook of Oil Palm Trunk Plywood Manufacturing. Malaysian Timber Industrial Board, pp. 89–109. Chapter 5.

Prociak, A., Malewska, E., Kurańska, M., Bąk, S., Budny, P., 2018. Flexible polyurethane foams synthesized with palm oil-based bio-polyols obtained with the use of different oxirane ring opener. Ind. Crop Prod. 115, 69–77. Available online from https://doi.org/10.1016/j.indcrop.2018.02.008.

Qi, Y., Li, N., Wang, D., Sun, X.S., 2013. Physicochemical properties of soy protein adhesives modified by 2-octen-1-ylsuccinic anhydride. Ind. Crop Prod. 46, 165–172.

Ratnam, C.T., Gunasunderi, R., Wan, M.Z.W.Y., 2007. Oil palm empty fruit bunch (OPEFB) Fiber reinforced PVC/ENR blend—electron beam irradiation. Nucl. Inst. Methods Phys. Res. B 265, 510–514.

Risanto, L., Hermiati, E., Sudiyani, Y., 2014. Properties of lignin from oil palm empty fruit bunch and its application for plywood adhesive. Makara J. Technol. 18 (2), 67–75. 2014.

Rokiah, H., Wan Nor Aidawati, W.N., Othman, S., Masatoshi, S., Salim, H., Fumio, K., Tokoko, S., Tay, G.S., d Ryohei T., 2012. Properties of Binderless particleboard panels manufacture from oil palm biomass. Bioresources 7 (1), 1357–1365.

Russita, M., Bahruddin, B., 2018. Production of palm frond based wood plastic composite by using twin screw extruder. IOP Conf. Ser.: Mater. Sci. Eng. 345. https://doi.org/10.1088/1757-899X/345/1/012039, 012039.

Suthon, S., Nirundorn, M., Frank, L., 2018. Property gradients in oil palm trunk (*Elaeis guineensis*). J. Wood Sci. 64, 709–719. https://doi.org/10.1007/s10086-018-1750-8. 2018.

Theapparat, Y., Chandumpai, A., Faroongsarng, D., 2018. Physicochemistry and Utilization of Wood Vinegar from Carbonization of Tropical Biomass Waste. Intechopen. https://doi.org/10.5772/intechopen.77380. (ISBN: 978-1-78923-562-3).

Wang, Z., Gu, Z., Hong, Y., Cheng, L., Li, Z., 2011. Bonding strength and water resistance of starch-based wood adhesive improved by silica nanoparticles. Carbohydr. Polym. 86 (1), 72–76.

Wang, Z., Gu, Z., Hong, Y., Cheng, L., 2012. Preparation, characterization and properties of starch-based wood adhesive. Carbohydr. Polym. 88 (2), 699–706.

Wang, Z., Chen, Y., Chen, S., Chu, F., Zhang, R., Wang, Y., Fan, D., 2019. Preparation and characterization of a soy protein based bio-adhesive crosslinked by waterborne epoxy resin and polyacrylamide. RSC Adv. 9, 35273–35279. 2019.

Yeoh, F.H., Lee, C.S., Kang, Y.B., Wong, S.F., Cheng, S.F., Ng, W.S., 2020. Production of biodegradable palm oil-based polyurethane as potential biomaterial for biomedical applications. Polymers 12 (8), 1842. https://doi.org/10.3390/polym12081842. 2020.

Zaini, M.J., Fuad, M.Y.A., Ismail, Z., Mansor, M.S., Mustafah, J., 1996. The effect of filler content and size on the mechanical properties of polypropylene/oil palm wood flour composites. Polym. Int. 40 (1), 51–55.

Zheng, P., Chen, N., Mahfuzul Islam, S.M., Ju, L.-K., Liu, J., Zhou, J., Chen, L., Zeng, H., Lin, Q., 2019. Development of self-cross-linked soy adhesive by enzyme complex from Aspergillus niger for production of all-biomass composite materials. ACS Sustain. Chem. Eng. 7 (4), 3909–3916. 2019.

PART 2

Processing and treatment based on oil palm biomass type

7

Processing of oil palm trunk and lumber

Edi Suhaimi Bakar[a,b], S.H. Lee[a], Wei Chen Lum[c], S.O.A. SaifulAzry[a], and Ching Hao Lee[a]

[a]Institute of Tropical Forestry and Forest Products (INTROP), Universiti Putra Malaysia, Serdang, Selangor, Malaysia, [b]Faculty of Forestry and Environment, Universiti Putra Malaysia, UPM, Serdang, Selangor, Malaysia, [c]Institute for Infrastructure Engineering and Sustainable Management (IIESM), Universiti Teknologi MARA, Shah Alam, Selangor, Malaysia

7.1 Introduction

Oil palm, rubber, rice, cocoa, and coconut are the main agricultural commodities grown in Malaysia. According to Griffin et al. (2014), 17 Mt of agricultural residues have been estimated. More than three-quarter of these residues are dominated by oil palm, while the remaining amount consists of rice, forestry, and other residues such as rubber, cocoa, and coconut. Oil palm biomasses could be classified into two general types based on their generation sites. Oil palm trunks (OPTs) and oil palm fronds are readily available in the planting sites. These two biomasses amount to 75% of the total oil palm biomasses. On the other hand, empty fruit bunches, mesocarp fibers, palm shell kernel, and palm oil mill effluent (POME) are generated at the mill sites after the extraction of fresh fruit bunch for palm oil. These biomasses make up to 25% of the total oil palm biomasses (Griffin et al., 2014). Annually, in Malaysia alone, the oil palm industry had generated a substantial amount of at least 30 million tons of underutilized residues in the form of trunks, fronds, empty fruit bunches, and leaves (Khalil et al., 2006).

From 2013 to 2019, the total oil palm area planted in Malaysia saw a steadily growing trend. In 2019, the overall planted area for oil palm was 5.90 million hectares, a rise of 13.8% over 5.23 million hectares in 2013 (MPOB, 2020; Mohammad Padzil et al., 2020). After an economic life span of approximately 25 years, the oil palm trees are felled for replanting activity, and OPTs are available as biomass. According to Sulaiman et al. (2012), replanting of oil palm trees usually takes place when the

Oil Palm Biomass for Composite Panels. https://doi.org/10.1016/B978-0-12-823852-3.00004-0
Copyright © 2022 Elsevier Inc. All rights reserved.

height of the palm trees has reached more than 13 m and annual yield of bunches has fallen below 10–12 tons per hectare. The felled trunks normally measure up to 45–65 cm in diameter at breast height. During replanting, OPT is the second most generated biomass after oil palm fronds. It is estimated that the generation rate of OPTs is 62.8 ton-dry/ha-replantation/year (Aljuboori, 2013). Therefore, it can be estimated that around 12.8 million tons of OPT are generated every year.

Bakar et al. (2013a) reported that an estimated of 230–250 m^3 of oil palm stems per hectare could be obtained during replanting. Utilization of oil palm wood (OPW) as a raw material could reduce the environmental burden of wood consumption. Since OPW is a potential ligno-cellulosic material, it is therefore possible to utilize it as an alternative for the declining supply of timber. Measures have to be taken to enhance the low quality of OPW and transform it into useful by-products that meet market demands. There has been a wide range of products made from OPW. Currently, the use of OPW as a raw material for certain applications has been a focus in the plywood industry in Malaysia.

However, before OPW can be utilized effectively, drying of the materials is of the utmost importance. Optimization in the drying process for any woody material aims at minimizing production costs and avoiding losses in productivity and quality. Studies found that the moisture content (MC) increased toward the core of the oil palm stem (Choo et al., 2011). Previous studies revealed that the MC of oil palm stems can exceed 500% (Lim and Khoo, 1986; Bakar et al., 2008). The variation of MC and density of the trunk inhibit its full utilization (Choo et al., 2011). Green MC of up to 300%–500% (Bakar et al., 2008) and density of 200–700 kg/m^3 (Anis et al., 2007) can be found mainly from the trunk. This abnormally high MC extends drying time and consumes more energy to dry. Another significant finding was a marked difference in the MC between lumbers obtained from the outer and the inner portion of the oil palm stem. As a consequence, the drying for these oil palm lumbers (OPL) needs to be adjusted specifically according to lumbers obtained from different parts of the OPT. The extremely hygroscopic nature of the OPW causes a higher rate of shrinking and swelling compared to wood (Loh et al., 2011a). OPW becomes susceptible to insect and fungal attacks when coupled with high sugar content (Bakar et al., 2013b). To date, an optimum drying schedule has yet to be developed for OPL; 30–35 days is needed for OPL dried using a conventional drying method in a kiln drier, which resulted in many drying defects. A drying method for OPL, called Oil Palm Lumber Drying Method and System Thereof, has been developed by Bakar et al. (2016). The concept of the method was to increase the evaporation rate of water from the material by using a holing technique on the OPL. This method successfully reduced the drying time of OPL from 30 to 35 days to approximately 3 h.

This chapter discusses the sawing, drying, and quality enhancement of OPT and lumber where different sawing patterns, drying methods, and quality enhancement procedures are reviewed.

7.2 Imperfections of oil palm trunks and lumber

The variations of MC and density of the trunk inhibit its full utilization (Choo et al., 2011). Green MC of up to 300%–500% (Bakar et al., 2008) and density of 200–700 kg/m^3 (Anis et al., 2007) can be found mainly from the trunk. One of the apparent characteristics of OPT is its high MC. Studies found that the MC increased toward the core of the oil palm stem (Choo et al., 2011). Previous studies revealed that the MC of oil palm stems can exceed 500% (Lim and Khoo, 1986; Bakar et al., 2008). Lim and Gan (2005) discovered that green OPT contains high MC, which ranged from 120% to more than 500%. Bakar et al. (2013a) reported that the MC of OPT increased from the bottom to middle part but declined slightly as the height increased. Bakar et al. (2008) justified that the lower MC is because relatively higher pressure is needed to allow the liquid to move upward against the gravitational force, therefore the amount of water attained in the top region is relatively lower (Bakar et al., 1998). On the other hand, Lim and Khoo (1986) explained that the MC variation at the radial direction is due to the relative quantity of vascular bundles and parenchymatous cells. Parenchymatous cells stored a higher volume of water compared to vascular bundles, thus a greater number of parenchymatous cells toward the inner section showed higher MC.

Density of the OPT was reported in the range from 110 to 900 kg/m^3 as reported by many studies. Prayitno (1995) reported that the OPT density is in the range of 280–750 kg/m^3. Lim and Khoo (1986) found that the average density value in different zones of the trunk ranged from 200 to 600 kg/m^3. They also reported that the density of oil palm stem shows considerable variability over the stem, both radially and vertically. The density at the peripheral region is over twice the values of the central region.

OPW is a hygroscopic material that absorbs and desorbs moisture from surrounding tissues and results in changes in the dimension of OPW (Sulaiman et al., 2012). The extremely hygroscopic nature of the OPW causes a higher rate of shrinking and swelling compared to wood (Loh et al., 2011a). Poor mechanical properties are one of the unfavorable properties of OPW (Ebadi et al., 2016). Abdullah et al. (2013) stated that the weak properties are a result of density variations inside the trunk itself and the anatomical structure of the trunk that comprises a solid vascular bundle and loose parenchyma cells. Another drawback that limits the utilization of OPW is that it requires a very long drying period since it possesses very high MC. Furthermore, as a consequence of a marked difference existing in the MC between lumbers obtained from the outer and inner portion of the oil palm stem, the drying for these OPLs needs to be adjusted specifically according to lumbers obtained from different parts of the OPT. OPW becomes susceptible to insect and fungal attacks when coupled with high sugar content (Bakar et al., 2013b).

7.3 Sawing of oil palm trunk and lumber

OPL has a wide range of green MC and density. There can be two different methods in segregation of the OPL, which is by density and MC. Segregation by density can be categorized into three different parts: the inner zone (IZ), central zone (CZ), and peripheral zone (PZ). Segregation by the MC can be differentiated between inner and outer portions. Characteristics of the inner part have the MC above 200% while the outer part has an MC below 200%. When sawing OPT, it is advisable to take this matter into consideration. Therefore, processing needs to be strategic by using a different cutting pattern than tropical lumber. From inner part to the outer part, density, number of vascular bundles, and MC varies significantly. Furthermore, Killmann (1983) proposed that the core of OPL cannot be used because of its poor drying properties, and because of that, he suggested the necessity of OPL segregation according to the potential uses. Similarly, Fauzi et al. (2012) also recommended segregating lumber made from OPT by densities prior to lumber manufacturing. In addressing this problem in terms of segregation of the lumber, a strategic sawing pattern must be selected. Sawing technique of OPL is not like other timber products. There is a need to segregate the OPL into inner and outer part, as the highest quality is the outer layer that only covers 30% (Bakar et al., 2006). Conventional sawing techniques such as life or plain sawing are designed based on the hardwood and softwood species. Due to production priority, the heartwood resides in the core while the sapwood is found at the periphery of the lumber (Bowyer et al., 2005; Desch and Dinwoodie, 1996). Contrarily for OPT, the sawing pattern intends for the outer portion as it is the best properties. Therefore, a different sawing pattern is needed to fully utilize the OPL for the inner and outer portion of the trunk. Addressing the sawing pattern issue, Bakar et al. (2006) have found that a polygon sawing pattern is the best method to adjust the inner and outer portion of the OPL.

Two modified sawing patterns (polygon sawing and cobweb sawing) plus one ordinary sawing pattern (life sawing) were compared in the sawing of oil palm stems (Fig. 7.1). Out of them, the most suitable

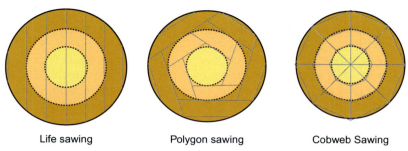

Life sawing   Polygon sawing   Cobweb Sawing

Fig. 7.1 Three sawing methods for oil palm lumber.

cutting pattern for sawing OPW is polygon sawing. It can produce wide-outer lumbers that are high in quality with a recovery of 27% (Bakar et al., 2005). As can be seen from Fig. 7.1, good quality lumber from polygon sawing needs for OPT to be on the carriage many times. Since the angle of rotation is not 90 degrees (a common rotation angle in sawing), polygon sawing is difficult to operate and requires a skilled operator. Therefore, a simpler sawing pattern needs to be constructed that can be used by moderately skilled workers without any modification to the existing sawing machine.

For that reason, Bakar et al. (2014) proposed a new sawing pattern, called reversed cant sawing. This sawing pattern is designed especially to maximally obtain the best outer portion of OPW, while shortening the sawing time with higher recovery since waned sawn timber from the outer part of OPT can be directly used. Because of that, the logs must be debarked first with a peeler before the trunk is sawn. Fig. 7.2 shows the reverse cant sawing method.

7.4 Drying of oil palm trunk and lumber

Initial drying is an essential step for OPT to be used effectively as lumber, pulp and paper, reconstituted boards, biocomposites, animal feed, and fuel (Anis et al., 2008). Based on a study by Anis et al. (2008), initial drying of OPT could offer several advantages such as lighter weight for cheaper material transportation cost, better durability against staining, reduced defects of the resultant products, and easier trimming due to better dimensional stability. However, the extreme moisture variant within the OPTs requires the trunk to be dried using a drying schedule dissimilar to that of normal wood. Therefore, drying of OPT and lumber is a very challenging task.

Normally, air drying and kiln drying are employed to dry OPT and lumber. Nevertheless, air drying and kiln drying of OPL resulted in unsatisfactory results. The different drying properties between inner and outer parts of OPL proved to be a difficult aspect to tackle. In this

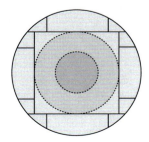

Fig. 7.2 Reversed cant sawing method.

section, four different drying methods are discussed. The drying methods include: (i) air drying, (ii) forced-air drying; (iii) super-fast drying, and (iv) modified super-fast drying. Generally, air drying refers to the drying process in which wood is exposed to open air, while forced-air drying involves the employment of a propeller fan to provide a constant flow of air. On the other hand, the super-fast drying method involves drilling holes in the OPL and applying light compression to produce the defect-free board. This process requires high precision as the method for the inner portion and outer portion are different. In super-fast drying, inner and outer portions of OPL are segregated by the MC. Segregation of the inner and outer portions is needed as both portions need different drying processes. The inner portion requires two higher compression cycles while the outer portion undergoes one cycle with lower compression. Meanwhile, modified super-fast drying is a modified version of the super-fast drying method where the initial MC of OPL is reduced to an intermediate MC using forced-air drying before super-fast drying takes place. The advantages of the modified super-fast drying method are the elimination of the holing process and the segregation of the inner and outer parts of OPL.

7.4.1 Air drying

Air drying is the drying of timber by exposing it to air. The technique of air drying consists mainly of making a stack of sawn timber (with the layers of boards separated by stickers) on raised foundations, in a clean, cool, dry, and shady place. The rate of drying largely depends on climatic conditions and on air movement (exposure to the wind). For successful air drying, a continuous and uniform flow of air throughout the pile of the timber needs to be arranged (Desch and Dinwoodie, 1996). This is a method of drying timber by exposing it to natural atmospheric conditions. As such, there is no control over drying rate as this will be determined by the prevailing weather (temperature, relative humidity, rainfall, and wind speed). The obvious advantage of air drying is its low capital cost in comparison to kiln drying procedures.

7.4.2 Forced-air drying

Drying of wood first involves the evaporation of the moisture in the drying process. As the wood becomes dry or has less MC, more energy is needed to release the hygroscopic forces that bounds in the wood. Green wood has a large amount of free water; as the moisture decreases, no changes occur on its physical aspect. As the wood continues to dry below 30%, physical changes occur as bound water decreases. Drying of wood requires optimum conditions in the surrounding atmosphere as it will affect the drying rate of the wood. The lower the relative humidity, the faster the drying, but it should not be too low as to reduce the drying defect. Air movement is an important factor to include when drying wood using air. Stacks of lumber require adequate space to bring heat energy into the stack, which draws in moisture and exits the stack. This flow is crucial to control the drying rate to be the same.

Air velocity has several important implementation reasons: to transfer moisture away from the lumber, to transport heat to the lumber, and to give uniform relative humidity and temperature throughout the kiln (Bachrich, 1980; Eckelman and Baker, 1976). Other factors that can relate to this process are the boundary layer, wood MC, moisture gradient, and variations in both MC and air velocity.

Water transfer to air is the primary medium to drying. When air passes across the wood stack, the air will absorb the moisture from the lumber. The statured moisture in the air will release to the atmosphere with lower velocity and humidity (Esping, 1996). Thus, the process is started again as the air will be flow through the fan and repeat the process. Three factors (temperature, humidity, and air velocity) can be manipulated to control the drying process and minimize drying defects, while accelerating the drying process. To utilize this concept in OPW, accelerated air drying or air-force drying was chosen.

Accelerated air drying can be defined as drying wood using accelerated air flow to reduce the MC of the lumber. The air flow can be accelerated using a fan, which creates air movements through the spaces between wood. When lumber use does not require a low MC, air drying is sufficient. Lumber used for outdoor furniture and other outdoor exposures, or for building structures such as barns, pole sheds, and garages that are not heated, can usually be air dried to a low enough MC. Rough sawn hardwood are often air dried at the producing sawmills to reduce weight so shipping costs are reduced. In addition, air drying reduces subsequent kiln drying time. High MC of wood of 60%–90% usually requires 25–30 days to reach a MC of 19% but may differ as different thicknesses may change the drying time.

7.4.3 Super-fast drying of OPL

Bakar et al. (2016) introduced an OPW super-fast drying method (PI 2016702162) with a reduced drying time to dry OPL. This method involves drilling holes into the lumber for faster evaporation rate of the lumber. It was reported that the distance between holes does not create a significant difference on the properties, but the sizing of the hole can affect the result. The next step is direct contact drying or hot pressing at 180°C; the duration depends on the portion of the lumber. The outer portion of the OPW, which has a MC of lower than 200%,

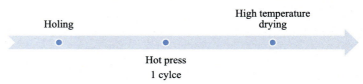

Fig. 7.3 Outer portion of the OPL with low initial MC (high density) Bakar et al. (2016).

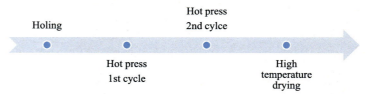

Fig. 7.4 Inner portion of the OPL with high initial MC (low density) Bakar et al. (2016).

only requires one cycle (40 min) of hot press (Fig. 7.3); the inner portion, which has a MC of higher than 200%, requires two cycles (80 min) (Fig. 7.4). The next process is high temperature drying to a target MC, which took a maximum of 3 h.

It is therefore necessary to note that, for an effective drying of OPW, holing and drying method of the OPW needs to be uniform. In other words, both the inner and outer portion of the OPW would need to dry at the same time, and any holing would need to be discarded. It is recommended to introduce forced-air drying and a modified super-fast drying method into the methodology. The methodology would need to separate into two segments. The first segment is to dry OPW from green to targeted MC. The second segment would dry OPW from the intermediate to final target MC with super-fast drying method. The end result would have the OPL reach the final target MC of less than 10% without any holing.

7.4.4 Modified super-fast drying method for OPL

Super-fast drying method mentioned earlier involves a holing process that resulted in unattractive surfaces for further application of the dried OPL. Therefore, an improved drying method, called modified super-fast drying method, has been proposed to solve the holed surfaces caused by the initial holing process (Muhammad Nadzim et al., 2021). The process has eliminated the holing process or has employed a semiholing process where the holes are not fully drilled through the OPL samples. In that case, at least one surface of the samples remained clear and hole-free, while another surface with holes could be hidden during lamination process. The elimination of the holing process could be achieved when the initial MC of the OPL samples was reduced to lower than 200% using forced-air drying. Practically, modified super-fast drying method is a combination of forced-air drying and super-fast drying methods.

In the study reported by Muhammad Nadzim et al. (2021), the MC of OPL samples were first reduced to an intermediate MC of 60%, 80%, and 100%, respectively, using forced-air drying. After that, the OPL was subjected to a direct contact drying method in a hot press set at a temperature of 180°C for 40 min. A compression ratio of 20% was applied during the process. Next, the OPL samples were dried in an oven at 100 ± 3°C for 3 h. As a result, modified super-fast drying method proved to be able to reduce the MC of the outer part of OPL to less than 10% within 3 h. Meanwhile, the inner part of OPL required slightly longer duration (3–4 h) to achieve a similar MC as that of the outer part. The study also reported that the thickness swelling and water absorption of OPL samples that had undergone modified super-fast drying method is lower compared to that of the OPL samples subjected to the super-fast drying method. Better dimensional stability could be attributed to the inexistence of holes and subsequently lower exposure area in the modified super-fast dried OPL samples. Furthermore, superior mechanical properties were also recorded in the OPL samples dried using the modified super-fast drying method.

While Muhammad Nadzim et al. (2021) included a complete removal of the holing process throughout the drying process, a study by Rais et al. (2021), on the other hand, investigated the feasibility of a semiholing process and its effects on the OPL samples. In the study by Rais et al. (2021), the effects of incising parameters, namely incising depths and distance, on the drying performance and properties of the dried OPL were explored. Different incising depths (1/3, 1/2, 2/3, and 100% of the total thickness of OPL) and distance (38 and 50 mm) were made on the OPL before subjecting to the super-fast drying procedures. Similar to the procedures employed by Muhammad Nadzim et al. (2021), the incised OPL samples were subjected to hot pressing at 180°C for 40 min followed by high temperature drying at 100°C for 3 h. Compression ratio of 10%–20% was employed during the hot pressing process. The results revealed that the OPL samples could be dried to less than 10% MC within 3 h accompanied with minimum drying defects. Overall, hole depth of 1/3 of the original thickness of OPL samples is the optimum incising depth, while hole distances (38 or 50 mm) did not exert a significant effect on the properties of OPL.

7.5 Quality enhancement of oil palm lumber and its products

OPW extracted from the outer part of oil palm stem was identified with four main imperfections. According to Bakar et al. (2013a), OPW has very poor strength and durability, is dimensionally instable, and has bad machining characteristics. Furthermore, the density variant of the trunk is very high due to its high radial density gradient. In addition, OPW is very high in MC, which makes it very susceptible to an attack of biodeterioration agents. All of these imperfections have limited the final application of the OPW. However, the properties could be improved using a compregnation method where the OPW is impregnated with phenol formaldehyde (PF) resin and compressed under high temperature. Treatment with PF resin (especially low-molecular-weight resin, or LMwPF) is reported to be effective in improving OPW strength, dimensional stability, and machinability (Bakar et al., 2008; Chong et al., 2010; Amarullah et al., 2010). Once improved, OPW has the potential to be utilized as a raw material for the production of engineered timber products.

OPT with inherently inferior properties was proven able to be converted into materials with superior properties comparable to wood once treated properly. Works on enhancement of the OPT has been conducted by Bakar et al. (2001) and Bakar (2003) using impregnation and compregnation techniques with phenolic resin, respectively. They found that the properties of the treated OPWs were greatly enhanced. The mentioned studies have been succeeded by a series of works focusing on various variables to improve the properties of OPW. These works are summarized in Table 7.1.

Table 7.1 Series of works on oil palm trunk enhancement.

Products	Method—variables	Findings	References
Oil palm wood	Impregnation and compregnation—0%, 25%, 50% compression level	– 65%–88% increment in resistance against termite attack – 47%–93% increment in resistance against white-rot fungi	Bakar et al. (2013a)
Oil palm wood	Impregnation and compregnation	– 30% antiswelling efficiency was attained – 15-fold increment in Young's modulus – 7-fold increment in shear strength - 66% improvement in resistance against termites	Bakar et al. (2013b)
Oil palm wood	Impregnation and compregnation	– Planning performance of treated OPW improved from Grade 3 (average) to Grade 1 (Excellent) – Surface roughness improved from Grade 3 to Grade 1	Chong et al. (2010)

Chapter 7 Processing of oil palm trunk and lumber **123**

Table 7.1 Series of works on oil palm trunk enhancement—cont'd

Products	Method—variables	Findings	References
Oil palm wood	Impregnation followed by semicuring—oven only, oven followed by microwave, microwave followed by oven	– Dual heating system took shorter time to dry the treated OPW, 4–11 h vs 24–30 h in single heating system	Nur Khairunnisha et al. (2014)
Laminated compregnated oil palm wood	Soaked in 30% LmwPF resin for 24 h followed by diffusion process for 2, 4, or 6 days. Assembled in three layers and compressed using hot press	– significant improvement in water absorption and thickness swelling – two to threefolds increment in bending strength	Aizat et al. (2014)
Oil palm wood	Steaming—120°C for 1 h. Compression—0%, 20%, 30%, and 40% compression level	– Steaming facilitated the compression process – Higher compression led to higher physicomechanical properties	Abare et al. (2014)
Oil palm stem plywood	Veneer pretreatment—soaking in phenol formaldehyde resin for 20 s	– 44% increment in resistance against termite attack – 69% increment in resistance against white-rot fungi	Loh et al. (2011a)
Oil palm stem plywood	Veneer from outer and inner layer of oil palm trunk—100% inner, 100% outer, and mixed	– Plywood made from outer oil palm stem had better mechanical properties	Loh et al. (2011b)
Oil palm trunk plywood	Impregnation—resin solid content of 15%, 23%, 32%, and 40%	– Higher mechanical properties in plywood made with oil palm veneer treated using higher resin content – Formaldehyde emission increased along with increasing solid content	Yeoh et al. (2013)
Oil palm trunk plywood	Impregnation—hot pressing time of 14, 16, 18, and 20 min	– Mechanical strength improved along with increasing hot press time	Yeoh and Paridah (2013)
Oil palm trunk plywood	Impregnation—press pressure of 20 bar, 30 bar, 2 stage 20 bar + 30 bar, and 2 stage 20 bar + 50 bar	– Two stage pressure resulted in better mechanical properties of the produced plywood	Yeoh et al. (2012)

7.5.1 The five-step processing method of oil palm wood

A series of works has been conducted to enhance the properties of OPW through bulking treatment with PF resin. Bakar (2003) began by inventing a method called impregnation and densification, which involved a five-step process involving sawing, drying, impregnation, re-drying, and hot pressing as shown in Fig. 7.5. As the result, the process improves strength, durability, dimensional stability, and machining characteristics of treated OPW.

The method was improved further by altering a few processes, which then involved drying, impregnation, semicure heating, and hot pressing (Bakar et al., 2005), which significantly enhanced treated OPW properties and appearance, making the raw material seem to be in the right path to enter the wood industry as a source of making high quality furniture. Unfortunately, both methods are time-consuming. The cutting technique of polygon sawing was really hard to conduct as only a skilled worker can do it (Bakar et al., 2006). Other than that, it took a very long time for the OPW to dry because generally the OPW contains higher MC. Furthermore, the reason lays on the OPW properties itself as there are various technical problems in the current processing technology to convert OPTs to lumber and other useful products. Sawing and impregnation process of OPW are generally considered difficult by the industry (Abare et al., 2014). Therefore, further improvement was made by Bakar et al. (2014) resulting in a new method called "Quality enhancement of OPW with integrated objective approach" that involves a six-step processing method as shown in Section 7.5.2.

7.5.2 The six-step processing method

The six-step processing method consists of six processes which are: Sawing→Steaming and Compression→Drying→Impregnation→Semicuring→Densification as shown in Fig. 7.6.

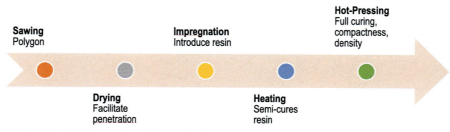

Fig. 7.5 The five-step processing method.

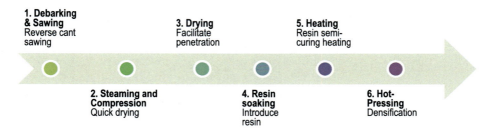

Fig. 7.6 The six-step processing method.

i. Sawing

Being a monocotyledon, oil palm stems have a very different characteristic to conventional hardwoods and softwoods, and thus the sawing patterns suitable for hardwoods and softwoods are not suitable for the oil palm stems. Two modified sawing patterns (polygon sawing and cobweb sawing) plus one ordinary sawing pattern (life sawing) were compared in the sawing of oil palm stems. Out of all these, the most suitable cutting pattern for sawing OPW is polygon sawing. It can produce wide outer lumbers that are high in quality with recovery of 27% (Bakar et al., 2005). But skilled operators and good carriage are needed to run the polygon sawing efficiently as it involves an angle that is not 90 degrees, thus difficult to run in a mill. For that reason, Bakar et al. (2014) proposed a new sawing pattern, called reversed cant sawing. This sawing pattern is designed to maximally obtain the best outer portion of OPW, while shortening the sawing time with higher recovery since waned sawn timber from the outer part of OPT can be used directly. Because of that, the logs must be first debarked with a peeler before the trunk is sawn.

ii. Steaming, compression, and drying

The six-step processing method involves the compression method; before compression, the OPW can be optionally steamed to reduce the cracking extent during the compression. The compression in this method, however, is not intended for density increment. It is intended to reduce the MC of OPW by force and create microcracks needed to hasten the drying process (Bakar et al., 2014). In this step, the OPW was steamed at 150°C for 15–30 min depending on the thickness of the samples. After that, the OPW was compressed at a compression ratio of 50% in a hot press set at 150°C until MC of 15% was achieved. Using the old five-step process method, the drying of OPW could take a very long time of 4–5 weeks. In this new six-step processing method, the drying time can be shortened. Instead of 4–5 weeks, the drying can be done for only 3–4 days. This is mainly due to the positive effect of compression before drying. Lower MC and existence of microcracks resulted in rapid drying (Bakar et al., 2014).

iii. Soaking

Resin inclusion can be carried out by a simple soaking method due to the presence of microcracks. As a result of the compression, the resin can penetrate easily into the OPW. Therefore, resin inclusion in the new six-step processing method can be carried out by just a simple soaking process instead of impregnation (Bakar et al., 2014). In this step, the OPW samples were soaked in phenolic resin for 30–45 min.

iv. Heating or semicuring

After soaking with phenolic resin, the OPW samples were semicured in an oven at 70°C until MC of 70% was attained.

v. Hot-pressing or densification

Lastly, the OPW samples were compressed for full curing in a hot press at 150°C for 30 to 60 min until a final desired thickness was achieved.

The advantages of the procedures involved in the six-step processing method are explained in detail in Table 7.2. Table 7.3 shows the comparison between the two "compreg" OPW processing methods.

Table 7.2 Advantages of every step in the six-step processing method.

Step	Details
1	Sawing involves a reverse cant sawing pattern that fully utilizes the best outer portions of the oil palm trunk, which makes the process much easier and faster without having to be supported by a perfect carriage system
2	Compression of wood will forcibly remove the water within the oil palm wood and lower the moisture content of the wood without the use of high inputs of energy (drying). Additionally, compression will also cause cracks in the wood, which facilitates drying and resin impregnation. Steaming can reduce the extent of cracks (microcracks).
3	Drying time is reduced because of the microcracks. Warping no longer occurs and cracking is not an issue because they will be eliminated in the densification process.
4	Resin introduction becomes easier due to the microcracks, and therefore the time taken to fully coat it is reduced. This opens up the possibility of using various coating methods and also resin that is more viscous/higher molecular weight.
5	Semicuring time is reduced because of the microcracks.
6	Densification allows irregularly shaped samples to be utilized. This step also aids in the curing of the resin.

Table 7.3 Comparison between five-step processing and six-step processing method of oil palm lumber.

	The five-step processing method	The six-step processing method
Diagram of process		
Characteristics of process	**Sawing**: Use polygon sawing that needs a perfect carriage system and highly skilled operator. **Drying**: Long drying time (30–35 days) due to very high MC of OPW. **Resin Introduction**: Needs impregnation method due to solid and thicker OPW. **Heating**: Very long heating time (30–45 h) since OPW must be heated at low temperature (70°C) to avoid full curing.	**Debarking and Sawing**: Debarking with peeler is needed before sawing. Use reverse cant sawing that produces waney timber, which is much easier and needs no high-skilled operator to run. **Compression**: Involves compression before drying that reduces initial MC and creates microcracks. **Drying**: Fast drying (3 days) due to much lower initial MC and the existence of microcracks. **Resin Introduction**: Can be carried out by a simple soaking method due to the existence of microcracks. **Heating**: Faster heating (5–8 h) because of the existence of microcracks.
Economic aspect	Long and impractical process that may not economically viable to the industry.	Fast and practical process and economically viable to the industry.
Final product	High performance "compreg" OPW suitable for high-end applications: • high strength • high durability • high dimensional stability (water resistant) • good machining • esthetic	

7.6 Conclusion

This chapter deals with the processing of OPT and lumber as well as quality enhancements of the OPL. To overcome its imperfections, several sawing patterns have been introduced such as life sawing, polygon sawing, and cobweb sawing. Polygon sawing pattern is the best sawing technique among the three. However, it requires skilled workers and some modifications on the existing sawing machine. Therefore, reverse cant sawing technique was introduced. Having a unique structure and very high MC, the drying has been one of big challenges in the utilization of OPL. However, several drying methods have been proposed that greatly reduced the drying time as well as minimized the drying defects. In conclusion, utilization of OPT and lumber effectively is currently on the right track. However, there is still a long way to go before the OPT and lumber could be accepted more widely by the industries.

References

Abare, A.Y., Bakar, E.S., Zaidon, A., 2014. Effect of steaming and compression on the physico-mechanical properties of compreg OPW treated with the 6-step processing method. In: Proceedings of the 6th International Symposium of IWoRS, 12–13 November, Medan, Indonesia.

Abdullah, A., Abdullah, M.M.A.B., Kamarudin, H., Ghazali, C.M.R., Salleh, M.A.A.M., Sang, P.K., Muhammad Faheem, M.T., 2013. Study on the properties of oil palm trunk fiber (OPTF) in cement composite. Appl. Mech. Mater. 421, 395–400.

Aizat, G., Zaidon, A., Nabil, F.L., Bakar, E.S., Rasmina, H., 2014. Effects of diffusion process and compression on polymer loading of laminated compreg oil palm (*Elaeis guineensis*) wood and its relation to properties. J. Biobaased Mater. Bioenergy 8, 1–7.

Aljuboori, A.H.R., 2013. Oil palm biomass residue in Malaysia: availability and sustainability. Int. J. Biomass Renew. 2 (1), 13–18.

Amarullah, M., Bakar, E.S., Ashaari, Z., Sahri, M.H., Febrianto, F., 2010. Reduction of formaldehyde emission from phenol formaldehyde treated oil palm wood through improvement of resin curing state. Jurnal Ilmu dan Teknologi Kayu Tropis 8 (1), 9–14.

Anis, M., Kamarudin, H., Lim, W.S., 2007. Challenges in drying of oil palm wood. In: Paper Presented at the 2007 PIPOC International Palm Oil Congress: Empowering Change, Kuala Lumpur, 26–30 August 2007.

Anis, M., Kamarudin, H., Astimar, A.A., Mohd Basri, W., 2008. Treatment of Oil Palm Lumber. MPOB Information Series. Malaysian Palm Oil Board, Bangi, Selangor. MPOB TT No. 379.

Bachrich, J.L., 1980. Dry Kiln Handbook. H.A. Simons (international) Ltd., Vancouver, ISBN: 0-87930-087-6.

Bakar, E.S., 2003. Improvement of Oil Palm Wood Quality Using "Kompress" Technique. IBP Press, Bogor, Indonesia.

Bakar, E.S., Rachman, O., Hermawan, D., Karlinasari, L., Rosdiana, N., 1998. Pemanfaatan batang kelapa sawit (*Elaeis guineensis* Jacq) sebagai bahan bangunan dan furniture (I): Sifat fisis, kimia dan keawetan alami kayu kelapa sawit. Jurnal Teknologi Hasil Hutan XI (1), 1–11.

Bakar, E.S., Hadi, Y.S., Surnadi, I., 2001. Quality improvement of oil-palm wood: impregnated with phenolic resin. Indones. J. For. Prod. Technol. 14 (2), 26–31.

Bakar, E.S., Tahir, P.M., Sahri, M.H., 2005. Properties enhancement of oil palm wood through the compreg method. J. Trop. Wood Sci. Technol. 2, 91–92.

Bakar, E.S., Febrianto, F., Wahyudi, I., Ashaari, Z., 2006. Polygon sawing: an optimum sawing pattern for oil palm stems. J. Biol. Sci. 6 (4), 744–749.

Bakar, E.S., Mohd Hamami, S., H'ng, P.S., 2008. A Challenge from the Perspective of Functional Wood Anatomy. Penerbit Universiti Putra Malaysia, Universiti Putra Malaysia, Serdang.

Bakar, E.S., Hao, J., Zaidon, A., Adrian, C.C.Y., 2013a. Durability of phenolic-resin-treated oil palm wood against subterranean termites a white-rot fungus. Int. Biodeter. Biodegr. 85, 126–130.

Bakar, E.S., Paridah, M.T., Sahri, M.H., Mohd Noor, M.S., Zulkifli, F.F., 2013b. Properties of resin impregnated oil palm wood (Elaeis Guineensis Jack). Pertanika J. Trop. Agric. Sci. 36 (S), 93–100.

Bakar, E.S., Ashaari, Z., Choo, A.C.Y., Abare, A.Y., 2014. A Method of Producing Compreg Oil Palm Wood. Patent number: PI2014700947.

Bakar, E.S., Soltani, M., Paridah, M.T., Choo, A.C.Y., 2016. Oil Palm Lumber Drying Method and System Thereof. Patent number: PI2016702162.

Bowyer, J.L., Febrianto, F., Ashaari, Z., 2005. The optimum sawing pattern for oil palm wood. In: Proceedings of IATC, p. 8.

Chong, Y.W., Bakar, E.S., Zaidon, A., Hamami Sahri, M., 2010. Treatment of oil palm wood with low-molecular weight phenol formaldehyde resin and its planing characteristics. Wood Res. 1 (1), 7–12.

Choo, A.C.Y., Tahir, P.M., Karimi, A., Bakar, E.S., Abdan, K., Ibrahim, A., Loh, Y.F., 2011. Density and humidity gradients in veneers of oil palm stems. Eur. J. Wood Wood Prod. 69 (3), 501–503.

Desch, H.E., Dinwoodie, J.M., 1996. Timber: Structure, Properties, Conversion and Use, seventh ed. Macmillan Press Ltd., London.

Ebadi, S.E., Ashaari, Z., Naji, H.R., Jawaid, M., Soltani, M., San, H.P., 2016. Mechanical behavior of hydrothermally treated oil palm wood in different buffered pH media. Wood Fiber Sci. 48 (3), 1–9.

Eckelman, C.A., Baker, J.L., 1976. Heat and air requirements in the kiln drying of wood. Indiana Agric. Exp. Station Res. Bull. 933, 3–19.

Esping, B., 1996. Trätorkning 1b praktisk torkning. Trätek, Stockholm, ISBN: 91-88170-23-3.

Fauzi, F., Bakar, E.S., Ashaari, Z., Yamani, S.A.K., Sahat, S., Ansar, S., 2012. Analysis drying defect of oil palm trunk. In: Paper Presented at the 4th International Symposium of IWoRs, Makassar, Indonesia.

Griffin, W., Michalek, J., Matthews, H., Hassan, M., 2014. Availability of biomass residues for co-firing in peninsular Malaysia: implications for cost and GHG emissions in the electricity sector. Energies 7, 804–823.

Khalil, H.S.A., Alwani, M.S., Omar, A.K.M., 2006. Chemical composition, anatomy, lignin distribution, and cell wall structure of Malaysian plant waste fibers. Bioresources 1 (2), 220–232.

Killmann, W., 1983. Some physical properties of the coconut palm stem. Wood Sci. Technol. 17 (3), 167–185.

Lim, S.C., Gan, K.S., 2005. Characteristics and utilization of oil palm stem. Timber Technol. Bull. 35, 1–7.

Lim, S.C., Khoo, K.C., 1986. Characteristics of oil palm trunk and its potential utilisation. Malays. For. 49 (1), 3–23.

Loh, Y.F., Paridah, M.T., Yeoh, B.H., Bakar, E.S., Anis, M., Hamdan, H., 2011a. Resistance of phenolic-treated oil palm stem plywood against subterranean termites and white rot decay. Int. Biodeter. Biodegr. 65, 14–17.

Loh, Y.F., Paridah, M.T., Yeoh, B.H., 2011b. Density distribution of oil palm stem veneer and its influence on plywood mechanical properties. J. Appl. Sci. 11 (5), 826–831.

Malaysian Palm Oil Board (MPOB), 2020. Oil Palm Planted Area 2013 to 2019. Available online http://bepi.mpob.gov.my/images/area/2019/Area_summary.pdf. accessed on 18 October 2020.

Mohammad Padzil, F.N., Lee, S.H., Ainun, Z.M.A., Lee, C.H., Abdullah, L.C., 2020. Potential of oil palm empty fruit bunch resources in nanocellulose hydrogel production for versatile applications: a review. Materials 13, 1245.

Muhammad Nadzim, M.N., Edi, S.B., Mojtaba, S., Zaidon, A., Lee, S.H., 2021. Drying of oil palm lumber by combining air force drying and modified super-fast drying methods. J. Oil Palm Res. https://doi.org/10.21894/jopr.2020.0000.

Nur Khairunnisha, I.P., Bakar, E.S., Nurul Azwa, A., Adrian, C.C.Y., 2014. Effect of combination oven and microwave heating in the resin semicuring process on the physical properties of "Compreg" OPW. Bioresources 9 (3), 4899–4907.

Prayitno, E.A., 1995. Layanan Bimbingan dan Konseling Kelompok (Dasar dan Profil). Ghalia Indonesia, Jakarta.

Rais, M.R., Edi, S.B., Zaidon, A., Lee, S.H., Mojtaba, S., Ramli, F., Paiman, B., 2021. Drying performance, as well as physical and flexural properties of oil palm wood dried via the super-fast drying method. Bioresources 16 (1), 1674–1685.

Sulaiman, O., Salim, N., Nordin, N.A., Hashim, R., Ibrahim, M., Sato, M., 2012. The potential of oil palm trunk biomass as an alternative source for compressed wood. Bioresources 7 (2), 2688–2706.

Yeoh, B.H., Paridah, M.T., 2013. Development a new method for pilot scale production of high grade oil palm plywood: effect of hot-pressing time. Mater. Des. 45, 142–147.

Yeoh, B.H., Loh, Y.F., Nor hafizah, A.W., Paridah, M.T., Jalaluddin, H., 2012. Development of a new pilot scale production of high grade oil palm plywood: effect of pressing pressure. Mater. Des. 36, 215–219.

Yeoh, B.H., Loh, Y.F., Chuah, L.A., Juliwar, I., Pizzi, A., Paridah, M.T., Jalaluddin, H., 2013. Development a new method for pilot scale production of high grade oil palm plywood: effect of resin content on the mechanical properties, bonding quality and formaldehyde emission of palm plywood. Mater. Des. 52, 828–834.

8

Rotary veneer processing of oil palm trunk

M.T. Paridah

Institute of Tropical Forestry and Forest Products (INTROP), Universiti Putra Malaysia, Serdang, Selangor, Malaysia

8.1 Introduction

The development of products from oil palm (*Elaeis guineensis*) trunk (OPT) has been intensified for several decades, particularly in the wood-based industry. This material has potential as an alternative to traditional forest timber. The trunks, which are abundantly available from replanting programs, have shown great potential for a wide range of products such as lumber, plywood, laminated veneer lumber (non-structural), blockboard, flooring, and various panel products such as particleboard, cement-bonded particleboard, gypsum-bonded particleboard, and medium-density fiberboard. With a combination of research knowledge, technical know-how, and the "right" technology, some of these products are now commercially available. Almost all the OPT-based products that are available are manufactured based on machinery and production set-up designed for tropical hardwoods. Some minor modifications, however, exist specifically in the peeling of logs, drying, and machining of lumber.

Plywood has been the most successful product developed from OPT logs so far. Because the anatomical features and the bulk properties of OPT logs are different than those of wood, rotary processing of logs from OPT differs slightly, principally in peeling, drying, veneer assembly, adhesive spreading, and hot pressing. Success in manufacturing OPT-based panel products largely depends on knowledge of OPT's characteristics and the know-how of correct processing techniques with regard to the woody material's morphology, as well as quality control measures. Properties that may be affected by rotary processing techniques include uniformity of thickness, surface roughness, sheet buckle, depth of checks into the veneer, color, and figure. Uniform thickness and relative smoothness and flatness are important for all veneer uses. Checks in the veneer, color, and figure are important for decorative face veneer. An "ideal" piece of veneer may be defined as

---

Oil Palm Biomass for Composite Panels. https://doi.org/10.1016/B978-0-12-823852-3.00005-2
Copyright © 2022 Elsevier Inc. All rights reserved.

uniform in thickness, roughness not greater than the wood structure, flat, with no checks on either side, and having a pleasing color. Figure is desirable for decorative face veneer and straight grain for other uses.

The rotary processing of OPT described in this chapter is a general overview of the process adopted by several OPT plywood mills, as well as those reported in the literature. Such process is not limited to, and may vary from, one method to another. "Veneer" in this chapter is defined according to the Palm Plywood standard stipulated in the Malaysian Standards (MS) 2629 and ISO 2426-4 as being produced by rotary cutting of OPT with a knife into thin continuous ribbons with maximum thickness of 6 mm. Since the characteristics of OPT and wood are very distinct, it is important to review the anatomical structure of OPT in relation to its physical properties and processing needs.

8.2 Characteristics of oil palm trunk logs

8.2.1 Morphological features

The size of OPT logs varies depending on age, species variety, and location, as well as the rate of growth. The rate of oil palm growth is highly dependent on both environmental and hereditary factors. For example, the current planting material, the tenera variety, grows taller by 40–75 cm, and the bole width increases from 2 to 5 cm annually. Replanting at a 25-year rotation, the bole length ranges between 9 and 13 m long with a mean volume of $1.76\,m^3$. With regard to bole axis, the form curve of OPT is neiloid (convex) in the region of its buttress until a point of inflection that is located approximately 1.5–2.0 m above the ground. From there, the trunk forms a curve that is paraboloid (concave) and terminates more or less conically in the region to its apical height (Kamarudin et al., 2014). In general, oil palm has a cylindrical trunk with a minimal taper, i.e., diameter differential at the butt and the top. Fig. 8.1 illustrates a typical 25-year-old matured oil palm plantation stand.

In another study on 27–year old oil palm, Bakar et al. (2008) reported that the trunk can reach a diameter as large as 60 cm at the butt, 50 cm at the top, with the frond-free height of about 14 m. The taperness, i.e., the differentiation of the diameter at the two ends for each meter length of the trunk was to be in the range of 0%–12%, of which the first 1 m–length butts experienced the largest taper. It is a normal practice to cut off the first 1 m–length of the taper butt, so that the maximum taperness will be not more than 6% (Bakar et al., 2008). The log is considered cylindrical if the log roundness (i.e., percentage of the smallest to the largest diameter of a log cross-section at a certain height of the trunk) is close to 100%. On average, OPT has a roundness of about 95%, thus in terms of morphology, matured OPT is almost perfectly cylindrical, which is favorable in sawing and peeling processes (Bakar et al., 2008).

Fig. 8.1 A typical 25-year-old oil palm stand (left) and a near-cylindrical form of an oil palm stem (right).

8.2.2 Anatomical structure

Oil palm belongs to family *Arecaceae,* which is a monocotyledon (Corley and Tinker, 2003). Typically, the stems of the oil palm consist of vascular bundles and parenchymatous tissues with high moisture content (MC; 1.5–2.5 times the weight of the dry matter), low cellulose and lignin content, and high content of water solubles and NaOH solubles in comparison with rubberwood and bagasse (Mohamad, 1985). Physical properties of trunks show heterogeneity and vary depending on both radial and vertical directions (Killmann and Lim, 1985). Some difficulties in utilizing OPTs also lie in the extremely tough outer bark and high content of decayable parenchyma cells.

As with other monocotyledons, water and food are transported by a system of cells arranged in a long series known as sieve tubes and vessels. The conducting cells are accompanied by another cell type, the vascular fibers, which contribute strength and rigidity to the plant. Generally, the vascular bundles of oil palm are distributed at random across the stem, which has no definable pith. Of the total bole volume, the amount of parenchyma tissues, vascular bundles, and bark (cortex) are approximately 32%, 54%, and 14%, respectively (Kamarudin and Wahid, 1997).

Killmann and Lim (1985) described OPT as a nonwood; it does not have cambium, secondary growth, growth rings, ray cells, sapwood and heartwood, or branches and knots. The growth and increase in diameter are mainly due to the overall cell division and cell enlargement in the parenchymatous ground tissues, together with the fiber enlargement in the vascular bundles. Being a monocotyledon, the OPT woody section has marked structural differences from timbers. The most remarkable features of woody monocotyledons are they achieve their stature without secondary thickening. Thus, instead of secondary xylem, OPT consists of primary vascular bundles embedded in parenchymatous tissue (Fig. 8.2).

The woody parts of OPT are not homogenous. There is usually a very hard peripheral rind surrounding the soft central region. Anatomically, the hard peripheral zone is composed of a narrow layer of parenchyma and congested vascular bundles giving rise to a sclerotic zone, which forms the main mechanical support of the palm stem. The central zone consists of larger and widely scattered vascular bundles embedded in the thin-walled parenchymatous tissue. Each vascular bundle consists of a fibrous sheath, phloem cells, xylem, and parenchyma cells and is surrounded by spherical, druse-like silica bodies. The xylem is always sheathed with parenchyma cells and usually consists of one or two wide vessels.

8.2.3 Physical, chemical, and mechanical properties

8.2.3.1 Moisture content

OPT is a highly anisotropic material, and its strength and stiffness properties vary with MC. Based on its dryness, the MC in OPT can range from 100% to 500% (Killmann and Lim, 1985; Lim and Khoo, 1986). Kamarudin et al. (2011) conducted a comprehensive study on MC distribution in radial and longitudinal directions of an OPT and plotted a trend in MC from the periphery zone (west outer, middle, and inner) to pith, to the next periphery zone (east outer, middle, and inner) along the tree height (Fig. 8.3). They concluded that, regardless

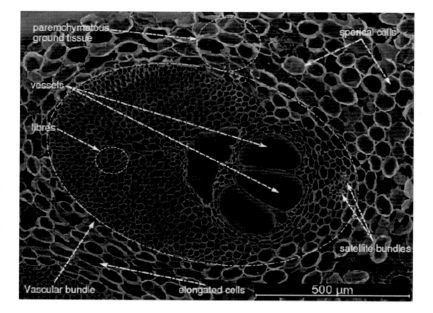

Fig. 8.2 A cross-section of an oil palm stem. Source: Dungani, R., Jawaid, M., Abdul Khalil, H.P.S., Jasni, J., Sri Aprilia, Hakeem, K.R., Sri Hartati, Islam, M.N., 2013. A review on quality enhancement of oil palm trunk waste by resin impregnation: future materials. Bioresources 8(2), 3136–3156.

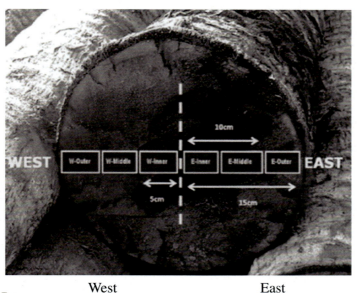

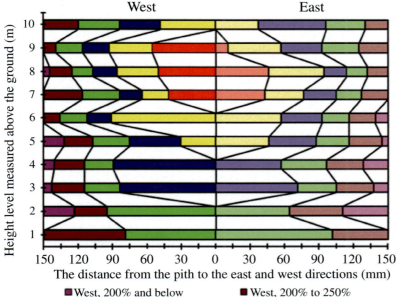

Fig. 8.3 Moisture content distribution in oil palm stem throughout radial-pith-radial and along the tree height. Source: Kamarudin, H., Jamaludin, K., Anis, M., Astimar, A.A., 2011. Characterisation of oil palm trunk and their holocellulose fibres for the manufacture of industrial commodities. Oil Palm Bull. 63, 11–23.

of height, the stem portion with MC of 300% and above appears to mainly be concentrated at the inner zone (pith) and partly contained at its intermediate middle zones. Of the total cross-sectional area, the woody portion accounted for about 47%–71% higher moisture than the periphery zone. Such extremely high and varied MC throughout the stem pose a challenge during handling and veneer drying. Choo et al. (2013, 2015) found that the MC increased toward the core of the oil palm stem.

8.2.3.2 Density

The average density of OPW shows an obvious decreasing trend from the outer zone toward the center and along the height (Fig. 8.4). Bakar et al. (2008) associated this trend with (1) the outer zone trunk is dominated by the high density vascular bundles by about 51%, while the center zone is dominated by low density parenchyma tissues by 70%; and (2) the cell walls of the parenchyma tissues at the outer zones of the trunk are relatively thicker than those present at the center. These two factors make woody parts from the outer zone much higher in density as compared to the center zone.

8.2.3.3 Shrinkage and swelling

Bakar et al. (2008) reported that the volumetric shrinkage of the woody part of oil palm stem increases from the outer to center, and from bottom to upper part of the stem. The center zones experienced higher shrinkage because they are dominated by thin-walled parenchyma tissues, which are easily ruptured during drying. The top center zones recorded the highest shrinkage, 74%, which is much higher than that recorded for hardwood or softwood species. Furthermore, this section appeared to be near collapse, whereby less than a quarter of its original volume was recovered after being dried (Fig. 8.5). In general, the radial swelling is higher than the tangential swelling, while its axial swelling is considered negligible. Therefore, it is not surprising that the density of veneer materials varies greatly within the same trunk, across the diameter, as well as along the tree height.

OPT generally shows serious raised grain, warping, and collapse when dried. These severe drying defects were also obtained by other researchers upon drying of the OPT (Ho et al., 1985; Anis et al., 2005; Haslett, 1990). This problem mostly occurred in the central region, which has low density and is virtually impossible to dry without excessive shrinkage and collapse. According to Abdullah (2010), microwave drying is effective in reducing the time and in the removal of moisture. It was observed in the case of microwave drying that increasing drying time led to higher moisture removal, while oven drying did not show any significant change within the first 12 min.

Chapter 8 Rotary veneer processing of oil palm trunk **137**

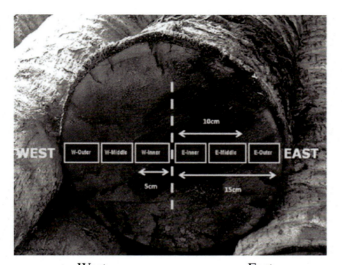

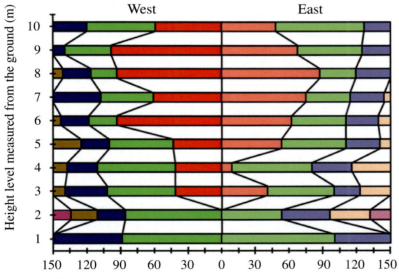

Fig. 8.4 Density distribution in oil palm stem throughout radial-pith-radial and along the tree height. Source: Kamarudin, H., Jamaludin, K., Anis, M., Astimar, A.A., 2011. Characterisation of oil palm trunk and their holocellulose fibres for the manufacture of industrial commodities. Oil Palm Bull. 63, 11–23.

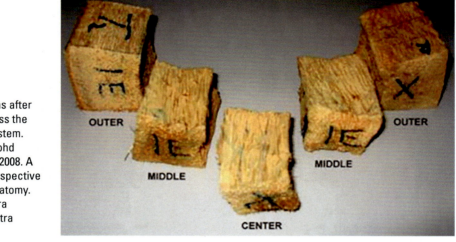

Fig. 8.5 Woody sections after drying taken from across the radial zone of oil palm stem. Source: Bakar, E.S., Mohd Hamami, S., H'ng, P.S., 2008. A Challenge from the Perspective of Functional Wood Anatomy. Penerbit Universiti Putra Malaysia, Universiti Putra Malaysia, Serdang.

8.2.3.4 Chemical properties

Tomimura (1992) characterized the chemical content of OPT and reported that starch content in OPT was remarkably high in parenchyma cells and vascular bundles. Xylose and glucose were the main sugar components in these two components. The lignin content was < 20%, which mainly belongs to p-hydroxybenzoic acid.

8.2.3.5 Mechanical properties

Sellers and veneer makers can make more money from a high quality log that is sliced into veneer than they can from sawing it into boards. While most wood species can be cut into veneer by suitable manipulation of the cutting conditions, it is more difficult to cut wood at the two extremes of the range of specific gravity (Lutz, 1971). Very lightweight species tend to cut with a fuzzy surface. Dense species require more power to cut and tend to develop deep cracks in the veneer as it passes over the knife. Basswood, with a specific gravity (based on green volume and oven dry weight) of about 0.32, is toward the low end of the gravity range for species that are successfully cut into veneer, and hickory with about 0.65 is near the high end.

One of the primary criteria in veneer processing is wood strength as it is directly proportional to specific gravity and density. Bakar et al. (1999) reported that mechanical properties of OPT decrease close to the center of the base and top of the trunk because the influence of the depth (the diameter) is greater than the effect of altitude. In the radial section, all the mechanical properties decreased sharply from the edge to the center and sloping down from the center to the center trunk due to differences in specific gravity and density of vascular bundles in each section. Compared to wood, the modulus of elasticity (MOE), modulus of rupture (MOR), compressive strength, shearing strength, and hardness of OPT's sliding exterior is almost equivalent to the strength of batai (*Paraserianthes falcataria*), a strength class IV–V (Bakar et al., 1999).

OPT has been seen as the most potential raw material for the manufacture of plywood (Abdullah, 2010; Anis et al., 2005; Bakar et al., 1999; Haslett, 1990). As such, OPT should possess some desirable mechanical properties such as stiffness, strength, toughness, and creep, which generally influence both the processing behavior and the product quality. Like MC and density, the strength of OPT varies greatly within the same trunk. Studies by Kamarudin et al. (2011) and Killmann and Lim (1985) revealed that the mechanical strengths of OPT are relatively lower compared to wood and coconut. In general, the woody portion from the periphery zone (outer section) is stronger than both the intermediate middle and inner zones, which were 28.5, 15.0, and 10.4 MPa, respectively (Kamarudin et al., 2014).

The sources of variation in the physical and mechanical properties of OPT were highly related to position in the radial direction with height. The results on basic density displayed an important density variation in pith to periphery zone. This information is a useful index of the product quality to which all end users can relate to. For example, to a sawmiller and a plywood mill, a high density of OPT materials indicates that the sawn lumber and oil palm veneer produced will be strong and stiff. In the case of wood-based panel industry, the use of dense OPT materials would produce more fiber per unit volume for the manufacture of biocomposites such as particleboard and medium-density fiberboard (MDF). Besides the production throughput, the basic density may also influence other properties such as strength and workability.

The woody portion at the periphery section has been reported by many studies as stronger compared to the middle and the inner (pith) sections. However, the demarcation of increasing or decreasing in strength is not well defined. Nonetheless, since wood density is used as an indicator for strength of wood, the density distribution pattern (Fig. 8.4) established by Kamarudin et al. (2011) can be used to estimate the strength distribution within the OPT. In the OPT plywood mills, sorting of veneers according to outer and inner types was done before the drying process to minimize the heterogeneity in the properties within the OPT.

8.3 Rotary OPT veneer processing

8.3.1 Definition of rotary veneer

Traditionally, a veneer is a thin layer of wood, usually less than 4 mm in thickness, removed from a log using either a rotary peeling or slicing process. Sliced veneers are usually much thinner and normally used as decorative layers, while rotary peeled veneers are relatively thicker, which fits its purpose as strengthening elements in plywood panel. All the veneers produced from OPT so far are rotary peeled and denoted as OPT veneers or palm veneers. Other forms of veneer, such as sliced or sawn veneers, are not discussed in this chapter.

The process of rotary veneer production is to remove a continuous thin ribbon of OPT veneer from a peeler billet periphery using a knife positioned parallel to the grain (Lutz, 1971). The billet is rotated against the knife using a drive mechanism that varies in design and approach, depending on the technology being used. Rotary-peeled veneers can be used to manufacture products suitable for structural and appearance applications. Common structural applications for products manufactured from rotary-peeled veneers include structural plywood, formwork plywood for concrete construction, packaging, marine ply, and laminated veneer lumber.

8.3.2 OPT plywood manufacturing process

There are several methods of manufacturing OPT plywood, depending on the advancement of machineries, and knowledge and experience of the millers. Anis et al. (2005) were the first to report on a typical manufacturing process of plywood from OPT, which are slightly different than that of normal plywood from timber logs. Paridah and Choo (2014) proposed additional steps to maintain good quality veneers by introducing a two-stage roller press predrying of veneers, sorting into outer (higher density) and inner (lower density) veneers, and a resin treatment process to produce veneers with consistent density. They also proposed several ways to improve the quality of OPT veneer by having (1) efficient log storage system, (2) veneer segregation according to density, (3) efficient drying method and MC control, and (4) use of machineries suitable for OPT.

The following sections describe methods of veneer processing from log end to dry veneer. This is followed by an additional section on quality control and troubleshooting to minimize veneer defects. The following sections discuss the process of rotary production of OPT veneers in a typical plywood processing mill that comprises (1) log end processing and (2) green veneer production.

8.3.3 Log-end processing

8.3.3.1 Felling and bucking

The oil palm trees are felled during the replanting operation (Fig. 8.6). The trunk is cross-cut into a length of either 2.7 m (9 ft) or 5.5 m (18 ft) in situ (Table 8.1). The shape of an OPT is usually tapered, thus it is a normal practice to make a cross-cut at a point above the buttress to maintain the cylindrical shape of the logs.

Fig. 8.6 Oil palm trunk after felling (left) and OPT logs being transported to the plywood mill (right). Extracted from Kamarudin, H., Anis, M., Yeoh, B.H., 2014. Properties of oil palm trunk. In: Loh, Y.F., Anis, M., Paridah, M.T., Choong, K.K., Hashim, W.S., Yeoh, B.H. (Eds.), Handbook of Oil Palm Trunk Plywood Manufacturing. The Malaysian Timber Industry Board, Kuala Lumpur, Malaysia.

Kamarudin et al. (2014) listed several crucial observations on the characteristics of the trunk and provided guidance in the selection of OPT logs as follows:

a. **Log attributes**: Preferably bigger and uniform diameter of logs is suitable with no presence of frond residues attached. The average diameter of OPT logs is about 35 cm (14 in.), and only the first 5.5 m (18 ft) is the best portion to be used for peeling purposes. Some OPT logs are larger in diameter, and in this case the logs are taken up to 8 m (27 ft) in length.

b. **Density and strength**: Density indicates the strength of the veneers, which are very much related to the origin of the oil palm trees. In general, trunks that are extracted from a normal dry soil to hilly land plantations produce acceptable good quality logs and the resulting veneers, whereas those obtained from watery or swampy plantation areas generate more porous, soggy, less dense woody material. OPT logs that are obtained from such conditions are significantly poorer in quality, so much so the resulting veneers are weaker and low in recovery.

8.3.3.2 Log storage

Logs intended for rotary processed veneers are called peeler or veneer logs. Peeler logs can be kept in good condition for some time providing the storage conditions are suitable. With poor storage conditions, logs can deteriorate by drying and cracking of the log ends and other exposed wood, development of blue stain, decay, and oxidation stain, attack by insects, development of undesirable odor, and increased porosity due to attack by bacteria (Lutz, 1974). The storage of peeler logs from OPT is similar to that of timber in the log yard. Unlike timber logs, OPT logs are more susceptible to insects and fungi, thus they need to be processed preferably within 3 weeks. Studies by Kamarudin et al. (2011) concluded that:

Table 8.1 A typical process flow in OPT plywood production.

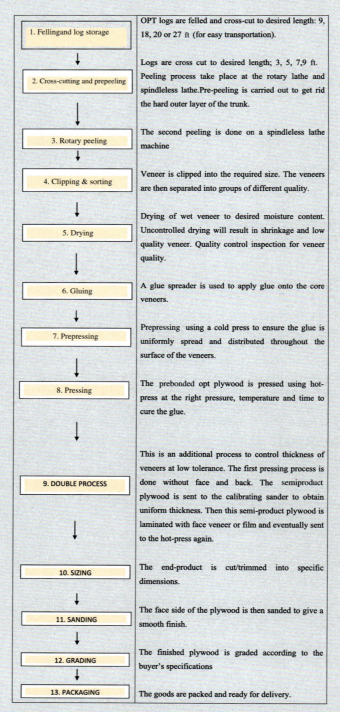

Source: Anis, M., Mohamad, H., Wan Hasamudin, W.H., Chua, K.H., 2005. Oil palm plywood manufacture in Malaysia. In: Proceedings of the 6th National Seminar on the Utilisation of Oil Palm Tree. Oil Palm Tree Utilisation Committee (OPTUC), Kuala Lumpur, Malaysia, p. 51–55.

a. OPT logs would not degrade for 3–4 weeks if stored under the shed. Depending on the harshness of weather, OPT logs would only last about 2–3 weeks if stored under an open environment. Continuous exposure to direct sunlight and rain will accelerate the rate of deterioration physically, mechanically, and chemically.
b. Some notable observations during log storage include apparent symptoms of heavy infestation of stains, molds (blue or black in color), and fungi. The logs smell sour (acidic) due to chemical reaction, i.e., fermentation of sugars and starch. Insect infestation on the OPT logs, however, is negligible due to the thick bark and high MC of the trunk that is not favorable to borers.
c. The degradation of OPT logs can be controlled by end-coating of both sides of logs' cross-sections with bitumen or paint to increase the storage time,
d. The implementation of a "first-in-first-out" concept would significantly help to control the quality of the logs.

8.3.3.3 Cross-cutting (bucking)

As mentioned earlier, the density in OPT varies along the height of the tree, thus it is a good practice if the logs can be segregated as early as possible based on the density range by distinguishing logs from the bottom, middle, and top of the tree stem. Depending on the manufacturers' preference, the length of the logs is cut into 3, 5, 7, or 9 ft (Fig. 8.7).

Fig. 8.7 Cross-cutting to required length at the factory.

8.3.4 Green veneer production

8.3.4.1 Peeling equipment

Lathe machine

Eighty to 90% of all wood veneer is cut by the rotary method (Lutz, 1974). The rotary method gives the maximum yield, it results in the widest sheets, knots are cut to show the smallest cross-section, and most juvenile wood and splits are left in the core. The veneer lathe works based on the principle that the wood is compressed with a nosebar, while the veneer knife cuts the blocks into veneers that are typically 3 mm (1/8 in.) thick (Fig. 8.8). In the case of OPT, there is a need to modify the spindle chuck pattern/design due to the soft structure of the trunk to create a firm hold during the "round up" (debarking) process. Such modification is crucial to avoid incidence of log spin off. This is one of the factors that many mills have turned into a spindleless lathe machine. It has the advantage in the ability to peel the logs down to 75 mm (3 in.) in diameter with acceptable uniform thickness. Nonetheless, a spindleless lathe requires a slight modification, especially on the opening (i.e., the gap seen in Fig. 8.8).

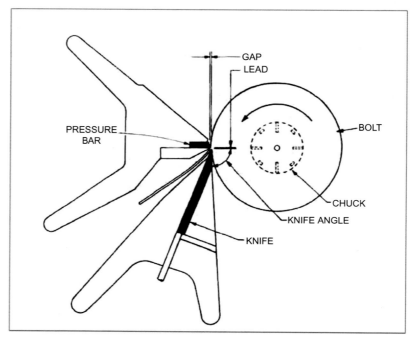

Fig. 8.8 A typical set-up for a rotary lathe machine.

Conventional spindleless lathe opening is about 203 mm (8 in.). These lathes should be rugged, able to peel OPT uniformly, and low in maintenance. Since OPTs possess very high MC, the lathe is expected to wear and tear faster. Most of the spindleless lathes are modified to accommodate bigger openings, e.g., > 330 mm (> 13 in.) so as to accommodate bigger diameter round up logs. The use of a log centering device would help to obtain a better log recovery.

Conveyor

Unlike wood veneer, OPT veneer breaks easily, particularly that obtained from the inner section of the trunk. Thus, manual handling of the veneer is difficult as it breaks easily. As an alternative, a conveyor attached to the peeler can be installed. This conveyor should have sufficient length and speed before clipping to optimize production time.

Clipper

To facilitate handling and sorting, a high throughput and rugged clipper should be used. The speed of peeling, conveying, and clipping should be synchronized for optimized production. The ribbon of veneer passing from the lathe is directed to an automated clipping machine, which cuts to the desired size. In general, veneer materials with standard widths are stacked automatically, while those veneers of random widths are manually stacked.

8.3.4.2 Peeling

Being a monocot, the woody palm trunk has a different anatomical structure than wood. The presence of parenchymatous tissues and vascular bundles and their distributions throughout the palm stem give rise to different MC and density patterns. For these reasons, there are basically three ways of processing OPT logs into veneers, which involve prepeeling or debarking, peeling of the outer sections, and peeling of the inner sections. These are:

Method 1: Involves two types of lathe machines. (1) Rotary lathe—peeling of logs to obtain a round cross-section by removing bark to obtain a uniform diameter, also known as "round up," and (2) Spindleless lathe—subsequent peeling of logs down to the smallest diameter, normally down to 100 mm (4 in.) depending on the type of lathe machine used.

Method 2: Involves two types of lathe machines. (1) Rotary lathe—peeling of logs to obtain a round cross-section by removing bark followed by a continuous peeling until the diameter reaches about 300 mm (9 in.), and (2) Spindleless lathe—subsequent peeling of logs down to the smallest diameter, normally down to 100 mm (4 in.) depending on the type of lathe machine used. The diameter of chucks (usually 9 in.) is the limiting factor for the peeling process. Further peeling using these chucks will invite spin-off incidence. Subsequently, a second stage peeling is carried out using a spindleless lathe.

Method 3: Involves only a spindleless lathe machine. Peeling is carried out to round up the logs (i.e., debarking) and continued until reaching the smallest diameter (about 100 mm or 4 in.) depending on the type of lathe machine. This type of lathe machine requires a bigger opening to cater for larger diameter OPT. It is modified with a rugged structure and parts to conduct both tasks, i.e., "round up" and peeling. In addition, the peelers must be able to remove the bark residues effectively during the peeling process. A continuous peeling process is possible for some mills set up with a conveyor and clipper.

Veneer thickness

Unlike wood veneer, OPT veneer is relatively thicker, particularly if the veneers are cut from the inner section of the trunk. The regular thickness of OPT veneer varies between 2.3 mm (from the outer section) to 6 mm (from the inner section). The average thickness varies from 4.0 mm to 5.0 mm. Thicker veneer is needed in OPT to compensate for "loss in volume/thickness" after drying and pressing due to the extreme anatomical variations within the OPT. Nonetheless, it is not advisable to peel too deep (final diameter < 100 mm or 4 in.) as the veneer quality becomes poorer and not acceptable for plywood manufacturing.

8.3.4.3 Veneer—Sorting and repairing

The OPT veneer sorting process is to sort inner and outer zones veneer into two different groups for drying purposes. Because of the different MC and density distributions throughout the stem, veneers obtained from the outer and inner sections of the trunk should be sorted prior to drying. The designation of outer and inner boundaries is done through experience where the outer section of the stem has a density of $\geq 250 \, \text{kg/m}^3$ and MC of < 200%.

8.3.5 Veneer drying

8.3.5.1 General drying principle

Optimization in the drying process for any woody material aims at minimizing production costs and avoiding losses in productivity and quality. Veneer drying is the process with the highest cost in the production of veneers or plywood (Walker, 2006). Furthermore, it is the most time-consuming process since it determines the speed of the overall production process (Baldwin, 1981). Veneer drying aims at reducing MC by 4%–12%, depending on the different wood species and type of applied glue. In general, 10%–15% of commercially produced veneers need to be redried for efficient use (Walker, 2006). Recent studies found that the MC increased toward the core of the oil palm stem (Choo et al., 2011, 2013). Previous studies revealed that the MC of oil palm stems can exceed 500% (Lim and Khoo, 1986; Lim and Gan, 2005; Bakar et al., 2008). This abnormally high MC extends drying time and consumes more energy to dry. Another significant finding was a marked difference in the MC between veneers obtained from the outer and inner portion of the oil palm stem. As a consequence, the drying system for these oil palm is differ to that of conventional drying system of hardwood.

8.3.5.2 Drying of OPT veneers

The natural characteristics of palm wood are the main challenge in a plywood processing line. As mentioned earlier, OPT exhibits vast variations in MC, density, and fiber morphology from bottom to top and vertically and horizontally (within the trunk). Such differences considerably influence the rate of drying, percent of shrinkage, and extent of drying defects of the palm veneers. The anatomical features of trunks give rise to the much softer tissues near the core, thus maintaining consistent veneer thickness tolerance is most challenging during peeling as well as drying.

Earlier reports revealed that MC of the oil palm stem varies between 150% and 500%. A gradual increase in MC occurs along the stem height and toward the central region, with both the outer and lower zones having far lower values than the other parts. Based on current drying practices, the dried palm veneers generally contain high variation in MC, e.g., > 40% in the inner or core region, and 10%–15% at the periphery region. The same trend exists from the top of the palm toward its basal, but in this case, the MC was found to vary from 25% to 40%.

Although palm veneer is usually thin (ranging from 3 mm to 6 mm), it possesses high initial MC. In some cases, it contains as high as 400% MC. In plywood manufacture, the green veneers are normally dried to an average of 8% (depending on adhesives). Due to its anatomical structures, the mechanism and rate of drying of palm veneers differ to that of wood. Depending on the drying system, veneers thickness, initial MC, and location where the veneers were taken, the drying time of OPT veneers varies from 50 to 90 min. Much shorter drying cycles are possible because the drying time varies exponentially with thickness. These are feasible if the capability of the dryer system suits the characteristics of palm veneers. For example, the drying operations can be carried out in a continuous production system where heat is transferred directly to the veneer surfaces rather than using an internal moisture flow.

In the actual drying process, an efficient drying rate is important to minimize drying time, production cost, energy, and safety of the workers. Basically, there are three types of dryers commonly used in the plywood industry throughout the whole world. The most popular and widely used in Malaysia is the air-circulation jet dryer. Most of these dryers were designed for temperate veneer species such as birch, maple, beech, aspen, etc. Later, these dryers were used to dry veneer of local tropical species with slight adjustment on the feeding speed, ventilation, and some proper drying planning due to variability on species used. In principle, the veneer characteristics such as thickness and MC (range about 50%–100%) are almost similar either for the temperate or tropical species. However, when handling with palm veneers features, one can justify that the same dryers are not designed specifically to perform such an extremely high MC during the drying process. Plywood manufacturers still have to dry palm veneers at a relatively longer period as compared to tropical hardwood veneers. The green palm veneers normally take more than 50 min per batch to be dried to 8% MC. In addition, those veneers located at the center (of the same batch) are still wet and require further drying. With the present drying system (net and roller dryers) and boiler capacity, any attempt to conduct "predrying" can reduce drying time. Though the practice may look as a major down-time, labor intensive etc., it does contribute to a significant reduction in drying time. Grouping veneers according to inner and outer (according to the trunk section where it is being peeled) would significantly expedite the drying process as well as minimize drying defects. Drying by batch is recommended.

8.3.5.3 Dryer and requirements

Kamarudin et al. (2011) conducted a series of study on OPT veneer drying and summarized the findings as follows:
1. Dryer
   Two types of dryers, roller and net dryers, can be used to dry the OPT veneers. Both dryers can be used to dry green OPT veneer down to about 8% MC. However, a roller dryer is favorable. It gives better veneer quality.
2. Energy source
   Since OPT veneer possesses extremely high MC, greater amount of energy is required to remove this moisture. Hence three options are recommended:
   - Use of a larger capacity boiler system—e.g., 25 tons and above (depends on the number of dryers and their capacity).
   - Use of larger capacity and longer haul dryer—more than 20 sections.
   - Use of an efficient and effective dryer system—optimized and retrofitted system for the vent opening, fan speed, temperature consistency, and requirements in evaluating any heat loss.

Both conventional net and roller dryers can be used for drying green OPT veneer. However, due to the higher MC, both dryers have shown a lower capacity in terms of output production. Normal drying time for OPT veneer is between 40 and 50 min. In some cases, the drying process has to be carried out twice.

8.3.5.4 Predrying and drying

Predrying: Due to the extremely high MC of OPT veneer and the undercapacity of conventional dryers, predrying is carried out. The aim is to reduce the MC of the OPT veneer to as low as possible prior to using the conventional dryer. The normal practice is by "sun-drying" (Fig. 8.9) and/or by mechanical pressure. The latter practice is by means of unused cold press. A pile of green veneer will be pressed (or squeezed) under constant pressure for a certain period. Both practices have shown a significant reduction of MC (down to about 70%).

A predrying stage of palm veneers is an essential sequence in veneer processing as it significantly affects the total drying time as well as the drying cost. By adding this sequence, millers could save between 20 and 40 min of drying time when using the current drying system such as those of net and roller dryers.

8.3.6 Veneer jointing/composing

The jointing of veneers is carried out to provide square and straight edges. This would allow two or more veneer sheets to be spliced to form a wider sheet. The narrower widths of dried veneer are edge

Fig. 8.9 OPT veneer drying: (left) "under the sun-drying" and (right) a typical steam-heated roller dryer for drying palm veneer. Extracted from Kamarudin, H., Jamaludin, K., Anis, M., Astimar, A.A., 2011. Characterisation of oil palm trunk and their holocellulose fibres for the manufacture of industrial commodities. Oil Palm Bull. 63, 11–23.

jointed into standard desired widths. This processing is only to be carried out by experienced manufacturers with the presence of splicing machines suitable for oil palm veneer.

The dried, graded veneers are usually assembled in two bundles in preparation for the glue spreading operation. In one bundle, the graded faces and long bands are assembled, and the other consists of the crossbands or, in the case of three ply, the cores. It is these crossbands or cores that are run through the glue spreader.

Longband: Core veneer that runs parallel to the grain of the face veneer in a 2440 mm by 1220 mm (8 ft by 4 ft) panel.

Crossband: Core veneer that runs across the panels at right angles to the face veneers is termed "crossband." In a 2440 mm by 1220 mm (8 ft by 4 ft) panel, the crossband can be produced by a smaller lathe (to produce short length veneers) or by cutting full sheets of veneer by two.

8.4 Challenges in rotary processing of OPT

When processing of OPT, it is best to understand some features and characteristic aspects about the trunk itself. The oil palm, being a monocotyledon, has marked anatomical differences from the timbers. The natural characteristics of OPT, i.e., anatomical structure, and physical and mechanical properties are the main challenges in making it a manageable raw material for plywood production. Handling and processing of OPT in commercial production requires proper and suitable equipment, as well as extensive knowledge on machine-material processing and equipment maintenance. Some of the limitations include:

a. OPT possesses a steep gradient over the physical characteristics, i.e., high in MC, varied in basic density, and fiber distributions suggesting the rate of drying of OPT veneer and shrinkage would vary considerably.

b. The freshly cut OPT is susceptible to fungal and insect infestation, thus it easily deteriorates and degrades when stored without proper treatment after felling.

c. The anatomical characteristics make the trunks much softer, especially near the core. In the case of plywood making, these would be a challenge to quality control to retain acceptable thickness tolerance of the plywood. Thus, the soft tissues (parenchyma) and surface roughness of OPT veneer tend to absorb more adhesives as compared to tropical wood.

d. The presence of silica causes the cutting knife to blunt prematurely. This is implied for both saw blades and peeler knifes.

Rationally, any attempt to utilize OPTs must consider the anatomical, physical, and mechanical properties whether for processing, product development, or marketing. Only then can the maximum potential of OPTs as an alternative source for veneer production be realized.

References

Abdullah, C.K., 2010. Impregnation of Oil Palm Trunk Lumber (OPTL) Using Thermoset Resins for Structural Applications. Master Thesis, Universiti Sains Malaysia, Penang, Malaysia.

Anis, M., Mohamad, H., Wan Hasamudin, W.H., Chua, K.H., 2005. Oil palm plywood manufacture in Malaysia. In: Proceedings of the 6th National Seminar on the Utilisation of Oil Palm Tree, Oil Palm Tree Utilisation Committee (OPTUC), Kuala Lumpur, Malaysia, pp. 51–55.

Bakar, E.S., Rachman, O., Darmawan, W., Hidayat, I., 1999. Utilization of oil palm trunk (*Elaeis guineensis* Jacq.) as construction materials and furniture (II): mechanical properties of oil palm wood. J. For. Prod. Technol. XII, 10–20.

Bakar, E.S., Mohd Hamami, S., H'ng, P.S., 2008. A Challenge from the Perspective of Functional Wood Anatomy. Penerbit Universiti Putra Malaysia, Universiti Putra Malaysia, Serdang.

Baldwin, R.F., 1981. Plywood Manufacturing Practices, second ed. Miller Freeman Publications Inc, San Francisco, USA.

Choo, A.C.Y., Tahir, P.M., Karimi, A., Bakar, E.S., Abdan, K., Ibrahim, A., Loh, Y.F., 2011. Density and humidity gradients in veneers of oil palm stems. Eur. J. Wood Wood Prod. 69 (3), 501–503.

Choo, A.Y.C., Paridah, M.T., Karimi, A., Bakar, E.S., Khalina, A., Azmi, I., Balkis, F.A.B., 2013. Study on the longitudinal permeability of oil palm wood. Ind. Eng. Chem. Res. 52, 9405–9410.

Choo, C.Y., Paridah, M.T., Karimi, A., Bakar, E.S., Khalina, A., Azmi, I., Juliana, A.H., 2015. Pre-drying optimization of oil palm veneers by response surface methodology. Eur. J. Wood Wood Prod. 73, 493–498.

Corley, R.H.V., Tinker, P.B., 2003. The Oil Palm, fourth ed. Blackwell Science Ltd, Oxford, UK.

Haslett, A.N., 1990. Suitability of oil palm trunk for timber uses. J. Trop. For. Sci. 2 (3), 43–51.

Ho, K.S., Choo, K.T., Hong, L.T., 1985. Processing, seasoning and protection of oil palm lumber. In: Proceedings of the National Symposium on Oil Palm By-Products for Agro-Based Industries, 5–6 November, Kuala Lumpur, Malaysia, pp. 43–51.

Kamarudin, N.H., Wahid, M.B., 1997. Status of rhinoceros beetle, *Oryctes rhinoceros* (Coleoptera: Scarabaeidae) as a pest of young oil palm in Malaysia. Planter 73 (850), 5–21.

Kamarudin, H., Jamaludin, K., Anis, M., Astimar, A.A., 2011. Characterisation of oil palm trunk and their holocellulose fibres for the manufacture of industrial commodities. Oil Palm Bull. 63, 11–23.

Kamarudin, H., Anis, M., Yeoh, B.H., 2014. Properties of oil palm trunk. In: Loh, Y.F., Anis, M., Paridah, M.T., Choong, K.K., Hashim, W.S., Yeoh, B.H. (Eds.), Handbook of Oil Palm Trunk Plywood Manufacturing. The Malaysian Timber Industry Board, Kuala Lumpur, Malaysia, pp. 16–33.

Killmann, W., Lim, S.C., 1985. Anatomy and properties of oil palm stem. In: Proceedings of the National Symposium of Oil Palm By-Products for Agro-based Industries. vol. 11. Palm Oil Research Institute of Malaysia Bulletin, pp. 18–42.

Lim, S.C., Gan, K.S., 2005. Characteristics and utilization of oil palm stem. Timber Technol. Bull. 35, 1–7.

Lim, S.C., Khoo, K.C., 1986. Characteristics of oil palm trunk and its potential utilisation. Malays. For. 49 (1), 3–23.

Lutz, J.F., 1971. Wood and log characteristics affecting veneer production. In: USDA Forest Service Research Paper Fpl 150. Available from https://www.fpl.fs.fed.us/documnts/fplrp/fplrp150.pdf.

Lutz, J.F., 1974. Techniques for Peeling, Slicing and Drying Veneer. Forest Products Laboratory, USDA Forest Service Forest Products Laboratory, Madison, Wisconsin, USA.

Mohamad, H., 1985. Potentials of oil palm by-products as raw materials for agro-based industries. In: Proceedings of Symposium on Oil Palm By-Products for Agrobased Industries. vol. 1. Palm Oil Research Institute of Malaysia Bulletin, Kuala Lumpur, pp. 7–15.

Paridah, M.T., Choo, A.C.Y., 2014. Drying of oil palm veneer. In: Loh, Y.F., Anis, M., Paridah, M.T., Choong, K.K., Hashim, W.S., Yeoh, B.H. (Eds.), Handbook of Oil Palm Trunk Plywood Manufacturing. The Malaysian Timber Industry Board, Kuala Lumpur, Malaysia, pp. 66–87.

Tomimura, Y., 1992. Chemical characteristics and utilization of oil palm trunks. JARQ 25, 283–288.

Walker, J.C.F., 2006. Primary Wood Processing: Principles and Practices, second ed. Springer, Dordrecht, Netherlands.

9

Pretreatment of empty fruit bunch fiber: Its effect as a reinforcing material in composite panels

Kit Ling Chin[a], Chuan Li Lee[a], Paik San H'ng[a,b], Pui San Khoo[a], and Zuriyati Mohamed Asa'ari Ainun[a]

[a]*Institute of Tropical Forestry and Forest Product, Universiti Putra Malaysia, Serdang, Selangor, Malaysia,* [b]*Faculty of Forestry and Environment, Universiti Putra Malaysia, Serdang, Selangor, Malaysia*

9.1   Introduction

Increasing environmental awareness has prompted changes in the market demand for product material, design, and manufacturing. This has motivated attempts toward advancing product innovation and development that must comply well with these demands and have a minimal effect on the environment. Due to the demand from various industries such as locomotive, agronomy, and construction, the biocomposite panel industry in Malaysia is advancing with new materials, formulations, and designs. Lignocellulosic fiber has been widely studied as a reinforcement material and was reported to contribute to an increase in the thermal and mechanical properties of composite panels (Pickering et al., 2016). The usage of biomaterial as a reinforcing substance focuses on lignocellulosic waste that is abundantly available and presently underused. The concept of lignocellulosic waste utilization for composite panel production is potentially feasible due to the low material cost, which reduces overall processing cost. Besides, composite panels from lignocellulosic waste also minimizes our dependence on nonrenewable materials such as synthetic fibers (Vaisanen et al., 2017).

One lignocellulosic waste that has obtained much interest in Malaysia is oil palm empty fruit bunch (EFB). EFB is sustainable and renewable as it is generated from the crude palm oil process as a by-product. As the whole fruit bunch undergoes a steaming process in the palm oil mill, EFB is normally in a wet condition with very high moisture content. If it is not processed immediately, EFB is prone to

Oil Palm Biomass for Composite Panels. https://doi.org/10.1016/B978-0-12-823852-3.00008-8
Copyright © 2022 Elsevier Inc. All rights reserved.

degradation, which affects the quality of the fiber. Basically, EFB consists of cellulose, hemicellulose, and lignin. EFB fibers are partially separated by shredding the bunches. Some of the palm oil mills double shred the bunch and send it out to their plantations as mulching material. However, many of the mills just pile it at the plantation, letting it degrade slowly to create organic stockpile for the soil. Despite utilizing EFB for this low-end application, EFB fiber has the potential to produce higher end products and has been recognized as an important lignocellulosic biomass. Due to its abundant availability in Malaysia, it is potentially considered as the main candidate as a substitute for wood and to assist in dealing with wood shortage problems.

The conventional manufacturing process utilizing wood or synthetic fiber as a raw material should never be directly replaced by EFB. Lignocellulosic waste such as EFB has different properties than wood or synthetic fiber as well as some major weaknesses such as lack of dimensional stability and presenting with a hydrophilic behavior. These issues will cause poor bonding of fiber with the matrix, which will affect the interface of the composite due to ineffective stress transfer (Gurunathan et al., 2015). Some modifications on the EFB are certainly necessary through selected pretreatments so that the properties of the fiber itself can be enhanced, which also leads to overall property improvement on the produced composite panel (Prabhakar et al., 2015). In dealing with EFB, pretreatment is the most important element that needs extensive research. The EFB fiber characteristics and the potential pretreatments to prepare the fiber as a reinforcement material for composite panel production are further discussed in the following sections.

9.2 The complex nature of EFB fibers

9.2.1 Morphological characteristics

EFB fibers are tough multicellular fibers, and each fiber encompasses a wide vessel in the core section of the vascular bundle (Jinn et al., 2015). The vascular bundle is fixed in the parenchyma tissues and encircled by porous cylindrical structure-like fibers. EFB is comprised of 75% fibers and 25% of vascular bundles in volume (Jinn et al., 2015). Vascular bundle and parenchyma tissues are very porous materials with high moisture absorption properties. Due to the high volume percentage of vascular bundle, EFB is capable of holding a high moisture content of up to two times the weight of the dry EFB (Lani et al., 2014; Ngo et al., 2014).

EFB fibers length range from 30 to 140 mm, while the kenaf fibers have a consistent length around 6 mm (Nafu et al., 2015). EFB fiber length is affected by the shredding process used to separate them from the bundle. EFB fiber has a diameter of 22.3 μm and lumen diameter of 15.7 μm

(Law et al., 2007). Aspect ratio is defined as the length of a fiber divided by its width, which has an influence on the tensile strength of a reinforced composite panel (Ghosh et al., 1995). EFB has an aspect ratio of 52–87.

EFB fiber is shorter, with a smaller diameter, thicker cell wall, and smaller lumen when compared with softwood. However, it has comparable fiber length and cell wall thickness with hardwood but bigger lumen and diameter (Jinn et al., 2015). Differences in fiber morphologies affects the physicomechanical properties of the fiber such as stiffness, tensile, and bending strength, which reflects fiber usability (H'ng et al., 2009; Benazir et al., 2010; Fidelis et al., 2013). The thickness of the cell wall reflects the fiber stiffness. EFB has long fibers that contribute to higher flexibility, thinner cell walls that gives a low Runkel ratio, and larger lumen area that creates low slenderness ratio (Jinn et al., 2015). EFB fiber with thin cell walls is beneficial in fiberboard production as it provides additional surface contact and better fiber adhesion (Dutt et al., 2004).

Runkel ratio is the thickness ratio of lumen and cell wall, which indicates the elasticity of the fiber. Fiber with the Runkel ratio below one is associated with good mechanical strength properties, and EFB has a Runkel ratio of 0.34 (Istek, 2006; Jinn et al., 2015). A higher slenderness ratio reflects a higher resistance of the fiber against tearing force. EFB has a lower slenderness ratio compared to wood but is still in an acceptable range for fiberboard production, which is 40.36 (Xu et al., 2006). Fibers are classified into different classes based on their flexibility ratio: (1) high elastic fibers with flexibility ratio higher than 75, (2) elastic fibers with flexibility ratio between 50 and 75, (3) rigid fibers with flexibility ratio between 30 and 50, and (4) high rigid fibers with flexibility ratio less than 30 (Bektas et al., 1999). The EFB fiber has a flexibility ratio of 70.40% and is classified under the type 2. Fibers under this category can be easily flattened and used to produce paper with high tensile strength properties (Bektas et al., 1999). Benazir et al. (2010) reported that fibers with high flexibility ratio are said to be mobile, crease easily, and have good interface and interfiber bonding.

EFB fiber surface were covered with silica bodies and bounded between the fibers. Silica bodies in nature act as a shield against fungal attack and give support to the plant's structure (Abdullah et al., 2016). The silica bodies observed on fiber strands have a round, spiky shape. EFB fibers exhibit a rough surface with border pits filled with silica (Law et al., 2007).

9.2.2 Chemical characteristics

Fresh from the palm oil mill, EFB comprises 30%–35% lignocellulose, 1%–3% residue oil, and roughly 60% of moisture (Gunavan et al., 2009). The viability of the lignocellulosic fiber for composite panel production was significantly affected by the chemical composition. EFB contains 74.9% holocellulose, 44.2% cellulose, 12.7% lignin, 4.8% ash, and 2.4% oil and grease. Cellulose forms both inter- and intramolecular hydrogen bonds. The fibrous structures and strong hydrogen bonds of cellulose creates a high tensile strength, and the fibers are insoluble in most solvents (Law et al., 2007). Low lignin content in EFB will facilitate the pretreatment process with less chemicals required. Due to the fertilization program at the plantation, elements such as calcium, potassium, magnesium, and phosphorus are provided throughout the plantation period and are mostly accumulated in the fruit stalk (pedicel) (Bondada and Keller, 2012). This may explain the reason for higher inorganic elements in EFB than wood. It is common in plantation practices to boost the yield and quality of fruit crops by applying excessive fertilization, especially potassium-rich fertilizer. Among the eight major inorganic elements presented by Chin et al. (2012), potassium content was found to be the predominant inorganic element in EFB.

9.2.3 Mechanical characteristics

EFB has a specific tensile stress with a mean of value of 1100 N/m, which is much lower than wood materials (3600–11,000 N/m) (Bodig and Jayne, 1993). EFB requires only one-third the force to attain the same elongated length as wood, thus EFB is much more brittle than wood. Tensile strength is an important property as it reflects the quality of the fiber and composite panels. The tensile strength of lignocellulosic fibers highly depends on their chemical composition and morphological structure. Low lignin content and high hemicellulose content of EFB lead to poor fiber stiffness. Voelker et al. (2011) reported that a drastic reduction of wood strength and stiffness was clearly shown when the lignin of poplar was reduced up to 40%, although the density was unchanged.

9.3 Constraints in producing composite panels from untreated EFB fiber

Regardless of the advantages of utilizing EFB fibers in composite panel production, there are a few complications that must be altered. Numerous wood composite production companies in Malaysia tried to substitute wood with EFB fiber to produce composite panels such as particleboards, fiberboards, and wood-plastic composites. However, most of these projects were produced in a small scale or halted due to various restrictions initiated by the natural characteristics of the EFB fiber itself and the inconsistency of the material's quality.

The major problem in the usage of EFB in the composite panel industry is the high content of grease, oil, and minerals in the EFB fibers. Oil consists of ester components that may disturb the bonding proficiency between EFB fibers and the matrix during composite panel production (Rozman et al., 2001). Oily constituents and impurities stick on the surface of EFB causing poor surface wetting and decreasing coupling efficiency between fiber and matrix. This could explain why composite panels produced from EFB are substandard in comparison with composite panels made from wood (Ratnasingam et al., 2008; Norul Izani et al., 2012). In addition, a high amount of silica bodies that attach on the EFB fibers create high abrasiveness on the machining tools. In comparison to particleboard made from rubberwood, EFB particleboard is nearly double the abrasivity in terms of the efficiency of the machining tool (Ratnasingam et al., 2008; Norul Izani et al., 2012), which incurs higher maintenance cost of the machinery.

Most of the EFB fiber bundles with low specific tensile stress are due to deficiencies of the fiber bundles, which consist of 75% of fiber and 25% of vascular bundle (v/v) (Jinn et al., 2015). Vascular bundle has a thin cell wall and is easier to collapse in comparison to fibers with thicker cell walls. This causes the fiber structure to have numerous mechanically weak points and inhomogeneous characteristics of the fiber structure. Besides, due to the high vascular bundle/fiber ratio in terms of volume, EFB fiber is a naturally porous material with high compaction ratio. EFB composite panels cannot be produced by applying a similar formula or method as in producing panels from wood. It is so much more difficult to attain a targeted density during the board pressing processes because of the high volume reduction and compaction ratios.

The large composition of vascular bundles also causes the composite panels from EFB to have a high hygroscopic range, which causes the fiber to swell and ultimately rot due to fungi infestation. Dimensional stability and moisture absorption properties are the main criteria for a composite panel to be applicable in the furniture industry or in structural material applications. Applying untreated EFB in such industries or applications could be an issue due to the high moisture absorption characteristics and low dimensional stability. Dimensional stability is the capability to uphold the initial dimensions when affected by an extraneous constituent. Low dimensional stability properties of composite panels will affect its durability, strength, screw-holding ability, and stiffness.

EFB are hydrophilic, which contains strongly polarized hydroxyl groups, thus EFB is characteristically incompatible with hydrophobic matrices. Utilizing EFB as reinforcements in such matrices will cause poor interfacial adhesion between polar-hydrophilic fiber and the nonpolar-hydrophobic matrix. Besides, poor wetting of the fiber with the matrix creates a problem during mixing, which eventually leads to composite panels with a weak interface (Mäder et al., 1996). These major drawbacks connected with EFB must be solved before utilizing them in composite panels. Thus, the EFB fiber needs to be pretreated before applied as a reinforcing material in composite panels. Pretreatments that alter the EFB fiber physically or chemically must be applied to reinforce the poor characteristics of the EFB fiber by (1) removal of contaminations, (2) altering the chemistry and crystallinity formation, (3) refining the interface of fiber and matrix, and (4) achieving good bonding between fiber and matrix (Fiore et al., 2015).

9.4 Types of lignocellulosic fiber pretreatment method

9.4.1 Physical pretreatment

Physical pretreatments can be utilized to treat lignocellulosic fiber to improve thermal properties and bonding strength of the composite panel by altering the physical characteristics without altering the chemical composition of the fiber (Cordeiro et al., 2011). These pretreatments are applied to the lignocellulosic fiber for separation of individual filaments from the fiber bundles and enhancement of fiber surfaces for composite panel implementations (Mukhopadhyay and Fangueiro, 2009), Physical pretreatments can be separated into three processes: (1) mechanical pretreatment, (2) solvent extraction pretreatment, and (3) electric discharge pretreatment. Each physical treatment creates different effects on the treated lignocellulosic fiber such as an increase in mechanical properties and surface area, high crystallinity, and enhanced durability.

9.4.1.1 Mechanical process

Conventional mechanical pretreatments such as stretching, calendaring, or rolling are applied to treat the fiber surface to improve the bonding between the lignocellulosic fibers and the matrix. Tensile strength can be increased through a prestretching method; however, this method may cause fiber extension in which the lignocellulosic fiber could slide across one another during prestretching causing expansion and further lengthening (Rozali et al., 2019). Pressure applied by the rollers in the calendaring method transforms the lignocellulosic fiber into an even continuous layer that can be cut and attached into a preferred mold. Through this calendaring method, the surface smoothness and density of the fibers are improved by reducing the pore size, and larger particles are bound together as a sheet (Gupta and Gupta, 2019). Fiber bundle separation was generated using a rolling method to enhance dispersibility and bonding with matrices (Potluri, 2019). It was reported that these mechanical pretreatments can lower the loss of fibers and improve the contact surface area between fibers and matrix in the composite panels (Satyanarayana et al., 2009). However, Varshney and Naithani (2011) found that such mechanical pretreatments lead to high energy consumption as the lignocellulosic fibers easily tangle together. Throughout the entire method, there will be no changes to the chemical composition of the lignocellulosic fiber, yet the pretreated fibers increase the characteristics of the composite panel as a reinforcement (Fuqua and Ulven, 2008). With pretreated fibers as reinforcement material, it was reported that the modulus of rupture (MOR) for the composite panels were increased, while no significant results were observed on the modulus of elasticity (MOE) (Rana et al., 2018).

Beating or refining is a physical surface modification pretreatment that is generally used in the pulp and paper industry. Thus, this method also increases the properties of the composite panels reinforced with fibers obtained from the beating process (Mohr et al., 2007). This pretreatment creates a standardized quantity of reduced fibrils, enhances fiber adhesion, and creates an optimal mechanical process in formulating the pulp for the paper production method. Generally, beating or refining pretreatment increases the elasticity of fibers by the collapse of fiber walls into isolated lamellae, exposing fibrils on the surface of the fibers, and generating fines from fibers that are unable to sustain compressive and/or shear forces during the pretreatment (Bhardwaj et al., 2004). Li et al. (2019) reported that beating at 60°SR (degree of beating) leads to greater porosity and a higher fibrillation degree. The microstructure and mechanical characteristics of the lignocellulosic fiber can be controlled through the fibrillation degree attained from the beating method (Jiang et al., 2019).

9.4.1.2 Solvent extraction

Solvent extraction is one of the simplest mechanical separation techniques to enhance the surface area and remove contaminations on lignocellulosic fiber (DeSimone, 2002). Lignocellulosic fibers can be separated to obtain fibers with high content of cellulose using selective solvent action. This pretreatment is an efficient method to remove a compound based on its solubility in a selected solvent (Li et al., 2016). But this pretreatment was not commonly applied due to the degradation effect on the lignocellulosic fiber, which decreases the fiber aspect ratio. Besides, fiber pretreatment using the solvent extraction method will produce harmful vapor and cause pollution if the by-products leach into the waterways or water supplies (Ramamoorthy et al., 2015). Many new solvent extraction methods were discovered, but due to the stringent condition in applying the selective solvents, many are not viable or applicable in the industry (Płotka-Wasylka et al., 2017).

9.4.1.3 Thermal process

Thermal pretreatment is applied with the aim of improving mechanical characteristics and increasing the melt viscosity of lignocellulosic fiber by separating the cellulose (Varshney and Naithani, 2011). This is a suitable pretreatment process to improve the interfacial properties between hydrophilic fiber and matrix by creating a coarse surface on the fiber (Varshney and Naithani, 2011). The thermal pretreatment has a lower environmental effect when compared with solvent extraction (Song et al., 2010). Thermal pretreatment alters the physical characteristics of the fiber while preserving its chemical structure (Cao et al., 2007). When the heating temperature is increased to between 100°C and 200°C, the fiber bundles are dried out and separated into single filaments (Mishra et al., 2004). Lignin and other chemical elements with lower glass-transition temperatures will decompose into simpler compounds (Ayrilmis et al., 2011). It was also reported that when lignocellulosic fibers are subjected to lower ranges of temperature, the fiber stiffness and compatibility between fiber-matrix increase as the fiber crystallinity increases. Temperature of 150°C was declared as the precise temperature to increase crystallinity of lignocellulosic fiber resulting in higher strength and modulus (Naveen et al., 2018).

9.4.1.4 Steam explosion

As concern about environmental pollution is increasing, steam explosion is an effective pretreatment to obtain high fiber yield that was invented with the aim of reducing the usage of chemicals, human toxicity, and environmental pollution (Deepa et al., 2011). This pretreatment helps in removing hemicelluloses prior to alkaline pretreatment. Saturated steam at a certain temperature and reaction time is used to treat the lignocellulosic fiber by shear force activated by the moisture expansion and through hydrolysis of acetyl groups in hemicelluloses (Tanpichai et al., 2018). This method physically breaks down the fibers from inside to release contaminants and produce fibrils without altering the chemical structure. Disintegration may happen, varying by severity degree of pretreatment in terms of temperature and time; however, with the strength and crystallinity structure naturally possessed by lignocellulosic fibers, no significant damage is observed through steam explosion.

9.4.1.5 Ultrasonication

Ultrasonication pretreatment for lignocellulosic fibers is still being studied under a laboratory scale. Imai et al. (2004) evaluated the effect of the ultrasonication pretreatment on pure cellulose (carboxyl methyl cellulose, CMC) prior to enzymatic hydrolysis. The conversion rate of the following enzymatic hydrolysis is almost doubled. Apparently, hydrogen bonds of the cellulose crystalline structure break when sufficient energy is applied. Nevertheless, significantly higher energy (130 kJ/g CMC) was supplied in this case, whereas the energy needed to break the hydrogen bond is only 0.12 kJ/g cellulose (Bochek, 2003). Chin et al. (2020) and Lee et al. (2019) have used ultrasonic treatment to increase the pore volume of palm kernel and coconut shell. Most of the impurities clogging the pores in palm kernel shell were successfully removed. However, in the case of coconut shell, the pore walls shattered and collapsed. This is due to the stronger structure of palm kernel shell compared to coconut shell.

9.4.1.6 Electron radiation

Electron radiation method is another pretreatment utilized on lignocellulosic fibers as a surface modification treatment to develop fiber-matrix interactions to improve fiber characteristics (Han et al., 2007). This pretreatment condition varies on (1) type of fiber/matrix; (2) additives; (3) reaction temperature; (4) pressure; (5) application rate; (6) morphological structure; (7) crystallinity; and (8) surface volume ratio (Huber et al., 2010). Electron radiation process generates free radicals on the surface of the lignocellulosic fiber and activates functional group's graft polymerization. The pretreated fiber can be applied in composite panels as the adhesion between fiber and matrix has been improved. This electron radiation pretreated fiber can also be used as independent functional materials such as an absorbent in a water purification application (Wu, 2019). This pretreatment substantially alters the composition, reactivity, and the mechanical and physicochemical characteristics of lignocellulosic fiber (Takacs et al., 2000).

9.4.2 Chemical pretreatment

Chemical pretreatment is applied to alter and stimulate the fiber structure by utilizing a hydroxyl group that can modify the structure of the lignocellulosic fiber through the exposure of additional elements to react with the substance (Valadez-Gonzalez et al., 1999). Fiber modification through the application of chemical reagents will enhance the mechanical characteristics of the fiber, thus increasing the mechanical characteristics of fiber-reinforced composite panels as the bonding between the fiber-matrix surface is improved.

9.4.2.1 Alkaline

Alkali pretreatment is efficient, cost-effective, and the most popular chemical pretreatment for lignocellulosic fiber modification. From this pretreatment, lignin, oil, and wax that cover the exterior surface of the cell wall will be partly removed, which activates cellulose disintegration and reveals short length crystallites (Chandrasekar et al., 2017). This pretreatment increases the surface roughness by breaking the internal hydrogen bonding that alters the surface structure, water absorption, pore formation, crystallinity, and alignment of fibrils, improving the mechanical characteristics of the fiber (Cai et al., 2015). Few researches have been conducted to evaluate the effect of sodium hydroxide (NaOH) on the characteristics of the lignocellulosic fibers and composite panels reinforced with the pretreated fibers (Mwaikambo and Ansell, 1999; Wei and Meyer, 2014; Tian et al., 2016). Depending on the concentration of alkali applied in the treatment, the fiber strength can be enhanced. However, a more severe condition of more than 10% alkali concentration will result in the deterioration of the fiber, which causes a reduction of tensile strength. Lower alkali concentration of around 6%–9% will increase the fiber tensile strength by nearly 30%. The alkali pretreatment helps remove impurities and creates a rough fiber surface, which directly affects the thermal and mechanical characteristics of the lignocellulosic fiber. It was proven by Barreto et al. (2011) that composite panels reinforced with alkali pretreated fibers have better mechanical characteristics compared to the composite panels reinforced with untreated fibers. The efficiency of alkali pretreatment on lignocellulosic fibers can be further improved by using higher reaction temperature as it would create further catalytic reaction in shattering the hydrogen bonds in the fibrils (Saha et al., 2016). Types of lignocellulosic fibers, the alkali concentration, reaction temperature, reaction time, and types of additives are the important parameters that should be considered when applying this pretreatment as these will directly affect the physical and mechanical characteristics of the pretreated fiber.

9.4.2.2 Coupling agent

Coupling agents are used to enhance the interfacial adhesion, which directly affects the strength properties of composite panels. The coupling agents have high compatibility with the fiber-matrix and help reduce moisture absorption, reducing the leaching effect and improving the fiber wettability (Saha et al., 2016). Coupling agents such as silanes, acetylation, and graft copolymerization is applied on lignocellulosic fibers to enhance the chemical bonding of the fiber surface, which contains oxide groups with the polymer molecules. These coupling agents help connect the hydrophobic functional polymers with hydrophilic fibers to enhance the strength and dimensional properties of the composite panel (Arrakhiz et al., 2012). Chemical pretreatment via coupling agents produces composite panels with notable enhancements in the mechanical and physical properties, which generally rely on the fiber, matrix, and type of surface treatment used (Bledzki and Gassan, 1999).

9.4.2.3 Bleaching

Bleaching pretreatments are environmentally friendly and applied on lignocellulosic fibers to separate individual microfibers by dispersing and removing lignin and hemicellulose (Arrakhiz et al., 2012). A majority of the fibers are bleached with hydrogen peroxide before being applied to textile fabrics. When applying hydrogen peroxide, temperature and pH are crucial parameters as extreme conditions (reaction time and temperature) and alkaline concentration will break the fibers. Bleaching contributes to the modification of appearance and aesthetic; however, it may also cause an unintentional deficiency of fiber tensile strength (Flitsch et al., 2013). This is mainly caused by the removal of lignin, which is a fortifying substance that leads to a reduction in the mechanical properties of the composite panel. Commonly, bleaching pretreatments is separated into two types: oxidative and reductive.

Reductive bleaching is a method using sodium dithionite ($Na_2S_2O_4$) applied to fibers comprising high cellulose by reducing chromophores (colored fibers) to leucochromophores (uncolored fibers). Reduction bleaching may cause considerable reduction in durability, so it is seldom studied for surface alteration of lignocellulosic fiber for composite panel applications (Fuqua et al., 2012). Oxidative bleaching is a more preferred pretreatment method utilizing hydrogen peroxide (H_2O_2), sodium hypochlorite (NaClO), or sodium chlorite ($NaClO_2$). NaClO reduces the lignin content by attacking the hydroxyl groups to produce aldehyde groups (CHO), which will affect the reduction of cellulose in lignocellulosic fiber. Whereas $NaClO_2$ bleaching creates a reduction in lignin and pectin but with only a slight degradation effect on cellulose. $NaClO_2$ bleaching is also more economically viable in comparison to NaClO and H_2O_2 bleaching. Nevertheless, H_2O_2 bleaching is more environmentally friendly and is being increasingly substituted for other oxidative bleaching methods; however, the cost is higher than most of the bleaching agents available with comparable functionality (Adeleye et al., 2015).

9.4.2.4 Biological pretreatment

Biological pretreatment is an efficient technique that requires low energy input to modify the fiber surface by removing lignin and hemicellulose. Microorganisms such as bacteria fungus release enzymes that release the cellulosic fibers by degrading or removing the substance (pectinase glue) that acts as an adhesive between fibers forming a bundle. This pretreatment enhanced the mechanical properties of the composite panels by increasing the surface hydrophilic interface between the fiber and matrix. The crystallinity and thermal properties increased after the lignocellulosic fiber bundles are split into separated fibers with a pretreatment using a mixture of enzymes (Li and Pickering, 2008). Biological pretreatment affects the composition, chemical characteristics, and final fiber properties (George et al., 2014). Biological pretreatment is increasingly common due to its environmentally friendly character (Ahmad et al., 2014) and the specific catalytic reaction from this pretreatment aimed toward a specific expected performance. After each usage, it can be recycled (Akil et al., 2015), but the release may create an environmental problem (Khalil et al., 2012).

9.5 Pretreated EFB fibers for fiber-reinforced composites

Fiber pretreatment is necessary and important to prepare EFB fibers for application in superior biocomposite panels. In theory, the objective of fiber pretreatment is to increase the interfacial adhesion and thus optimize the properties and structure of the fibers before they are used as reinforcements in composite panels. Numerous physical and chemical pretreatments on EFB have been reported, which include using ultrasonication, steam treatment, autoclave, nanoparticle impregnation, bleaching, and acidic and alkaline pretreatments (Table 9.1).

Theoretically, the outer layer of the EFB cell wall contains of few undesirable compounds such as extractives, hemicellulose, pectin, lignin, oil, and waxy constituents. These unwanted compounds will be removed through chemical pretreatment, exposing the fibrils. This pretreatment caused physical features of the fiber surface to have an uneven arrangement. Chemical pretreatment such as alkaline, acidic, bleaching, and saline pretreatments are reported to remove the outer layer of oil/wax, lignin, and hemicellulose and detach silica from the EFB surface. This increases the effective surface area available for contact with the matrix due to the cleaner and rougher fiber surface. Alkaline pretreatment has been the most studied pretreatment for the modification of EFB (Suradi et al., 2010; Chowdhury et al., 2013; Ibrahim et al., 2015, 2019; Akasah et al., 2019; Latip et al., 2019; Sawawi et al., 2020). Mwaikambo and Ansell (2002) and Islam et al. (2010) stated that alkaline is one of the best chemical pretreatments for modifying natural fiber. Ibrahim et al. (2016) found that the detachment of hydroxyl groups from chemical treatment will reduce the hydrophilic behavior of fiber. Composite panels from pretreated EFB displayed lower water intake compared to untreated EFB. This will eventually improve dimensional stability along with improvement in the mechanical properties of EFB-reinforced composite panels.

Table 9.1 Physical and chemical pretreatments on EFB.

Pretreatment method	Pretreatment effects	Pretreatment condition/ process	Type of composite panel	Reference
Ultra-sonication and alkaline peroxide	**i.** Disassemble the lignocelluloses and opens up for the penetration of reagents **ii.** High cellulose yield **iii.** Fibre with rough surface	Three steps involved **i.** EFB fibres were suspended in 2% NaOH (30 min; 250 rpm). **ii.** Addition of 30% H_2O_2 and sonication (40 kHz; 2 h; room temperature) **iii.** Treated with H_2O_2 at (pH 11; 90 min)	Fibre reinforced thermoplastic composites (polypropylene)	Abdullah et al. (2016)
Steam treatment	**i.** Improvement on thermal stability **ii.** Better hydrophobicity **iii.** Silica bodies were partially removed making the strand to have a rough surface **iv.** Fibre surfaces were cracked **v.** Improve the tensile strength of produced panel **vi.** Better adhesion of EFB fibre with the polypropylene polymer matrix	Steam treatment was conducted using an autoclave at 120 °C and 21 psi for 30 min.	Fibre reinforced thermoplastic composites (polypropylene)	Bujang and Nordin (2020)
Autoclave and alkali pretreatment	**i.** Disassemble the lignocelluloses and opens up for the penetration of reagents **ii.** Complete removal of cementing wax, hemicelluloses, and lignin **iii.** Large crystalline region and small amount of amorphous region **iv.** Fibre with rough surface **v.** High cellulose yield	Three steps involved **i.** EFB washed with washing detergent **ii.** Treating fibres in 70% ethanol using a Soxhlet system (6 h) **iii.** fibre suspended in 10% NaOH followed by 10% H_2O_2 and autoclaved for 1 h	–	Nazir et al. (2013)
Acetic acid	Improvement in dimensional stability	Immersed in 0.4% acetic acid for 24 h	Medium density fibreboard	Ibrahim et al. (2016)
Acetic acid	**i.** Fibre surface becomes rougher and more silica bodies were removed **ii.** Fibre surface have clear cleavage lines and cracks formation **iii.** Fibre with higher tensile strength	Soaked in different concentration levels of 0.8% acetic acid for 24 h	–	Ibrahim et al. (2015, 2019)
Alkaline-NaOH/ KOH	**i.** Increasing in tensile strength due to improvement in cellulose structure and increase the rigidity of cellulose (increment in crystallinity index) **ii.** Higher thermal stability	EFB soaked in 15% NaOH or KOH (130 °C; 40 min, fibre to solution ratio 1:20)	–	Latip et al. (2019)

Continued

Table 9.1 Physical and chemical pretreatments on EFB.—cont'd

Pretreatment method	Pretreatment effects	Pretreatment condition/ process	Type of composite panel	Reference
Alkaline-NaOH	i. Removed oil on EFB ii. Improvement in mechanical property of the produced panel	Soaked in different concentration levels of 4% NaOH for 24 h	Cement bonded fibreboards	Akasah et al. (2019)
Alkaline-NaOH	i. Silica bodies are detached from the EFB surface and more voids are present ii. An increase in the effective surface area available for contact with the matrix due to the cleaner and rougher fibre surface iii. Increment in the flexural strength of the EFB/UF particleboard due to the improvement in the bonding strength	EFB fibres were soaked in 1.5% concentration of NaOH at room temperature for 24 h	Urea formaldehyde resin particleboard	Sawawi et al. (2020)
Alkaline-NaOH	i. Oil content was reduced from EFB ii. Fibre surface rougher with less amount of silica bodies iii. Fibre surface becomes uneven and contains some substances deposited on the surface iv. Fibre with higher tensile strength	Soaked in 0.8% NaOH concentration for 24 h	–	Ibrahim et al. (2015, 2019)
Nanoparticle impregnation and alkali treatment	i. Higher crystallinity index ii. Improvements in fibres strength due to the improved fibre-filler interfacial bonding iii. Increased thermal stability iv. Improvement in mechanical property and durability of the produced board	Eight steps involved i. EFB fibres were treated in 0.5% NaOH solution (24 h; 200 rpm) ii. Fibres washed with distilled water and dried in an oven (70 °C; 48 h) iii. EFB fibres were dipped in 36% 3-chloro-hydroxypropyltrimethyl ammonium chloride solution (solid liquor ratio of 1:30; 60 °C; continuous agitation) iv. 15% NaOH was added in three steps at an interval of 5 min, and the mixture was further stirred for 15 min v. Fibres rinsed with water, neutralized and dried at the ambient temperatures vi. The fibres were introduced in the synthesized nanocopper solution of copper concentration 250 mg/L (fibre to liquid ratio of 1:30; continuous shaking; 12 h) vii. Fibres rinsed with water and dried in a dark place at room temperature	Copper nanoparticles reinforced cationized EFB	Chowdhury et al. (2013)

Alkaline peroxide	i. Improve adhesion characteristics between fibre-PP matrix by removing surface impurities such as lignin content ii. Higher thermal stability	Three steps involved i. EFB soaked in distilled water (70 °C; 4–6 h) ii. EFB soaked in 15% NaOH (75 °C; 3 h) iii. EFB soaked in 2% H2O2 (45 °C; 8 h, 50 rpm)	fibre reinforced thermoplastic composites (polypropylene)	Suradi et al. (2010)
Bleaching	i. 95% reduction in fibre size ii. Removal of impurities, waxes, and lignin part in the natural fibre iii. Increased surface roughness of the OPEFB fibres	Four steps involved i. Treating fibres in ethanol:nhexane:acetone solvent using a Soxhlet system (24 h) ii. Bleached in NaClO solution at 70 °C for various different times iii. Fibres were soaked with 6% KOH solution (20 °C; 24 h) iv. Fibres were hydrolyzed under strong agitation (stirring) in 64% H_2SO_4 solution (45 °C; 30 min)	fibre reinforced thermoplastic composites (vinylester resin)	Nabinejad et al. (2017)
Seawater	i. Tensile strength of the single fibre increased ii. Improve the interfacial bonding between the matrix due to the removal of the outer layer of hemicellulose and lignin	EFB immersed in seawater for 30 days	fibre reinforced thermoplastic composites (poly (vinyl alcohol)	Sarjadi et al. (2018)

EFB surface is covered with silica bodies. Physical pretreatment such as ultrasonication, steam pretreatment, and autoclave have been reported to be effective in removing silica bodies from the EFB surface (Nazir et al., 2013; Abdullah et al., 2016; Bujang and Nordin, 2020). The removal of silica on the fiber surface is good for fiber-matrix adhesion because it provides a physical interlocking between the fiber and matrix and increases the tensile strength of the produced panel. These physical pretreatments also create crack formation on the fiber surface, which creates a better adhesion of EFB fiber with the matrix. Besides, these physical pretreatments were also conducted in combination with chemical pretreatments (physicochemical pretreatment). Physical pretreatment disassembles the lignocelluloses and opens up the EFB fiber for better penetration of chemical reagents during chemical pretreatment.

9.5.1 Conclusion

The advantages of applying EFB fiber as a reinforcing material in composite panels, as well as its weaknesses, have been discussed in this chapter. Due to the complex characteristics of EFB fiber in nature, EFB is a material with high stiffness, which can be obtained for little cost. However, the major drawbacks are high sensitivity to moisture and inconsistent fiber quality. Most of these issues can be overcome via appropriate pretreatments before converting them into higher end products. Although the additional treatments required to increase the competency of EFB as a reinforcing material raises the overall processing cost, it may still be an affordable production process due to the low material cost, which reduces the overall cost. Recognizing the fundamental properties of this natural fiber is an important element in unlocking higher application possibilities of EFB in the composite panel industry and to make the process economically feasible. The composite panels reinforced with EFB will be commercially feasible only if the fibers possess higher value-in-use for the similar application as the current in-use materials, which are aimed to be replaced. A new product has a better value-in-use than the current existing product if it is more economical and eco-friendlier, besides having similar features. The composite panel industry players should have a comprehensive understanding to develop an appropriate production system utilizing EFB. This is an important aspect of a successful deployment of new products, especially from materials with complex characteristics such as EFB. The successful development of an EFB pretreatment process can be valuable to local manufacturers by providing larger application value to the fiber itself to be produced into numerous products eligible for commercialization.

References

Abdullah, M.A., Nazir, M.S., Raza, M.R., et al., 2016. Autoclave and ultra-sonication treatments of oil palm empty fruit bunch fibers for cellulose extraction and its polypropylene composite properties. J. Clean. Prod. 126, 686–697. https://doi.org/10.1016/j.jclepro.2016.03.107.

Adeleye, A.S., Conway, J.R., Garner, K., et al., 2015. Engineered nanomaterials for water treatment and remediation: costs, benefits, and applicability. Chem. Eng. J. 15, 640–662. https://doi.org/10.1016/j.cej.2015.10.105.

Ahmad, F., Choi, H.S., Park, M.K., 2014. A review: natural fiber composites selection in view of mechanical, light weight, and economic properties. Macromol. Mater. Eng., 1–15. https://doi.org/10.1002/mame.201400089.

Akash, Z.A., Dullah, H., Soh, N.M.Z.N., Guntor, N.A.A., 2019. Physical and mechanical properties of empty fruit bunch fibre-cement bonded fibreboard for sustainable retrofit building. Int. J. Mater. Sci. Eng. 7, 1–9. https://doi.org/10.17706/ijmse.2019.7.1.1-9.

Akil, H., Zamri, M.H., Osman, M.R., 2015. The use of kenaf fibers as reinforcements in composites. In: Faruk, O., Sain, M. (Eds.), Biofiber Reinforcements in Composite Materials. Woodhead Publishing, Cambridge, pp. 138–161.

Arrakhiz, F.Z., El, A.M., Kakou, A.C., et al., 2012. Mechanical properties of high density polyethylene reinforced with chemically modified coir fibers: impact of chemical treatments. Mater. Des. 37, 379–383. https://doi.org/10.1016/j.matdes.2012.01.020.

Ayrilmis, N., Jarusombuti, S., Fueangvivat, V., Bauchongkol, P., 2011. Effect of thermal-treatment of wood fibres on properties of flat-pressed wood plastic composites. Polym. Degrad. Stab. 96, 818–822. https://doi.org/10.1016/j.polymdegradstab.2011.02.005.

Barreto, A.C.H., Rosa, D.S., Fechine, P.B.A., Mazzetto, S.E., 2011. Properties of sisal fibers treated by alkali solution and their application into cardanol-based biocomposites. Compos. Part A 42, 492–500. https://doi.org/10.1016/j.compositesa.2011.01.008.

Bektas, I., Tutus, A., Eroglu, H., 1999. A study of the suitability of Calabrian pine (PinusBrutia ten.) for pulp and paper manufacture. Turk. J. Agric. For. 23, 589–598.

Benazir, J.A.F., Manimekalai, V., Ravichandran, P., et al., 2010. Properties of fibres/culm strands from mat sedge-Cyperus Pangorei Rottb. Bioresources 5, 951–967.

Bhardwaj, N.K., Duong, T.D., Nguyen, K.L., 2004. Pulp charge determination by different methods: effect of beating/refining. Colloids Surf. A Physicochem. Eng. Asp. 236, 39–44. https://doi.org/10.1016/j.colsurfa.2004.01.024.

Bledzki, A., Gassan, J., 1999. Composites reinforced with cellulose based fibres. Prog. Polym. Sci. 24, 221–274.

Bochek, A.M., 2003. Effect of hydrogen bonding on cellulose solubility in aqueous and nonaqueous solvents. Russ. J. Appl. Chem. 76, 1711–1719.

Bodig, J., Jayne, B.A., 1993. Mechanics of wood and wood composites. Krieger, Florida Bondada B, Keller M (2012) morphoanatomical symptomatology and osmotic behavior of grape berry shrivel. J. Am. Soc. Hort. Sci. 137, 20–30.

Bondada, B., Keller, M., 2012. Morphoanatomical symptomatology and osmotic behavior of grape berry shrivel. J. Am. Soc. Hortic. Sci. 137 (1), 20–30. https://doi.org/10.21273/JASHS.137.1.20.

Bujang, A.M.B., Nordin, N.I.A.B.A., 2020. Effect of steam treatment on the characteristics of oil palm empty fruit bunch and its biocomposite. Indones J. Chem. 20, 292–298. https://doi.org/10.22146/ijc.40906.

Cai, M., Takagi, H., Nakagaito, A.N., et al., 2015. Influence of alkali treatment on internal microstructure and tensile properties of abaca fibers. Ind. Crop Prod. 65, 27–35. https://doi.org/10.1016/j.indcrop.2014.11.048.

Cao, Y., Sakamoto, S., Goda, K., 2007. Effects of heat and alkali treatments on mechanical properties of kenaf fibers. In: Proceedings of 16th International Conference on Composite Materials (ICCM), pp. 1–4. Kyoto, Japan, July 2007.

Chandrasekar, M., Ishak, M.R., Sapuan, S.M., et al., 2017. A review on the characterisation of natural fibres and their composites after alkali treatment and water absorption. Plast., Rubber Compos. 46, 119–136. https://doi.org/10.1080/14658011.2017.1298550.

Chin, K.L., H'ng, P.S., Chai, E.W., et al., 2012. Fuel characteristics of solid biofuel derived from oil palm biomass and fast growing timber species in Malaysia. Bioenergy Res. 6, 75–82. https://doi.org/10.1007/s12155-012-9232-0.

Chin, K.L., Lee, C.L., H'ng, P.S., et al., 2020. Refining micropore capacity of activated carbon derived from coconut shell via deashing post-treatment. Bioresources 15, 7749–7769.

Chowdhury, M.N.K., Beg, H., Khan, M.R., Mina, F., 2013. Modification of oil palm empty fruit bunch fibers by nanoparticle impregnation and alkali treatment. Cellul. 20, 1477–1490. https://doi.org/10.1007/s10570-013-9921-7.

Cordeiro, N., Gouveia, C., John, M.J., 2011. Investigation of surface properties of physico-chemically modified natural fibres using inverse gas chromatography. Ind. Crop Prod. 33, 108–115. https://doi.org/10.1016/j.indcrop.2010.09.008.

Deepa, B., Abraham, E., Cherian, B.M., et al., 2011. Structure, morphology and thermal characteristics of banana nano fibers obtained by steam explosion. Bioresour. Technol. 102, 1988–1997. https://doi.org/10.1016/j.biortech.2010.09.030.

DeSimone, J.M., 2002. Practical approaches to green solvents. Science (80-) 297, 799–803. https://doi.org/10.1126/science.1069622.

Dutt, D., Upadhyaya, J.S., Malik, R.S., Tyagi, C.H., 2004. Studies on pulp and paper—making characteristics of some Indian non-woody fibrous raw materials: part 1. J. Sci. Ind. Res. (India) 63, 48–57.

Fidelis, M.E.A., Pereira, T.V.C., Gomes, O.D.F.M., et al., 2013. The effect of fiber morphology on the tensile strength of natural fibers. J. Mater. Res. Technol. 2, 149–157. https://doi.org/10.1016/j.jmrt.2013.02.003.

Fiore, V., Scalici, T., Di Bella, G., Valenza, A., 2015. A review on basalt fibre and its composites. Compos. Part B Eng. 74, 74–94. https://doi.org/10.1016/j.compositesb.2014.12.034.

Flitsch, A., Prasetyo, E.N., Sygmund, C., et al., 2013. Cellulose oxidation and bleaching processes based on recombinant Myriococcum thermophilum cellobiose dehydrogenase. Enzyme Microb. Technol. 52, 60–67. https://doi.org/10.1016/j.enzmictec.2012.10.007.

Fuqua, M.A., Huo, S., Ulven, C.A., 2012. Natural fiber reinforced composites. Polym. Rev. 52, 259–320. https://doi.org/10.1080/15583724.2012.705409.

Fuqua, M.A., Ulven, C.A., 2008. Characterization of polypropylene/corn fiber composites with maleic anhydride grafted polypropylene. J. Biobaased Mater. Bioenergy 2, 258–263. https://doi.org/10.1166/jbmb.2008.405.

George, M., Mussone, P.G., Bressler, D.C., 2014. Surface and thermal characterization of natural fibres treated with enzymes. Ind. Crop Prod. 53, 365–373. https://doi.org/10.1016/j.indcrop.2013.12.037.

Ghosh, S.K., Talukdar, M.K., Dey, P.K., 1995. Effect of number of passes on tensile and tear properties of nonwoven needle-punched fabrics. Indian J. Fibre Text Res. 20, 145–149.

Gunavan, F.E., Homma, H., Brodjonegoro, S.S., et al., 2009. Mechanical properties of oil palm empty fruit bunch fiber. J. Solid. Mech. Mater. Eng. 3, 943–951. https://doi.org/10.1299/jmmp.3.943.

Gupta, A., Gupta, R., 2019. Treatment and recycling of wastewater from pulp and paper mill. In: Singh, R.L., Singh, R.P. (Eds.), Advances in Biological Treatment of Industrial Waste Water and their Recycling for a Sustainable Future. Springer Nature, Singapore, pp. 13–49.

Gurunathan, T., Mohanty, S., Nayak, S.K., 2015. A review of the recent developments in biocomposites based on natural fibres and their application perspectives. Compos. Part A Appl. Sci. Manuf. 77, 1–25. https://doi.org/10.1016/j.compositesa.2015.06.007.

H'ng, P.S., Khor, B.N., Tadashi, N., et al., 2009. Anatomical structure and fibre morphology of new kenaf varieties.Pdf. Asian J. Sci. Res. 2, 161–166.

Han, Y.H., Han, S.O., Cho, D., Il, K.H., 2007. Kenaf/polypropylene biocomposites: effects of electron beam irradiation and alkali treatment on kenaf natural fibers. Compos. Interfaces 14, 559–578. https://doi.org/10.1163/156855407781291272.

Huber, T., Biedermann, U., Müssig, J., 2010. Enhancing the fibre matrix adhesion of natural fibre reinforced polypropylene by electron radiation analyzed with the single fibre fragmentation test. Compos. Interfaces 17, 371–381. https://doi.org/10.1163/092764410X495270.

Ibrahim, Z., Ahmad, M., Aziz, A.A., et al., 2019. Properties of chemically treated oil palm empty fruit bunch (EFB) fibres. J. Adv. Res. Fluid Mech. Therm. Sci. 57, 57–68.

Ibrahim, Z., Ahmad, M., Aziz, A.A., et al., 2016. Dimensional stability properties of medium density Fibreboard (MDF) from treated oil palm (Elaeis guineensis) empty fruit bunches (EFB) Fibres. Open J. Compos. Mater. 6, 91–99. https://doi.org/10.4236/ojcm.2016.64009.

Ibrahim, Z., Aziz, A.A., Ramli, R., et al., 2015. Effect of treatment on the oil content and surface morphology of oil palm (Elaeis guineensis) empty fruit bunches (EFB) fibres. Wood Res. 60, 157–166.

Imai, M., Ikari, K., Suzuki, I., 2004. High-performance hydrolysis of cellulose using mixed cellulase species and ultrasonication pretreatment. Biochem. Eng. J. 17, 79–83. https://doi.org/10.1016/S1369-703X(03)00141-4.

Islam, M.S., Pickering, K.L., Foreman, N.J., 2010. Influence of alkali treatment on the interfacial and physico-mechanical properties of industrial hemp fibre reinforced polylactic acid composites. Compos. Part A Appl. Sci. Manuf. 41, 596–603. https://doi.org/10.1016/j.compositesa.2010.01.006.

Istek, A., 2006. Effect of phanerochaete chrysosporium white rot fungus on the chemical composition of populus Tremula L. Cellul. Chem. Technol. 40 (6), 475–478.

Jiang, Y., Wu, Q., Wei, Z., et al., 2019. Papermaking potential of Pennisetum hybridum fiber after fertilizing treatment with municipal sewage sludge. J. Clean. Prod. 208, 889–896. https://doi.org/10.1016/j.jclepro.2018.10.148.

Jinn, C.M., San, H.P., Ling, C.K., et al., 2015. Agricultural biomass based potential materials - Google books. In: Hakeem, K., Jawaid, M., Alothman, O. (Eds.), Agricultural Biomass Based Potential Materials. Spinger, Cham, pp. 375–389.

Khalil, H.P.S.A., Bhat, I.U.H., Jawaid, M., et al., 2012. Bamboo fibre reinforced biocomposites: a review. Mater. Des. 42, 353–368. https://doi.org/10.1016/j.matdes.2012.06.015.

Lani, N.S., Ngadi, N., Johari, A., Jusoh, M., 2014. Isolation, characterization and application of nanocellulose from oil palm empty fruit bunch (EFB) fiber as nanocomposites. J. Nanomater. 2014, 1–9.

Latip, N.A., Sofian, A.H., Ali, M.F., et al., 2019. ScienceDirect structural and morphological studies on alkaline pre-treatment of oil palm empty fruit bunch (OPEFB) fiber for composite production. Mater Today Proc. 17, 1105–1111. https://doi.org/10.1016/j.matpr.2019.06.529.

Law, K.N., Daud, W.R.W., Ghazali, A., 2007. Morphological and chemical nature of fiber strands of oil palm empty-fruit-bunch (OPEFB). Bioresources 2, 351–362. https://doi.org/10.15376/biores.2.3.351-362.

Lee, C.L., H'ng, P.S., Chin, K.L., et al., 2019. Characterization of bioadsorbent produced using incorporated treatment of chemical and carbonization procedures. R. Soc. Open Sci. 6. https://doi.org/10.1098/rsos.190667, 190667.

Li, J., Yang, X., Xiu, H., et al., 2019. Structure and performance control of plant fiber based foam material by fibrillation via refining treatment. Ind. Crop Prod. 128, 186–193. https://doi.org/10.1016/j.indcrop.2018.10.085.

Li, Y., Pickering, K.L., 2008. Hemp fibre reinforced composites using chelator and enzyme treatments. Compos. Sci. Technol. 68, 3293–3298. https://doi.org/10.1016/j.compscitech.2008.08.022.

Li, Z., Smith, K.H., Stevens, G.W., 2016. The use of environmentally sustainable bioderived solvents in solvent extraction applications—a review. Chin. J. Chem. Eng. 24, 215–220. https://doi.org/10.1016/j.cjche.2015.07.021.

Mäder, E., Jacobasch, H.J., Grundke, K., Gietzelt, T., 1996. Influence of an optimized interphase on the properties of polypropylene/glass fibre composites. Compos. Part A Appl. Sci. Manuf. 27, 907–912. https://doi.org/10.1016/1359-835X(96)00044-9.

Mishra, S., Mohanty, A.K., Drzal, L.T., et al., 2004. A review on pineapple leaf fibers, sisal fibers and their biocomposites. Macromol. Mater. Eng. 289, 955–974. https://doi.org/10.1002/mame.200400132.

Mohr, B.J., Biernacki, J.J., Kurtis, K.E., 2007. Supplementary cementitious materials for mitigating degradation of Kraft pulp fiber-cement composites. Cem. Concr. Res. 37, 1531–1543. https://doi.org/10.1016/j.cemconres.2007.08.001.

Mukhopadhyay, S., Fangueiro, R., 2009. Physical modification of natural fibers and thermoplastic films for composites—a review. J. Thermoplast. Compos. Mater. 22, 135–162. https://doi.org/10.1177/0892705708091860.

Mwaikambo, L.Y., Ansell, M.P., 1999. The effect of chemical treatment on the properties of hemp, sisal, jute and kapok for composite reinforcement. Angew. Makromol. Chem. 272, 108–116. https://doi.org/10.1002/(sici)1522-9505(19991201)272:1<108::aid-apmc108>3.3.co;2-0.

Mwaikambo, L.Y., Ansell, M.P., 2002. Chemical modification of hemp, sisal, jute, and kapok fibers by alkalization. J. Appl. Polym. Sci. 84, 2222–2234. https://doi.org/10.1002/app.10460.

Nabinejad, O., Debnath, S., Taheri, M.M., 2017. Oil palm fiber vinylester composite: effect of bleaching treatment. Mater. Sci. Forum 882, 43–50. https://doi.org/10.4028/www.scientific.net/MSF.882.43.

Nafu, Y.R., Foba-Tendo, J., Njeugna, E., et al., 2015. Extraction and characterization of Fibres from the stalk and Spikelets of empty fruit bunch. J. Appl. Chem. 2015, 1–10. https://doi.org/10.1155/2015/750818.

Naveen, J., Jawaid, M., Amuthakkannan, P., Chandrasekar, M., 2018. Mechanical and physical properties of sisal and hybrid sisal fiber-reinforced polymer composites. In: Mohammad, J., Mohamed, T., Naheed, S. (Eds.), Mechanical and Physical Testing of Biocomposites, Fibre-Reinforced Composites and Hybrid Composites. Woodhead Publishing, Cambridge, pp. 427–440.

Nazir, M.S., Wahjoedi, B.A., Yussof, A.W., Abdullah, M.A., 2013. Eco-friendly extraction and characterization of cellulose from oil palm empty fruit bunches. Bioresources 8, 2161–2172. https://doi.org/10.15376/biores.8.2.2161-2172.

Ngo, W.L., Pang, M.M., Yong, L.C., Tshai, K.Y., 2014. Mechanical properties of natural fibre (Kenaf, oil palm empty fruit bunch) reinforced polymer composites. Adv. Environ. Biol. 8, 2742–2747.

Norul Izani, M.S., Paridah, M.T., Astimar, A.A., et al., 2012. Mechanical and dimensional stability properties of medium-density fibreboard produced from treated oil palm empty fruit bunch. J. Appl. Sci. 12, 561–567.

Pickering, K.L., Efendy Aruan, M.G., Le, T.M., 2016. A review of recent developments in natural fibre composites and their mechanical performance. Compos. Part A Appl. Sci. Manuf. 83, 98–112. https://doi.org/10.1016/j.compositesa.2015.08.038.

Płotka-Wasylka, J., Rutkowska, M., Owczarek, K., et al., 2017. Extraction with environmentally friendly solvents. Trends Anal. Chem. 91, 12–25. https://doi.org/10.1016/j.trac.2017.03.006.

Potluri, R., 2019. Natural Fiber-based hybrid bio-composites: Processing, characterization, and applications. In: Muthu, S.S. (Ed.), Green Composites, Textile Science Adn Clothing Technology. Springer Nature, Singapore, pp. 1–46.

Prabhakar, M.N., Shah, A.U.R., Song, J.-I., 2015. A review on the flammability and flame retardant properties of natural fibers and polymer matrix based composites. Compos. Res. 28, 29–39. https://doi.org/10.7234/composres.2015.28.2.029.

Ramamoorthy, S.K., Skrifvars, M., Persson, A., 2015. A review of natural fibers used in biocomposites: plant, animal and regenerated cellulose fibers. Polym. Rev. 55, 107–162. https://doi.org/10.1080/15583724.2014.971124.

Rana, S., Parveen, S., Pichandi, S., Fangueiro, R., 2018. Development and characterization of microcrystalline cellulose based novel multi-scale biocomposites. In: Fangueiro, R., Rana, S. (Eds.), Advances in Natural Fibre Composites. Spinger International Publishing, Cham, pp. 159–173.

Ratnasingam, J., Tek, T.C., Farrokhpayam, S.R., 2008. Tool wear characteristics of oil palm empty fruit bunch particleboard. J. Appl. Sci. 8, 1594–1596.

Rozali, S.N.M., Paterson, A.H.J., Hindmarsh, J.P., Huffman, L.M., 2019. Understanding the shear and extensional properties of pomace-fibre suspensions prior to the spray drying process. LWT 99, 138–147. https://doi.org/10.1016/j.lwt.2018.09.061.

Rozman, H.D., Tay, G.S., Kumar, R.N., et al., 2001. The effect of oil extraction of the oil palm empty fruit bunch on the mechanical properties of polypropylene-oil palm empty fruit bunch-glass fibre hybrid composites. Polym.-Plast. Technol. Eng. 40, 103–115. https://doi.org/10.1081/PPT-100000058.

Saha, P., Chowdhury, S., Roy, D., et al., 2016. A brief review on the chemical modifications of lignocellulosic fibers for durable engineering composites. Polym. Bull. 73, 587–620. https://doi.org/10.1007/s00289-015-1489-y.

Sarjadi, M.S., Aziz, S.A., Rahman, L., 2018. Effect of seawater treatment on the mechanical properties of oil palm empty fruit bunch fibre/poly(vinyl alcohol) composites. ASM Sci. J. 11, 87–94.

Satyanarayana, K.G., Arizaga, G.G.C., Wypych, F., 2009. Biodegradable composites based on lignocellulosic fibers-an overview. Prog. Polym. Sci. 34, 982–1021. https://doi.org/10.1016/j.progpolymsci.2008.12.002.

Sawawi, M., Mohammad, N.H., Sahari, S.K., et al., 2020. Effects of chemical treatment on mechanical properties of oil palm empty fruit bunch (EFB) with urea formaldehyde (UF) resin Particleboardtype. Int. J. Recent Technol. Eng. 8, 1330–1334. https://doi.org/10.35940/ijrte.e6094.018520.

Song, H., Ankerfors, M., Hoc, M., Lindström, T., 2010. Reduction of the linting and dusting propensity of newspaper using starch and microfibrillated cellulose. Nord. Pulp Pap. Res. J. 25, 495–504. https://doi.org/10.3183/npprj-2010-25-04-p519-528.

Suradi, S.S., Yunus, R.M., BMD, H., et al., 2010. Oil palm bio-fiber reinforced thermoplastic composites-effects of matrix modification on mechanical and thermal properties. J. Appl. Sci. 10, 3271–3276.

Takacs, E., Wojnarovits, L., Foldvary, C., et al., 2000. Effect of combined gamma-irradiation and alkali treatment on cotton-cellulose. Radiat. Phys. Chem. 57, 399–403.

Tanpichai, S., Witayakran, S., Boonmahitthisud, A., 2018. Study on structural and thermal properties of cellulose microfibers isolated from pineapple leaves using steam explosion. J. Environ. Chem. Eng. 7. https://doi.org/10.1016/j.jece.2018.102836, 102836.

Tian, H., Zhang, Y.X., Ding, Y., 2016. Recent advances in experimental study on mechanical behaviour of natural fibre reinforced cementitious composites. Struct. Concr. 17, 564–575. https://doi.org/10.1002/suco.201500177.Submitted.

Vaisanen, T., Das, O., Tomppo, L., 2017. A review on new bio-based constituents for natural fiber-polymer composites. J. Clean. Prod. 149, 582–596. https://doi.org/10.1016/j.jclepro.2017.02.132.

Valadez-Gonzalez, A., Cervantes-Uc, J.M., Olaya, R., Herrera-Franco, P.J., 1999. Chemical modification of henequen fibers with an organosilane coupling agent. Compos. Part B Eng. 30, 321–331.

Varshney, V.K., Naithani, S., 2011. Chemical functionalization of cellulose derived from nonconventional sources. In: Kalia, S., Kaith, B.S., Kaur, I. (Eds.), Cellulose Fibers: Bio- and Nano-Polymer Composites. Springer-Verlag, Berlin Heidelberg, Berlin, pp. 43–60.

Voelker, S.L., Lachenbruch, B., Meinzer, F.C., Strauss, S.H., 2011. Reduced wood stiffness and strength, and altered stem form, in young antisense 4CL transgenic poplars with reduced lignin contents. New Phytol. 189, 1096–1109.

Wei, J., Meyer, C., 2014. Improving degradation resistance of sisal fiber in concrete through fiber surface treatment. Appl. Surf. Sci. 289, 511–523. https://doi.org/10.1016/j.apsusc.2013.11.024.

Wu, G., 2019. Radiation sources and radiation processing. In: Wu, G., Wang, M., Zhai, M. (Eds.), Radiation Technology for Advanced Materials: From Basic to Modern Applications. Academic Press, Cambridge, pp. 1–11.

Xu, F., Zhong, X.C., Sun, R.C., Lu, Q., 2006. Anatomy, ultrastructure and lignin distribution in cell wall of Caragana Korshinskii. Ind. Crop Prod. 24, 186–193. https://doi.org/10.1016/j.indcrop.2006.04.002.

10

Microstructure, physical, and strength properties of compressed oil palm frond composite boards from *Elaeis guineensis*

Razak Wahab[a,b], Mohd Sukhairi Mat Rasat[c], Rashidah Kamarulzaman[a], Mohamad Saiful Sulaiman[a,b], Sofiyah Mohd Razali[b], Taharah Edin[a], and Nasihah Mokhtar[a,b]

[a]University of Technology Sarawak, Sibu, Sarawak, Malaysia, [b]Centre of Excellence in Wood Engineered Products, UTS, Sibu, Sarawak, Malaysia, [c]University Malaysia Kelantan (UMK), Jeli, Kelantan, Malaysia

10.1   Introduction

The supply chain of bioresources is causing problems for the wood-based sector in Malaysia and around the world. The maturation of timber species might take anywhere from 40 to 60 years depending on the species. Solid wood products are becoming increasingly difficult to come by. The industry needs easy-to-get alternative materials in large quantities.

Oil palm fronds are abundant in Malaysia, which has 6 million hectares of land cultivated with oil palm trees. Every year, these oil palm plantations generate a large number of oil palm fronds. Every month, a large number of fronds are dropped from trees, and they are regarded as one of Malaysia's most plentiful agricultural by-products. However, this biomaterial is often left to rot in the field and, after some time due to the biodegradation process, serves for soil conservation, erosion control, and the long-term benefit of nutrient recycling (Hassan et al., 1995; Pamin and Guritno, 1997).

Preliminary investigations suggest that the fronds might be used as an alternative raw material for the wood-based sector if collected and

Oil Palm Biomass for Composite Panels. https://doi.org/10.1016/B978-0-12-823852-3.00014-3
Copyright © 2022 Elsevier Inc. All rights reserved.

processed correctly. Composite boards and scrimber products are examples of high-value-added goods that can be created (Sreekala et al., 2002; Abdul Khalil et al., 2001). Several extensive research projects are presently underway to develop appropriate technology for turning oil palm fronds into commercially viable composite board products (Sulaiman et al., 2008; Laemsak and Okuma, 2000; Chew, 1987; Ho et al., 1987).

Bioresidues from oil palm fronds are used to maximize the usage of biomass agricultural by-products. Furthermore, it contributes to a greater understanding and knowledge of non-wood materials for future sustainability development (Bledzki and Gassan, 1999). However, non-wood bioresources, particularly oil palm fronds, have yet to contribute to commercialization (Gurmit, 1999). To utilize a considerable volume of agrowaste, novel technologies with higher efficiency and lower environmental implications are urgently needed (Yang et al., 2006).

10.2   Processing of oil palm composite boards

The oil palm fronds for the studies were provided by a private oil palm farm owner. They looked for and chose decay-free and defect-free items. According to their maturity, the oil palm fronds were divided into three sample groups. The first group of oil palm fronds (matured fronds) was removed from below the frond's crown. Next, the middle of the frond's crown was used to obtain the second (intermediate fronds) maturity fronds. Finally, the juvenile maturity fronds plucked from above the frond's crown formed the third group (young fronds). The difference between maturity groups is depicted in Fig. 10.1A.

Each maturity group was further segregated into the bottom, middle, and top (see Fig. 10.1B). The specimen was peeled off the epidermis layer manually by a scraper and then sliced in a longitudinal direction (Fig. 10.1C). A roller compress machine was used to loosen up the bonding between the cells in the sample fronds, particularly the fibers in the vascular bundles (Fig. 10.1D). The material then was air-dried under the shed for 12h before mixing with resin phenol and urea-formaldehyde at 12%–15% and hot-pressed into composite boards. A 1% hardener of ammonium chloride (NH_4Cl) was then added and compressed into a mold of size 350mm (length) × 350mm (width) × 20mm (thickness) boards. This was undertaken by transferring the material to a hydraulic hot-press machine with a platen temperature set at $125 \pm 5°C$ for resin phenol-formaldehyde or set at $100 \pm 5°C$ for resin urea-formaldehyde and pressed into a subsequent shape for testing. The preparation of the oil palm composite boards followed International Organization for Standardization (ISO) standards.

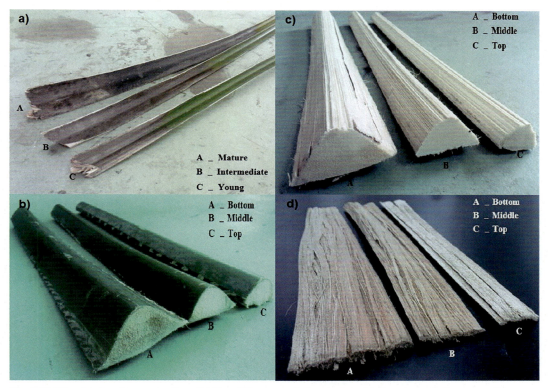

Fig. 10.1 Oil palm fronds preparation. (A) The difference of oil palm frond between maturity group. (B) The segregation of oil palm frond into bottom, middle and top part. (C) Oil palm frond after the removal of epidermis layer. (D) Oil palm frond after compression by roller compress machine. Source: Wahab, R., Rasat, M.S.M., Fauzi, N.M., Sulaiman, M.S., Samsi, H.W., Mokhtar, N., Ghani, R.S.M., Razak, M.H., 2021. Processing and properties of oil palm fronds composite boards from Elaeis guineensis. IntechOpen, https://doi.org/10.5772/intechopen.98222. [online first], Available from: https://www.intechopen.com/online-first/77099.

10.3 Physical properties of compressed oil palm fronds (COPF) composite

The fundamental physical qualities of composite boards are critical in determining their behavior and performance. Furthermore, it is thought that this has an effect on the specimen's strength qualities. As a result, the fundamental density and COPF composite board were investigated such as the parameters used in this study, which included parts of fronds, a maturity group, and the types of resin used.

10.3.1 Measurement of COPF composite density

As a portion of wood characteristics, density is thought to be a good predictor of substance (Abdullah, 2010). It was found that the parameter has a direct effect on the COPF composite board's results. Table 10.1 shows the mean value results for the density depending on the parameter employed in this investigation. The most significant density value is shown at the bottom, followed by each maturity group specimen's middle and top places. Also worth noting is that the matured age had the highest density value, followed by the intermediate and young ages.

Table 10.1 highlights that the internal structure influences the specimens' outcomes through the abundance of parenchymatous tissues and vascular bundles. Nonetheless, statistical analysis of an ANOVA (see Table 10.6) shows a significant difference between the portion and maturity groups against the density. Notwithstanding, the table also highlights no significant difference in the use of resin. Consequently, the resin types did not affect the density value of the COPF composite. Although the resin penetration in producing a composite directly affects the composite manufactured densities compared to the oil palm frond raw material, the phenomenon causes an increase in a material substance per unit volume in COPF composite.

Table 10.1 A summary of density value for composite board manufacturing.

Maturity groups	The different portion on density (g/cm^3)		
	Bottom	Middle	Top
Matured			
COPF with PF[a]	0.45	0.44	0.42
COPF with UF[b]	0.46	0.43	0.42
Intermediate			
COPF with PF	0.43	0.42	0.40
COPF with UF	0.44	0.42	0.41
Young			
COPF with PF	0.42	0.41	0.40
COPF with UF	0.42	0.41	0.40

[a] Phenol-formaldehyde (PF).
[b] Urea formaldehyde (UF).
Source: Rasat, M.S.M., Wahab, R., Kari, Z.A., Yunus, A.A.M., Moktar, J., Ramle, S.F.M., 2013. Strength properties of bio-composite lumber from lignocelluloses of oil palm fronds agricultural residues. Int. J. Adv. Sci., Eng. Inf. Technol. 3, 9–19.

10.3.2 Investigation of compressed oil palm frond's (COPF) composite basic density

The mean values of basic density for a COPF composite based on different groups' maturity, the position of portion, and types of resin are represented in Table 10.2. The decreasing basic density value is influenced by the higher concentration of fibrous vascular bundle in the COPF composite specimen from mature to young and bottom to the top portion in maturity groups (Mohamad et al., 1985; Rowell et al., 1994). The size, thickness, and relative amount of solid cell wall materials are considered when determining the factor that modifies the basic density value. The mature and thick cells toward the bottom of the tree, on the other hand, have higher basic density values than the top and center. The fundamental density correspondingly reduced the variations in anatomical cell maturity development from the bottom to the top locations (Haygreen and Bowyer, 1930). The excellent strength properties of wood come from a greater basic density.

The density had a significant difference between portions and groups of maturity (see Table 10.6), but there were no significant

Table 10.2 A summary of basic density value for composite board manufacturing.

Maturity groups	The different portion on basic density (g/cm^3)		
	Bottom	Middle	Top
Matured			
COPF with PF[a]	0.38	0.36	0.33
COPF with UF[b]	0.39	0.35	0.32
Intermediate			
COPF with PF	0.36	0.35	0.32
COPF with UF	0.37	0.34	0.31
Young			
COPF with PF	0.34	0.33	0.30
COPF with UF	0.34	0.32	0.30

[a] Phenol-formaldehyde (PF).
[b] Urea formaldehyde (UF).
Source: Rasat, M.S.M., Wahab, R., Kari, Z.A., Yunus, A.A.M., Moktar, J., Ramle, S.F.M., 2013. Strength properties of bio-composite lumber from lignocelluloses of oil palm fronds agricultural residues. Int. J. Adv. Sci., Eng. Inf. Technol. 3, 9–19.

differences between the types of resin used. The increasing value of the basic density in composite manufacturing was influenced by the effortless absorption of phenol and urea-formaldehyde resin. The results were supported by Paridah and Anis (2008) who asserted that the parenchyma acted like a sponge and quickly absorbed the moisture.

10.4 COPF composite board strength properties

Wood's strength qualities are defense mechanisms against external pressures that directly deform its bulk (Erwinsyah, 2008). The forces are usually determined by their computation and loading method, such as compressive strength, tension, flexural strength, etc. According to a prior work by Tsoumis (1991), wood exhibits variable strength properties in distinct growth routes, making it highly anisotropic. However, Bowyer et al. (2004) believe that the strength characteristics are crucial perspectives to describe the outcome utilized in structural purposes. Hence they are designated as fundamental criteria for choosing the substance. The strength element functions for many applications, either interior or exterior purposes (Erwinsyah, 2008).

The study determined the compressive strength, MOR, and MOE on flexural strength from the COPF composite. The specimen was tested by ISO standard for strength performance purposes. The prepared COPF composite board followed a few parameters such as different sample portions, groups maturity, and resin types.

10.4.1 Static bending strength of COPF composite board

Bending stress test was conducted to measure the MOE, and the MOR expresses the force needed until specimen failure. The bending strength is commonly expressed in MOR and is the most vital parameter infrequently used for engineered products (Walker et al., 1993). The static bending strength relates to the tests executed (Erwinsyah, 2008). Tables 10.3 and 10.4 highlight a summary of MOE and MOR results in static bending strength. It is indicated that the COPF composite had the highest MOE and MOR from the bottom portion. The results have been followed by intermediate and young age groups of maturity. The resin type's parameter is represented by a decreasing order of strength value from bottom to the top and young to mature, respectively. MOE value from the COPF composite bonded with phenol-formaldehyde highlighted around 999.61, 952.29, and 844.18 N/mm^2, while the urea-formaldehyde resin indicated the MOE value at 980.31, 949.40, and 840.40 N/mm^2 for the bottom, middle, and top parts of materials, respectively. Hence, the strength on MOE value was

Table 10.3 The mean of MOE value for COPF composite static bending.

Maturity groups	MOE value at different portion (N/mm^2)		
	Bottom	Middle	Top
Matured			
COPF with PF	999.61	952.29	844.18
COPF with UF	980.31	949.40	840.40
Intermediate			
COPF with PF	979.15	942.44	817.29
COPF with UF	953.93	928.34	776.04
Young			
COPF with PF	935.36	837.24	761.14
COPF with UF	936.24	836.67	666.30

Source: Rasat, M.S.M., Wahab, R., Kari, Z.A., Yunus, A.A.M., Moktar, J., Ramle, S.F.M., 2013. Strength properties of bio-composite lumber from lignocelluloses of oil palm fronds agricultural residues. Int. J. Adv. Sci., Eng. Inf. Technol. 3, 9–19.

Table 10.4 The mean of MOR value for COPF composite static bending.

Maturity groups	MOR value at different portion (N/mm^2)		
	Bottom	Middle	Top
Matured			
COPF with PF	16.66	12.55	11.72
COPF with UF	15.40	12.38	11.63
Intermediate			
COPF with PF	14.38	12.37	10.87
COPF with UF	12.62	12.07	10.51
Young			
COPF with PF	12.61	11.62	10.27
COPF with UF	12.25	11.19	9.10

Source: Rasat, M.S.M., Wahab, R., Kari, Z.A., Yunus, A.A.M., Moktar, J., Ramle, S.F.M., 2013. Strength properties of bio-composite lumber from lignocelluloses of oil palm fronds agricultural residues. Int. J. Adv. Sci., Eng. Inf. Technol. 3, 9–19.

designated in decreasing order from bottom to the top on the maturity groups for both types of resin used. The situation is repeated at an intermediate and young age, which concluded that the COPF bonded with PF had the highest value of MOE and MOR compared to manufacturing using a UF resin. COPF composite from the bottom portion and bonded with PF represented the MOE value at 999.61, 979.15, and 935.36 N/mm^2. Then, the COPF bonded with UF indicated the value around 980.31, 953.93, and 936.24 N/mm^2 for matured, intermediate, and young, respectively. On COPF composite manufacturing, the MOE value proportionally decreases from the matured to the young for the bottom portion on both resin types. Also, the MOR is defined when the specimen reaches the fracture point and cannot recover its original position. Table 10.4 summarizes the mean value for MOR on COPF composite at different resin types, portions, and maturity groups.

For the age group, the MOR of COPF composite boards increased strength from top to bottom, and for each component, from young to mature. On PF resins, the MOR strengths for the maturity groups were 11.72, 12.55, and 16.66 N/mm^2 for the top, middle, and bottom, respectively. Nonetheless, the portion side highlighted MOR for UF was 11.63, 12.38, and 15.40 N/mm^2. The bottom of each maturity group of COPF bonded with PF have highlighted results at 12.16, 14.38, and 16.66 N/mm^2 for the young, intermediate, and mature, respectively.

The bottom for each maturity group (mature, intermediate, and young) and the grouped section of the PF composite board were 16.66, 14.38, and 12.16 N/mm^2, respectively, whereas the MOR for the UF composite board was 15.40, 12.62, and 12.25 N/mm^2. For both types of resins employed in the maturity group, strength declined from the mature to the bottom. MOE values showed a similar pattern. In the maturity group, MOR ascends from top to bottom, and each part ascends from the mature to the young group.

From top to bottom, the MOE and MOR values for the COPF composite board increased. In the frond maturity group, the young, intermediate, and mature groups all made similar observations, which applied for both PF and UF resin composite boards. The decline could be explained by the variation trend in MOE and MOR, the height of trees in the maturity of wood, and the fiber length from the top to the bottom of the tree (Rulliarty and America, 1995). Wood materials with higher value density and basic density will have higher strength (Haygreen and Bowyer, 1930). The author also reported that the bottom of each division has a higher value for both MOE and MOR strength than the middle and upper parts of the mature group (Haygreen and Bowyer, 1930). The statement is supported by the descending vascular bonds from the bottom to the top, palm fronds, and the mature to young maturity group. A more significant number of vascular bundles in oil palm fronds with more fibrous cells resulted in higher density values and basic density in both composites. According to the study's findings on the effect of resin type in static bending, composite panels made of PF resin have higher MOE and MOR values than composite panels made of UF resin. In comparison to PF resins, the latter has a higher solid content (Desch, 1968).

The value of basic density and density highlighted the significant difference in static bending (Desch, 1968). In the manufacturing of COPF composite, results showed that the MOR and MOE strength value is affected from top to bottom. This outcome is indicated in Table 10.6. Also, the COPF composite bonded with PF shows higher MOR and MOE values than the one bonded with UF resin. It is because UF has a higher solid content than PF resin.

Furthermore, the PF resin dispersion is distributed irregularly within the composite board structure (Abdullah, 2010). Therefore, when stress is applied, it is not consistently transferred between the fiber and the matrix. Furthermore, high-viscosity penetration of UF resins may break the cell walls of COPF composite boards (Abdullah, 2010).

10.4.2 The COPF composite's compressive strength

Compressive strength can be defined as the maximum stress borne by compression of a specimen with the smallest length and dimension ratio (Thanate et al., 2006). Ronald and Gjinoli (1997) reported that the static bending strengths are similar to the characteristics of the compressive deformation load curve. The composite's compressive strength is highly dependent on the matrix's ability to support the fibers against buckling (Ronald and Gjinoli, 1997).

A statistical analysis was applied to investigate the effectiveness of selected parameters on compressive strength and static bending, which were specimen portion, group of maturity, and types of resin used. Table 10.5 highlights the compressive strength of the mean value and is based on the test results. According to the results in Table 10.5, the compressive strength values of the matured position for the PF composite board are 473.17, 395.93, and 260.22 N/mm^2, respectively, while for the UF composite board they are 459.52, 344.60, and 260.00 N/mm^2, respectively. For the old maturity group, the compressive strength decreases from the bottom to the middle and top. The intermediate and young maturity groups had similar decline distribution data toward the bottom, middle, and top.

Table 10.5 depicts each COPF composite board trends from the mature, intermediate, and young in the maturity group. For PF composite boards, the mature, intermediate, and young results were 473.17, 453.67, and 301.46 N/mm^2, respectively. The UF composite yielded 459.52, 431.88, and 312.94 N/mm^2 values, respectively. Thus,

Table 10.5 The mean of compressive strength value for COPF composite.

Maturity groups	Compressive value at different portion (N/mm^2)		
	Bottom	Middle	Top
Matured			
COPF with PF	473.17	395.93	260.22
COPF with UF	459.52	344.60	260.00
Intermediate			
COPF with PF	453.67	318.88	196.71
COPF with UF	431.88	274.90	190.70
Young			
COPF with PF	301.46	235.60	183.48
COPF with UF	312.94	198.79	181.06

Source: Rasat, M.S.M., Wahab, R., Kari, Z.A., Yunus, A.A.M., Moktar, J., Ramle, S.F.M., 2013. Strength properties of bio-composite lumber from lignocelluloses of oil palm fronds agricultural residues. Int. J. Adv. Sci., Eng. Inf. Technol. 3, 9–19.

it can be concluded that the data demonstrated in decreasing order from mature, intermediate, and through to the young age. Moreover, the equivalent pattern was found in the middle and top portions.

Variations in the number of vascular bundles and oil palm fronds, which influence density and basic density values, are thought to be the cause. The results of the compressive strength distribution for the sections and maturity groups are determined by the difference between the two. The bottom half of the structure has a higher compressive strength than the top part. The ANOVA findings are shown in Table 10.6, which reveals that there is a significant difference in compressive strength between the section and maturity groups. According to Nordahlia (2008), several wood properties, including compressive strength failure, are expected at low wood densities. The obtained results show that the PF composite board has a higher value than the UF composite. This phenomenon is because adequately cured PF composite resins are typically more complicated than bonded (Baldwin, 1995). However, as shown in Table 10.6, the difference in compressive strength is not as significant. As a result, we can conclude that resin is unimportant as long as it can be economically implemented to produce large quantities of COPF composite board.

Table 10.6 ANOVA analysis on strength and physical properties of COPF composite.

Source of variance	Dependent variable	Sum of square	df	Mean square	F-ratio
Maturity	Density	0.01	2	0.01	7.94**
	Basic Density	0.01	2	0.02	28.75**
	MOE Bending	155,675.00	2	77,837.50	57.05**
	MOR Bending	79.02	2	39.51	40.39**
	Compression	255,794.00	2	127,897.00	63.81**
Portion	Density	0.01	2	0.01	8.26**
	Basic Density	0.04	2	0.01	28.75**
	MOE Bending	507,856.00	2	253,928.00	186.12**
	MOR Bending	157.72	2	78.86	80.62**
	Compression	565,023.00	2	282,512.00	140.95**
Resin type	Density	0.00	1	0.00	0.20^{ns}
	Basic Density	0.00	1	0.00	1.28^{ns}
	MOE Bending	11,232.80	1	11,232.80	8.23^{ns}
	MOR Bending	8.23	1	8.23	8.41^{ns}
	Compression	7538.01	1	7538.01	3.76^{ns}

** $P \leq 0.01$.

Source: Rasat, M.S.M., Wahab, R., Kari, Z.A., Yunus, A.A.M., Moktar, J., Ramle, S.F.M., 2013. Strength properties of bio-composite lumber from lignocelluloses of oil palm fronds agricultural residues. Int. J. Adv. Sci., Eng. Inf. Technol. 3, 9–19.

10.4.3 ANOVA analysis on strength and physical properties of COPF composite

The ANOVA analysis for COPF composite board due to the strength and physical properties are represented in Table 10.6. The study focused on the significance level between strength and physical properties and also to determine the dependent variables on COPF composite, which are a portion of the specimen, group of maturity, and types of resin used. Results show there is a significant difference between strength and physical properties (Table 10.6). Nonetheless, there is a substantial difference between the group of maturity and the selected portion.

In conclusion, the selected parameter demonstrated that it affected and influenced the strength and physical properties of COPF composite, representing the p-value was significantly different at 0.01. However, there is no discernible difference in strength and physical properties between the types of resin used. Nevertheless, the test for the PF and UF resin to make a COPF board are very similar.

10.5 Microstructure of COPF composite board

High-performance microscopy was used to analyze the microstructure of the composite board made from COPFs, and scanning electron microscopy (SEM) demonstrated a more detailed structure. As a result, the microscopic resin penetration of the composite board from COPFs received more attention. The macroscopic structural characteristics of the composite board are the features that use high-performance microscopy to magnify from 0.75 to 8.0 times. Fig. 10.2A and B showed a roughly structural composite board from COPFs at the longitudinal sectional view for phenol and urea-formaldehyde composite board that had been observed under high-performance microscopy.

The reddish color shown in Fig. 10.2A in the composite board was affected by the presence of the resin phenol-formaldehyde. The resin urea-formaldehyde, on the other hand, shows whitish in

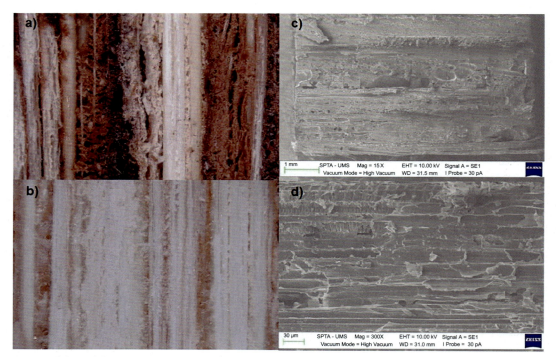

Fig. 10.2 (A) and (B) represent high-performance micrograph of the phenol-formaldehyde and urea-formaldehyde composite board at the longitudinal sectional view at 2 × magnification, respectively; (C) and (D) indicate the SEM of the composite board at the longitudinal sectional view at 15 × magnification and 300 × magnification, respectively. Source: Wahab, R., Rasat, M.S.M., Fauzi, N.M., Sulaiman, M.S., Samsi, H.W., Mokhtar, N., Ghani, R.S.M., Razak, M.H., 2021. Processing and properties of oil palm fronds composite boards from Elaeis guineensis. IntechOpen, https://doi.org/10.5772/intechopen.98222. [online first], Available from: https://www.intechopen.com/online-first/77099.

color (see Fig. 10.2B). The color between the two boards gave urea-formaldehyde resin a better appearance than the composite board from phenol-formaldehyde resin. This is why urea-formaldehyde resin is much preferred in mostly wood-based composites for internal applications. Fig. 10.2C and D show a more complex arrangement of the boards observed under SEM observation at different magnification.

10.5.1 Resin penetration on the COPF composite board from compressed oil palm fronds

The resins that bonded various cells and tissues improved the COPF composite boards from the oil palm fronds. The resins fill all void spaces in the boards and thus improve the physical and strength properties. The density and the basic density are improved by increasing the penetration of the resin in the board. This can be seen in Fig. 10.3A in which the resin penetrated through the intercellular cavities of the composite board from COPFs. The microscopic image of resin penetration on the composite boards from COPFs is presented in detail in Fig. 10.3B and C. The strength of the composite board increases as the density and basic density increase. The specimens tested show an increase in strength properties. This includes the MOE and MOR strengths in static bending strength and the compression tested in this study.

The resin applied to the compressed fronds to produce the COPF composite boards significantly improves their strength qualities. However, the use of the type of resin does not show significant differences in the properties or characteristics of the boards.

10.6 Conclusion

The fronds of the oil palm can be used as a biomaterial in the composites sector. The COPF composite boards manufactured from oil palm fronds had physical and strength attributes comparable to those of composites prepared from other tropical wood species. Furthermore, the abundant and year-round availability of oil palm fronds from oil palm trees could alleviate the burden of using wood from natural forests.

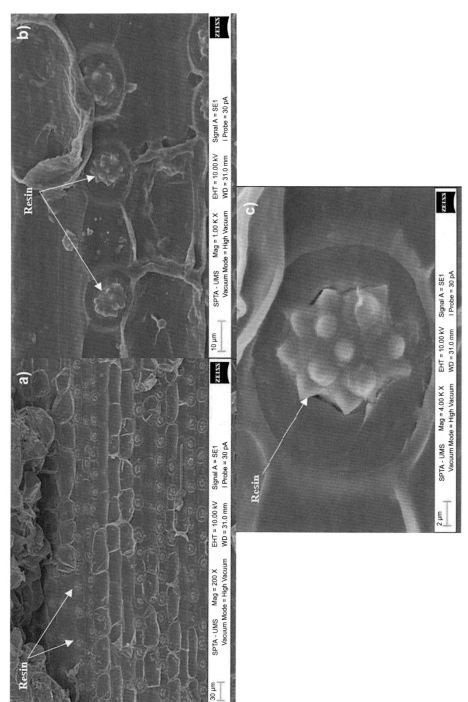

Fig. 10.3 SEM of resin penetration on composite board from compressed oil palm fronds at different magnifications. (A) SEM of resin penetration of COPF by 200x magnification. (B) SEM of resin penetration of COPF by 1000x magnification. (C) SEM of resin penetration of COPF by 4000x magnification. Source: Wahab, R., Rasat, M.S.M., Fauzi, N.M., Sulaiman, M.S., Samsi, H.W., Mokhtar, N, Ghani, R.S.M., Razak, M.H., 2021. Processing and properties of oil palm fronds composite boards from Elaeis guineensis. IntechOpen, https://doi.org/10.5772/intechopen.98222. [online first], Available from: https://www.intechopen.com/online-first/77099.

References

Abdul Khalil, H.P.S., Ismail, H., Ahmad, M.N., Ariffin, A., Hassan, K., 2001. The effect of various anhydride modifications on strength properties and water absorption of oil palm empty fruit bunches reinforced polyester composites. Polym. Int. 50 (4), 395–402.

Abdullah, C.K., 2010. Impregnation of Oil Palm Trunk Lumber (OPTL) Using Thermoset Resins for Structural Applications. Master Thesis, Universiti Sains Malaysia, Gelugor, Penang.

Baldwin, R.F., 1995. Adhesives and bonding techniques. In: Plywood and Veneer-Based Products Manufacturing Practices. Miller Freeman Inc., California.

Bledzki, A.K., Gassan, J., 1999. Composites reinforced with cellulose-based fibres. Prog. Polym. Sci. 24, 221–274.

Bowyer, J.L., Shmulsky, R., Haygreen, J.G., 2004. Forest Product and Wood Sciences—An Introduction, fourth ed. Blackwell Publishing Company, London.

Chew, L.T., 1987. Particleboard Manufacture from Oil Palm Stems a Pilot-Scale Study. FRIM Occasional Paper (Malaysia). Available from https://agris.fao.org/agris-search/search.do?recordID=MY19880080143.

Desch, H.E., 1968. Timber, its Structure, Properties, and Utilisation. London and Basingstoke Associated Companies, New York.

Erwinsyah, 2008. Improvement of Oil Palm Wood Properties Using Bio-Resin. Doctoral dissertation, Dresden University of Technology, Dresden, Germany.

Gurmit, S., 1999. The Malaysian oil palm industry: progress towards environmentally sound and sustainable crop production. Ind. Environ. 22, 45–48.

Hassan, O.A., Ishida, M., Mohd Sukri, I., 1995. Oil palm fronds (OPF) technology transfer and acceptance, a sustainable in-situ utilization for animal feeding. In: Proceedings of the 17th Malaysian Society of Animal Production (MSAP) Annual Conference, 28–30 May 1995, Penang, pp. 134–135.

Haygreen, J.G., Bowyer, J.L., 1930. Introduction to Forest Product and Wood Science. Subtitled by Suhaimi Muhammed and Sheikh Abdul Karim Yamani Zakaria. Ampang Press Sdn. Bhd, Kuala Lumpur.

Ho, K.S., Choo, K.T., Hong, L.T., 1987. Processing, seasoning, and protection of oil palm lumber. In: National Symposium on Oil Palm By-products for Agro-based Industries, Kuala Lumpur, 5–6 Nov 1985.

Laemsak, N., Okuma, M., 2000. Development of boards made from oil palm frond II: properties of binderless boards from steam-exploded fibers of oil palm frond. J. Wood Sci. 46, 322–326.

Mohamad, H., Zin, Z.Z., Abdul Halim, H., 1985. Potentials of oil palm by-products as raw materials for agro-based industries. In: Proceedings of National Symposium on Oil Palm by-Product for Agro-Based Industries, 5–6 November, Kuala Lumpur, Malaysia, pp. 34–42.

Nordahlia, A.S., 2008. Wood Quality of 10-Year-Old Sentang (Azadirachta Excelsa) Grown from Seedlings and Rooted Cuttings. Master Thesis, University Putra Malaysia, Serdang, Selangor.

Pamin, K., Guritno, P., 1997. Utilisation of oil palm empty fruit bunch fiber for oil palm seedling pot in pre-nursery. Jurnal Penelitian Kelapa Sawit 5, 179–190.

Paridah, M.T., Anis, M., 2008. Process optimisation in the manufacturing of plywood from oil palm trunk. In: Proceedings of 7th National Seminar on the Utilisation of Oil Palm Tree, Oil Palm Tree Utilization Committee, Kuala Lumpur, Malaysia, November, pp. 12–24.

Ronald, W.W., Gjinoli, A., 1997. The use of recycled wood and paper in building applications. In: Proceedings of Forest product society no. 7286, LaGrange, GA, United States.

Rowell, R.M., O'Dell, J.L., Rials, T.G., 1994. Chemical modification of agro-fiber for thermo-plasticization. In: Second Pacific Rim Bio-Based Composites Symposium, Vancouver, Canada, November 6-9, pp. 144–152.

Rulliarty, S., America, W.A., 1995. Natural variation in wood quality indicators of Indonesian big leaf mahogany (*Swietenia macrophylla*. King). In: Proceedings of XX IUFRO World Congress, Tampere, Finland, 6–12 August, pp. 76–83.

Sreekala, M.S., George, J., Kumaran, M.G., Thomas, S., 2002. The strength performance of hybrid phenol-formaldehyde-based composites reinforced with glass and oil palm fibres. Compos. Sci. Technol. 62 (3), 339–353.

Sulaiman, O., Hashim, R., Wahab, R., Samsi, H.W., Mohamed, A.H., 2008. Evaluation on some finishing properties of oil palm plywood. Holz Roh Werkst. 66, 5–10.

Thanate, R., Tanong, C., Sittipon, K., 2006. An investigation on the strength properties of trunks of palm oil trees for the furniture industry. J. Oil Palm Res., 114–121. Special issue:.

Tsoumis, G., 1991. Science and Technology of Wood—Structure, Properties, and Utilisation. Van Nostrand Reinhold, New York.

Walker, J.C.F., Butterfield, B.G., Langrish, T.A.G., Harris, J.M., Uprichard, J.M., 1993. Primary Wood Processing, first ed. Chapman and Hall, London.

Yang, H., Yan, R., Chen, H., Lee, D.H., Liang, D.T., Zheng, C., 2006. Pyrolysis of palm oil wastes for enhanced production of hydrogen-rich gases. Fuel Process. Technol. 87 (10), 935–942.

11

The processing and treatment of other types of oil palm biomass

Norul Hisham Hamid[a,b], Mohd Supian Abu Bakar[c], Norasikin Ahmad Ludin[d], Ummi Hani Abdullah[b], and Asmaa Soheil Najm[e]

[a]*Institute Of Tropical Forestry and Forest Products (INTROP), Universiti Putra Malaysia, Serdang, Selangor, Malaysia,* [b]*Faculty of Forestry & Environment, Universiti Putra Malaysia, Serdang, Selangor, Malaysia,* [c]*Advance Engineering Materials and Composites Research Center, Department of Mechanical and Manufacturing Engineering, Faculty of Engineering, Universiti Putra Malaysia UPM, Serdang, Selangor, Malaysia,* [d]*Solar Energy Research Institute (SERI), Universiti Kebangsaan Malaysia UKM, Bangi, Selangor, Malaysia,* [e]*Department of Electrical Electronic & Systems Engineering, Faculty of Engineering & Built Environment, Universiti Kebangsaan Malaysia, UKM, Bangi, Selangor, Malaysia*

11.1   Introduction

The fixed and operational costs influence investment in new processing and technology in almost all types of industries, including palm oil mills, as they will determine the profit and survival of the business. The British introduced the oil palm plant to Malaya (now Malaysia) as an ornamental plant in the 1870s. The oil palm is an important plant to overcome poverty in rural areas of Malaysia with a current coverage area of 5.87 million hectares. Around 19.14, 4.70, 2.20, and 2.5 million tons of crude palm oil (CPO), palm kernel, crude palm kernel oil, and palm kernel cake are respectively generated by the oil palm tree. Malaysia is one of the world's largest producers and exporters of palm oil. More than half a million workers are employed in the industry, and an approximate 1 million people depend on it for their survival (MPOB, 2021).

Oil generated from the oil palm tree can be categorized either from the flesh fruit or kernel/seed. Generally, 1 ton of kernel is produced for every 10 tons of palm oil. To generate finished palm oil that satisfies the specifications of customers, many processing operations are used. The first stage in the process is performed at the mill, as the CPO is extracted from the fruit. A second stage includes a further process of CPO to produce a variety of specified quality palm products. The partially

Oil Palm Biomass for Composite Panels. https://doi.org/10.1016/B978-0-12-823852-3.00020-9
Copyright © 2022 Elsevier Inc. All rights reserved.

and entirely processed grades need some additional processing prior to usage, providing the end user a saving in processing costs. Simple crystallization and separation methods may be used to fractionate palm oil into solid (stearin) and liquid (olein) fractions with different melting features. The fractions' various features make them appropriate for a broad range of food and nonfood products. Although the economic value of edible oil and fats is derived from palm fruits, the extraction process of each product generates wastes, namely empty fruit bunch, palm kernel endocarp, palm kernel press cake, and liquid palm oil mill effluent (POME). POME is the largest waste generated after obtaining the CPO. To produce 1 ton of CPO, at around 5–7.5 tons of water are needed. More than half of the water used in CPO processing becomes liquid waste (Ma et al., 1999).

The oily waste in POME needs to be removed to avoid interface in water treatments, problems in the biological treatment stages, and finally to fulfill water discharge requirements (Ahmad et al., 2005a, b). The POME residue is a hazardous pollutant for the aquatic ecosystem and toxic to aquatic organisms (Fig. 11.1).

Fig. 11.1 Examples of types of oil palm waste. Source: Author.

11.2 Source and characteristics of POME

Generally, there are three processing phases responsible for producing POME, which are listed in Table 11.1.

The purification of the extracted CPO contributes the highest POME (60%) followed by sterilization of FFB (36%) and hydrocyclone separation of a cracked mixture of kernel and shell (4%). Based on 50% of wastewater gains after the CPO process, it was assessed that Malaysia produced around 49.65–74.47, 48.79–73.18, and 47.85–71.87 million tons of POME in 2018, 2019, and 2020, respectively (Table 11.2).

The characteristic and content of POME is dependent on the season, raw material quality, and processing condition. The pH of POME is generally low, ranging from 4 to 5 due to the fermentation process of organic acid, and it largely contains solid (40,500 mg/L) and oil and grease (4000 mg/L). The waste from a palm oil mill can significantly affect the environment and water ecosystem if it is discharged without any treatment to neutralize it (Davis and Reilly, 1980; Ma, 2000; Singh et al., 2010).

Table 11.1 The source of POME during processing of crude oil palm.

No	Source of POME	Percent
1	Sterilization of fresh fruit bunch (EFB)	36
2	Clarification of the extracted crude palm oil (CPO)	60
3	Hydrocyclone separation of cracked kernel and shell mixtures	4

Source: Sethupathi, S., 2004. Removal of Residue Oil from Palm, Oil Mill Effluent (POME) Using Chitosan. Universiti Sains Malaysia.

Table 11.2 Malaysian crude oil palm production and estimation of POME.

Year	Million metrics tons		
	2018	2019	2020
Crude oil production	19.86	19.52	19.14
Estimation of used water	99.29–148.94	97.58–146.37	95.70–143.55
Estimation of POME production	49.65–74.47	48.79–73.18	47.85–71.78

Source: MPOB (2021).

Besides water, POME contains a significant amount of solids, which are either suspended or dissolved ranging from 18,000 and 40,500 mg/L. They are referred to as palm oil mill sludge (POMS). According to Rupani et al. (2010), the solid waste produced in POME after the extraction process include leave, trunk, decanter cake, empty fruit bunch, seed shells, and mesocarp fiber.

The oil and grease contain about 4000 mg/L, becoming the most challenging substance for water treatment. It also has about 25,000 and 50,000 mg/L of biochemical oxygen demand (BOD) and chemical oxygen demand (COD), respectively. The characteristics of raw POME are listed in Table 11.3.

POME is a nontoxic material because no chemicals have been used in the oil palm extraction. However, it is considered a main origin of aquatic pollution due to the reduction of dissolved oxygen when discharged into water without treatment (Khalid and Wan Mustafa, 1992).

Despite its negative perspective, the POME contains a significant amount of N, P, K, Mg, and Ca (Habib et al., 1997; Muhrizal et al., 2006), which makes it suitable as a nutrient fertilizer for plant growth. A high content of organic nitrogen in the POME increases the quality of the fertilizer (Agamuthu et al., 1986). POME also contains a higher amount of aluminum than chicken dung and wood sawdust (Muhrizal et al., 2006). A toxic metal like lead is similarly located in POME owing to contamination from plastic, metal pipe, paint, and glazing material, but the quantity is below a sublethal level (> 17.5 µg/g) per James et al. (1996) and Habib et al. (1997).

Table 11.3 The features of raw palm oil mill effluent (POME).

Parameter	Value
Temperature (°C)	80–90
pH	4.7
Biochemical oxygen demand BOD 3; 3 days at 30°C	25,000
Chemical oxygen demand (COD)	50,000
Total solids (T.S)	40,500
Total suspended solids (T.S·S)	18,000
Total volatile solids (T.V·S)	34,000
Oil and grease (O & G)	4000

Source: Ma, A.N., 2000. Environmental management for the oil palm industry. Palm Oil Dev. 30:1–10.

11.3 POME treatment methods

11.3.1 Biological treatment

The microorganism treatment system generally degrades the biodegradable substance and reduces the COD of the wastewater. Because of the capacity of anaerobic treatment to rapidly decrease COD and BOD without the use of oxygen, this treatment becomes a priority at the start of the POME process (Eddy, 2003). The biological treatment method is preferred by industries because it reduces operating costs, has high organic loading capacity, is simple, and the consumption of energy is low (Najafpour et al., 2006). An anaerobic process involves many reaction stages such as hydrolysis, acidogenesis, and methanogenesis (Gerardi, 2003).

The open pond concept is commonly used in anaerobic treatment to degrade POME, containing a de-oiling tank, acidification ponds, anaerobic ponds, and a facultative pond (Chan and Chooi, 1984). At the time being, 85% of POME treatment uses an anaerobic treatment and open pond system (Ma et al., 1999). One of the primary issues with ponding systems is their incapacity to achieve department of environment (DOE) discharge requirements after 80 days of retention. It is reported that the final COD and BOD for the final effluent of the anaerobic pond are 1725 mg/L and 610 mg/L, respectively (Chin et al., 1996).

An improvisation of the biological system was conducted by the Borja and Banks (1994a, b). They treated POME with an upflow anaerobic sludge blanket reactor (UASB), which gives 96% removal of COD. The UASB has an advantage over the pond system, which has a shorter hydraulic retention (1.5–3 days) and requires less operation area. One of the UASB system's limitations is that it can't sufficiently hold a high amount of microorganisms for the high loading treatment (Borja and Banks, 1994a, b).

The upflow anaerobic filtration (UAF) system to treat POME was introduced following the problematic gains by the UASB system (Borja and Banks, 1994a, b). The UAF system was introduced to retain the dense microorganisms in the reactor, and this give a more efficient treatment for a higher loading of POME; almost 90% of the substrate was oxidized and the reactor condition was stable under acidic and alkaline conditions (Borja and Banks, 1994a, b). The ability of both UASB and UAF to capture methane gas in the reactor is another one of their advantages. The methane gas can be used as a source of energy to power palm oil processing mills.

The methane is obtained by a methanogenesis step during the anaerobic reaction. It is reported that for every 1 g of COD removed from the POME, the UAF will produce about 0.69–0.79 dm^3 of methane gas (Borja and Banks, 1994a, b). Although it's beneficial in producing methane gas, the UAF suffers a reactor malfunction at high loading rates because of suspended solids in the POME (Najafpour et al., 2006).

The UAF problems are overcome by integrating the UASB with an upflow fix system (UFF). This new integration system (UASFF) successfully treated the POME in a laboratory scale and enhanced solid/liquid/gas separation in the reactor even at a very high solid content in POME. A COD removal of 97% within 3 days was reported (Najafpour et al., 2006). This technology only works on the laboratory scale and not in an industrial scale due to failure of the upscaling process.

Research on treating POME with an aerobic system is in progress with a focus to more toward decreasing hydraulic retention time of the anaerobic treatment. A 98% COD removal within 60 h was achieved by aerobic treatment of POME via an activated sludge reactor (Vijayaraghavan et al., 2007). The use of aerobic treatment to POME not only reduced carbon content but also the inorganic nitrogen; it also altered the pH from acidic to alkaline (Agamuthu et al., 1986).

However, the aerobic system proposed by Vijayaraghavan et al. (2007) faces a problem. Biomass in the effluent causes issues with the receiving water body. This problem is countered by the improvement of aerobic granules for POME treatment. Gobi et al. (2011) formed aerobic granules in POME and utilized them to effectively treat POME. Aerobic granules have a good settling ability and are robust in nature (Pijuan et al., 2009). The Gobi et al. (2011) system focuses on increasing the reactor's height over diameter (H/D) ratio to facilitate forming of aerobic granules; the sequencing batch reactor (SBR) system only uses a small space and handles POME with 90% efficiency in only 24 h.

11.3.2 Nonbiological method

In addition to the biological treatment, several methods have been identified that can be used to treat POME, including coagulation-flocculation, adsorption, membrane technology, and integrated technology. Chemicals such as polyacrylamide, aluminum sulfate, and polyaluminum chloride are used in a coagulation-flocculation treatment to destabilize the colloids in POME (Ahmad et al., 2006a). The mechanism of this treatment is to suspend the solids and separate it from the POME wastewater. In contrast to the biological treatment, the COD value is not remarkably reduced in the coagulation-flocculation treatment.

The adsorption technique is only capable of removing residual oil from POME and is used as a final polisher; it cannot eliminate COD in POME. Both coagulation-flocculation and adsorption techniques are technically ineffective to treat POME due to their high operating and maintenance costs (Zhang et al., 2008; Ahmad et al., 2006a, 2005a, b).

The membrane technique is reported to remove 99% of the influent COD from POME, but the technique is not autonomously operated and needs pretreatment with several techniques. Basically, before the POME undergoes membrane technology, it is subjected to three pretreatment processes to decrease the turbidity, COD, and BOD. In each of these stages, it requires the use of chemicals and extra energy, which increase the operational cost (Ahmad et al., 2006b). The membrane process is superior to biological treatment, but its implementation is not successful at an industrial level.

11.3.3 Integrated system

The integrated system was born by mixing both biological and nonbiological treatments. In the first stage of this integrated system, the POME undergoes a granular sludge bed (EGSB) treatment and then aerobic reactor. In the second stage, the POME goes through the ultrafiltration membrane and reverse osmosis membrane, which can also trap methane gas. The final effluent is almost clear crystal and can be used for boiler feed water (Zhang et al., 2008).

Chaiprapat and Laklam (2011) combined an ozone pretreatment method with an anaerobic treatment system in another system. The use of ozone to degrade the recalcitrant components in POME will open the way for the anaerobic system. The anaerobic sequencing batch reactor (ASBR) will treat POME at high organic loadings ($9.04 \, \text{kg} \, COD \, m^{-3} day^{-1}$) in just 12 h, with a longer hydraulic retention period (10 days).

The integration of anaerobic and aerobic methods using a single bioreactor was described by Chan et al. (2012). This system can replace the mill operation area and inefficient ponding system. It can remove about 99% of the COD and BOD in POME and trap methane gas in the reactor.

A single system to treat POME with a combination of anaerobic hybrid reactor (AHR), anaerobic baffled filter (ABF) reactor, and anaerobic downflow filter (ADF) reactor was conducted by Choi et al. (2013). This system aims to maintain a high effluent recycling ratio while reducing treatment retention time. At $13 \, \text{kg} \, COD \, m^{-3} day^{-1}$, this system is shown to be able to remove 95.6% of influent COD. Without biomass, the anaerobic and filtration systems cleans the effluent. The methane gas is generated at 0.171 and $0.269 \, L \, CH_4/g \, COD$ removed.

11.4 Products from POME treatment

Many by-products are generated through the treatment of POME either using biological or nonbiological treatments. Some of the by-products are economically valuable, while others are not valuable at all. To date, the economic value of the POME treatment by-products are not known. The by-products produced by treatment of the POME and its potential usage are listed in Table 11.4.

11.4.1 Production of gas

Methane gas is without a doubt one of the most popular by-products in a variety of processes. Methane collected is used to supply power to the mill, and it can reduce the burden of power cost in overall production of the CPO (Arthur and Glover, 2012). Besides methane gas, other gases including hydrogen can be generated from the anaerobic treatment plant. This biohydrogen is collected through the acidogenic fermentation of wastewater (Angenent et al., 2004).

11.4.2 Production of solid

Sludge is the second largest substance produced in a wastewater treatment plant. The waste activated sludge has been utilized as an adsorbent material to remove textile dye (Gobi et al., 2011; Li et al., 2011) and as a fertilizer (Chan and Chooi, 1984).

Table 11.4 The by-product and potential usage from POME treatment.

Treatment	By-product	Potential usage	Reference
Membrane technology	Biohydrogen	Power generation	Ahmad et al. (2006a, b)
	Water	Boiler feed	
	Methane gas	Power generation	
UASB	Waste anaerobic granular sludge	Activated carbon	Najafpour et al. (2006)
	Biohydrogen	Power generation	
	Methane gas	Power generation	
	Volatile fatty acid	PHA feedstock	
Sequencing batch reactor	Waste aerobic granules	Activated carbon	Gobi et al. (2011)
Aerobic digestion	Waste activated sludge	Activated carbon	Vijayaraghavan et al. (2007)

Table 11.5 The adsorption capacity of sludge.

Adsorbent	Adsorption capacity (mg/g)	Reference
Rhodamine	33.33	Zou et al. (2013)
Phenol	100.00	Zou et al. (2013)
Toluene	350.00	Anfruns and Martin (2011)
Chloroform	244.00	Tsai et al. (2008)
Methylene blue	130.69	Li et al. (2011)
Reactive black 5	93.00	Gulnaz et al. (2006)
Methylene blue	66.23	Gobi et al. (2011)

According to Tsai et al. (2008), the adsorption capacity of waste activated sludge is similar to that of industrial activated carbon (373 mg/g). This is attributed to the functional groups attached to the surface (Gobi et al., 2011). The waste sludge helps the adsorption process through the physical and chemical adsorption procedures. The adsorption capacity of waste activated sludge is shown in Table 11.5.

11.5 Waste aerobic granules (WAG)

WAGs differ in terms of morphology, robustness, and performance as compared to waste activated sludge. WAG is formed by a series of processes starting with inoculation, pellet formation, shear force, colonization of bacteria, oxygen limitation, and hydrolysis to the final forms of granule of bacteria colony. It inherits the microbial behavior of the activated sludge, is compact and dense, and has an efficient ability to settle compared to conventional activated sludge (Liu et al., 2011). A good settling capacity of aerobic granules led to the effluent free of residual biomass. The application of granulation technology has gained considerable interest owing to its ability to reduce the working area.

11.5.1 The action mechanism of aerobic granules used in wastewater treatment

For both growth and maintenance, the aerobic granules will biodegrade the organic contents of wastewater. As with typical activated sludge, some portion of the consumed organic content is stored as polyhydroxyalkanoate (PHA) within the microorganism's cells. PHA acts as an energy storage material that can be utilized during times of famine (Chakraborty et al., 2009). The accumulation of PHA is also reported in aerobic granules (Wu et al., 2012, 2010).

11.5.2 Aerobic granule formation in POME

The aerobic granules are formed within the SBR using POME as the substrate. The maturation period for aerobic granules is 120 days (Gobi et al., 2011) and 60 days (Abdullah et al., 2011). The difference in maturation time is attributed to the exchange ratio and reactor configurations, but both removed 90% of COD.

11.5.3 Waste aerobic granules generation

The presence of WAG occurs as a consequence of the aerobic granules' growth process. To preserve the sludge retention time (SRT) within the stable reactor, an excess of aerobic granules should be discarded. WAGs may be reused for other uses. The reuse of excess aerobic granules is mostly unexploited.

11.5.4 Application of waste aerobic granules

The inactive aerobic granules are used to adsorb Yellow 2G and reactive brilliant red K-2G, with the adsorption capacity of 58.50 and 66.18 mg/g for yellow 2G and reactive red K-2G (Gobi et al., 2011). The inactive aerobic granules are also used to adsorb residual COD when minimizing turbidity in POME treated in SBR. The remaining COD and turbidity have been removed to a percentage of 21% and 99%, respectively (Gobi et al., 2011). Aerobic granules are used as seed sludge to initiate aerobic granules in a new reactor. When crushed, aerobic granules are used as a seed sludge, and the reactor startup period is reduced to 18 days, as compared to 120 days when active sludge is used as the seed sludge (Pijuan et al., 2011; Gobi et al., 2011). The difference is clarified by the fact that certain crushed aerobic granules remain intact, causing the floccular sludge on the crushed aerobic granules to be enhanced. As a consequence of this phenomenon, aerobic granules have grown more rapidly; simultaneously, the reactor's startup period has decreased.

Wang et al. (2012) obtained a nitrobenzene-degrading bacterium from broken aerobic granular sludge (*Klebsiella ornithinolytica* NB1). With the use of that bacterium, the biodegradation rate of nitrobenzene was found to be about 9.29 mg/L 1 h^{-1}. Nevertheless, in almost all cases of reuse of aerobic granules, the production of PHA was the most attractive. The synthesis of PHA is not well exploited since PHA is an intermediate product.

11.6 Polyhydroxyalkanoate (PHA)

11.6.1 Method of production

PHA is an energy storage molecule created naturally within microorganisms (Hassan et al., 2013). The total accumulation of monomers are, for example, hydro- xybutrate (HB) and 3-hydroxyvalerate (3HB-3HV), 3-hydroxyvalerate (3HV), 3-hydroxy-2-methylvalerate (3H2MV), or 3-hydroxyhexanoate (3HHx), which are described as PHA (Sudesh et al., 2000). PHA is categorized into two categories: short carbon chain length (SCL) or medium carbon chain length (MCL).

PHA is a categorized as a biodegradable and biocompatible polymer with characteristics that are almost equivalent to those in conventional plastics (Albuquerque et al., 2010a). Several researchers have suggested that carbon-rich waste streams could be applied as a substrate for the production of PHA (Castilho et al., 2009). Other wastes, like olive oil mill effluent, sugar molasses, and food waste, could be utilized as a substrate for the processing of PHA (Albuquerque et al., 2010a; Venkateswar Reddy and Venkata Mohan, 2012).

PHA is accumulated within microorganisms by consuming available volatile fatty acids (VFA) in wastewater. As compared to unfermented carbon sources, fermented substrates have been shown to produce further PHA (Venkateswar Reddy and Venkata Mohan, 2012). The proportion of VFAs with shorter chain lengths is higher in the fermented substrate. Rather than a complex substrate, the PHA-accumulating microorganism will more simply take up the VFA. VFA would be produced by acidogenic fermentation of wastewater. The VFA is then fed into a reactor containing PHA-accumulating organisms. The organisms in the reactor then convert VFA into the appropriate acyl-CoA, which is then converted into PHA (Morgan-Sagastume et al., 2010). The operating conditions play a crucial role to accrete PHA within the microorganisms. The aerobic dynamic feeding (ADF) method is the most classical method to accumulate PHA within cells. In ADF, a feast and famine phase are settled in the reactor, during which the ADF suppresses internal growth of the microorganisms, causing them to adapt to the nutrient limitation caused by ADF. Throughout the adaptation phase, the substrate is stored as PHA within the cells of microorganisms (Beccari et al., 1998). Despite understanding PHA accretion within the microorganisms, the PHA accretion within the microorganisms is generally appropriate to pure culture methods only. Approximately 86% CDW PHA is accreted inside the pure culture with the current available knowledge (Mumtaz et al., 2010).

11.6.2 PHA production through mixed culture

As mentioned earlier, the production of PHA using pure culture method is expensive, and other ways such as the mixed culture method offers a better alternative (Johnson et al., 2010). With the mixed culture, another problem arises due to poor accretion of PHA inside the microorganism (Moita and Lemos, 2012). The microorganisms are able to function to accrete PHA within their cells. The solution to this problem is by enriching the PHA-accreting microorganisms in the mixed culture system (Albuquerque et al., 2011; Johnson et al., 2010). The microbials that are highly enriched in PHA-accreted organisms can be cultured simply by changing the reactor's operating conditions. Inhibition of the PHA-accumulation mechanism is a key challenge in a mixed culture method. Under conditions with excess ammonium, PHA accumulation in a mixed culture is just 69% after 4.4 h (Johnson et al., 2010).

The addition of 180 Cmmol/L acetate is one single pulse inhibitor of the accretion of PHA in 67% of cell dry weight of PHA (Serafim et al., 2004). A high level of substrate inhibition raises production costs, like the pure culture method, that conflicts with the mixed culture method's goal of low cost. Because of the low recovery of the PHA accumulated within the microorganisms, the inhibition process raises production costs. Operating conditions, excess nutrients, nutrient limitation, and thermodynamics are all factors that inhibit PHA (Albuquerque et al., 2010a; Johnson et al., 2010; Liu et al., 2011; Wen et al., 2010).

11.6.3 Current direction of generating PHA via mixed culture

An effort was made to reduce the cost of mixed culture method to produce PHA by utilizing waste streams rich with organic matter as a carbon substrate. Wastewaters derived from many sources are successfully converted to carbon substrate for PHA production:

i. Sugar molasses (Albuquerque et al., 2010a).
ii. Olive oil pomace (Waller et al., 2012).
iii. Food waste (Chen et al., 2013).
iv. POME (Din et al., 2012).

The lower level of PHA accumulation in the mixed culture method relative to the pure culture method is one of the major issues that needs to be addressed (Albuquerque et al., 2011). To maximize the PHA accumulation in the mixed culture method, a feast-famine period is introduced. Albuquerque et al. (2010b) used a mixed culture method with a feast-famine ratio of 0.5 and 0.22, yielding 0.18 and 0.59 Cmol PHA/Cmol VFA, respectively. This demonstrates that a lower feast-to-famine ratio will improve PHA yield. In a continuous mode reactor, this discovery improves PHA production. PHA yields between 0.213 and 0.257 g PHA/g acetate were produced in a continuous mode reactor (Chakravarty et al., 2010). The enrichment of the PHA-accretion organism and the efficient operation of the continuous mode reactor can open the direction for future commercialization of PHA production from wastewater, including the POME. Therefore, knowledge of the reaction between microorganism and POME for production of PHA is needed.

11.6.4 PHA accretion using POME

11.6.4.1 Composition of POME

POME is very rich with biodegradable organic content, and a few researchers have tried to produce PHA from POME. In an anaerobic process, the digested POME is converted to biodegradable organic content of VFAs. Generally, the VFA acts as the carbon substrate for PHA production. Complete anaerobic digestion of POME will produce methane and carbon dioxide. Generally, a complete anaerobic digestion comprises of three theoretical stages (hydrolysis, acidogenesis, and methanogenesis) to produce methane and carbon dioxide gases (Eddy, 2003).

The VFA and biohydrogen are formed once the anaerobic process stops at the acidogenesis stage. The VFs content could further produce other linear chain carboxylic acids like acetic, propionic, butyric, and isobutyric acids (Fang et al., 2011; Rasdi et al., 2012). The types of PHA from POME, either 3-hydroxybutyrate, 3-hydroxyvalerate (3HV), and 3-hydroxy-2- methyl valerate (3H2MV), depend on the constituents of VFA.

11.6.4.2 PHA-accumulating microorganisms

The PHA-accretion microorganisms collecting the VFA store it within the cells then turns the VFA into PHA. The PHA quantity is determined by the PHA content (% of cell dry weight). A cell dry weight with 90% PHA is obtained by using *Comamonas* sp. EB172 (microorganism) to generate PHA (Mumtaz et al., 2010). The PHA can be also generated by adding a photosynthetic of *Rhodobacter sphaeroides* to the VFAs (Hassan et al., 1997a).

Other bacteria such as FLP 1 and FLP 2 can also use the palm oil olein to synthesize PHA up the 18% of the dry cell wall (Alias and Tan, 2005). A significant difference in the accretion of PHA in a study by Mumtaz et al. (2010) and Alias and Tan (2005) is due to the different types of substrates. In the case of Mumtaz et al. (2010), a higher percentage of VFA causes a higher anaerobic procedure.

11.6.4.3 The quantity of PHA

The yield and PHA content obtained from different microorganisms in shown in Table 11.6. Generally, a low yield obtained in pure culture could be interpreted that the presenting organic content is not injected enough for PHA production. It is most probably because of the complexity of the VFAs formed in POME or a limitation of the microorganisms that use POME to accumulate PHA. One way to solve the problem is to use a strong microorganism that is highly selective for PHA accumulation.

11.7 Generation of biohydrogen by using POME

The POME can produce biohydrogen, either by using pure culture strains or mixed culture bacteria, with a maximum yield of 31.95 mL H2/g COD for pure culture (Chong et al., 2009a, b) and 4708 mL H2/L-POME mixed culture (Atif et al., 2005), respectively. Generally, mixed culture gives a better option for biohydrogen than those of pure culture. The quantity of biohydrogen from POME is listed in Table 11.7.

Biohydrogen is forecasted as a possible fuel in the future for a carbon-free energy system (Leaño and Babel, 2012). Biohydrogen has an advantage over fossil fuels in terms of a higher yield, and it has attracted much recent research (Zhang et al., 2008; Chong et al., 2009a, b). Biohydrogen can be generated in several ways: fermentation, photofermentation, a two-stage process (integration of dark and photofermentation), or biocatalyzed electrolysis (Manish and Banerjee, 2008).

Overall, it was reported that dark fermentation is the most possible method to yield biohydrogen since dark fermentation has a high rate of cell growth without light, has no oxygen limitation issues, and most importantly, its lower capital cost (Chong et al., 2009a, b). The fermentation process needs to be stopped at the acidogenesis

Table 11.6 The production of PHA in POME.

Microorganism	Type	PHA yield	PHA content (%)	Reference
Comamonas sp. EB172	Pure	0.31	85.8	Mumtaz et al. (2010)
Rhodobacter sphaeroides	Pure	0.5	67.0	Hassan et al. (1997a)
Burkholderia cepacia	Pure	NA	57.4	Alias and Tan (2005)
Alcaligenes eutrophus	Pure	0.32	45.0	Hassan et al. (1997b)
Heterotropic aerobic bacterium	Mixed culture	0.80	74.0	Din et al. (2012)

Table 11.7 The quantity of biohydrogen obtained from POME.

Medium	H$_2$ Production rate	Reference
Clostridium butyricum EB6	31.95	Rasdi et al. (2012)
Anaerobic sludge	4708	Chong et al. (2009a, b)
Microflora	102.6	Atif et al. (2005)
Thermophilic microflora	4.4[a]	Vijayaraghavan et al. (2007)
Thermoanaerobacterium		
Thermosaccharolyticum	4800	Thong et al. (2007)
Rhodopseudomonas palustris PBUM001		
Thermoanaerobacterium	1050	Jamil et al. (2009)
Anaerobic mixed microflora		
Thermotolerant consortia		
Mixed culture	4200	Mamimin et al. (2012)
Mixed culture	6700	Badiei et al. (2011)
Mixed culture	702.52	Yossan et al. (2012)

stage to enhance biohydrogen collection. The degradation of dark fermentation product will further affect the production of biohydrogen gas if methanogenesis is not immediately stopped.

The temperature influences the biohydrogen production. The microbial species are recognized with a temperature tag of either thermophilic or mesophilic. Biohydrogen production is optimum at the thermophilic region of 601 °C. At this temperature, the activity of the methanogens stops and this further improves the production of biohydrogen. In a nutshell, biohydrogen is a by-product that can be generated in significant quantities by utilizing POME.

11.8 PHA and biohydrogen processing issues

11.8.1 PHA from POME

The generation of PHA from POME has a few limitations that need to be overcome before it can be commercialized up to an industrial level. The first problem is that the pure culture method to produce PHA within the microorganisms is very fragile, requires a high cost, and is sensitive to environmental changes and other parameters in the operating reactor (Johnson et al., 2010). Although the pure culture method could provide higher PHA accumulation, it is constantly susceptible to failure on a large scale. However, the mixed culture method is in the very earlier stages for large scale commercialization. Very detailed studies are needed for a comprehensive understanding of PHA accretion via a mixed culture. The best inhibition features in the mixed culture must be clearly identified before making a step toward commercialization.

Next problem is to find a way for the generation of PHA with a continuous mode. To date, there is no research conducted to find the feasibility of PHA production with a continuous mode using POME. The main limitation is to determine the retention time for the POME at each stage of the operation and a suitable time to allow the organic acids to be synthesized at the acidogenesis process for PHA production. The production of PHA is also hindered by some nonsustainable extraction methods. The halogenated method is commonly used to extract of PHA (Mohammadi et al., 2012). However, many hazardous chemicals, including chloroform and sodium hypochlorite, that are compulsory for this process are not practicable for commercial usage. In addition, a complex treatment system is necessary to remove the hazardous solvents when PHA is extracted.

11.8.2 Biohydrogen from POME

The production of biohydrogen from POME is not in a commercialized stage due to a few reasons. The key reason is that the biohydrogen-producing microorganism should be controlled in the POME treatment plant's mixed culture medium to ensure optimum hydrogen generation from anaerobically digested POME (Yokoi et al., 1998). The second reason is that the recent methods for enriching bacteria are expensive and impractical to implement at an industrial level. As mentioned earlier, biohydrogen is produced by the acidogenesis process. The VFAs are formed at a higher proportion during the acidogenesis process. These VFAs causes a decrease of pH and thus the mixed liquor is acidic. This condition inhibits the production of biohydrogen (Chow and Ho, 2000). It is reported that the optimum pH for biohydrogen production in wastewater is around 5.5 (Thong et al., 2011). Therefore, a balance between adequate VFA production and the pH of the mixed liquor needs to be identified to ensure optimum biohydrogen production. VFA production is very important for both biohydrogen and PHA production. Another obstacle that must be overcome is the continuous generation of biohydrogen since it is critical to guarantee the chain of energy source for both the treatment plant and business orientation.

11.9 The by-product of oil palm kernel

The oil palm kernel is the second important oil palm mill waste after the POME. Several studies are reported to use oil palm kernel for activated carbon, bioabsorbent, feedstock, and others.

11.9.1 Activated carbon

Rugayah et al. (2014) explored the production of granular activated carbon from palm kernel shell (PKS) utilizing commercial scale carbonization and activation systems. Carbonization occurred in a commercial scale rotary kiln, while activation took place in a kiln earth system. During the activation process, steam is used as an oxidizing agent at temperatures varying from 900°C to 1000°C. PKS activated carbon produces activated charcoal with a low ash content (2.3%) and a high carbon and volatile content of 23% and 61.7%, respectively.

Maximum thermal stability is detected for PKS activated carbon and raw PKS up to 700°C, but only 600°C for PKS charcoal. PKS activated carbon has a Brunauer–Emmett–Teller (BET) surface area of $607.76 \, m^2g^{-1}$ with a micropore area of $541.76 \, m^2g^{-1}$, which is equal to commercial activated carbon, which has a BET surface area of $690.92 \, m^2g^{-1}$ with a micropore area of $469.08 \, m^2g^{-1}$. The PKS activated carbon will remove up to 80.7% of COD with an adsorption capacity of 8.83 mg/g, which is equivalent to the properties of commercial activated carbon obtainable on the market.

Nahrul et al. (2017) investigated the production of charcoal from PKS utilizing a microwave-assisted precarbonization system. At a carbonization temperature of 250–300°C, a high heating value (HHV) of 27.63 MJ/kg is obtained. The HHV is similar to those found in other studies that used a carbonization temperature of more than 450°C and microwave heating for the entire carbonization period. The gaseous emission was lower than the standard limits set by the Malaysian Ambient Air Quality Standards (2014). The high HHV of charcoal with low gaseous emission can be used for co-combustion for green power generation purposes.

To achieve a high yield and surface activated carbon, a double-insulated carbonization-activation reactor was designed (Nahrul et al., 2018). This reactor is double insulated with low cement castable and coated with a stainless-steel plate and a fiberglass jacket with a thermal insulation coating around the internal space of the reactor, enabling efficient heat transfer into the reactor's bed of material. The carbonization of oil palm kernel shell (OPKS) at 400°C, accompanied by steam activation at 500–1000 °C continuously in the same reactor, with a steam flow rate of 12.80–18.17 L/min, increases the activated carbon surface area from 305 10.2 to 935 36.7 m^2g^{-1} and gives a high yield of 30% during a 7-h retention period.

The activated carbon generated has been successfully used as a bioadsorbent for the treatment of POME final discharge, with reductions in total suspended solids (TSS), COD, color, and BOD of up to 90%, 68%, 97%, and 83%, respectively.

11.10 Conclusions

The production of PHA and biohydrogen from treatment of POME has a great prospective as another source of income for the oil palm mils. Commercialization of PHA and biohydrogen could provide a return of investment and zero waste discharge from the POME treatment plant. The PHA generated from POME should be considered as it has a high market value and sustainability, and the same is applied for biohydrogen. The SBR using aerobic granules should be carefully developed as it can produce both PHA and biohydrogen by maintaining its main function to treat the POME. Development of this technology up to an industrial scale will attract mill operators to invest in new technology for their wastewater treatment plants. The production of activated carbon from oil palm kernel shows a potential market to generate a local economy surrounding the palm oil mill area, but it may face a high competition with other commercial activated carbon made from mangrove trees and recycled wood waste materials. The demand for charcoal and activated carbon in Southeast Asia is not as important in European countries with four weather seasons.

References

Abdullah, N., Ujang, Z., Yahya, A., 2011. Aerobic granular sludge formation for high strength agro-based wastewater treatment. Bioresour. Technol. 102, 6778–6781. https://doi.org/10.1016/j.biortech.2011.04.009.

Agamuthu, P., Tan, E.L., Shaifal, A.A., 1986. Effect of aeration and soil inoculum on the composition of palm oil effluent (POME). Agric. Wastes 15, 121–132. https://doi.org/10.1016/0141-4607(86)90043-0.

Ahmad, A.L., Sumathi, B.H., Hameed, B.H., 2005a. Residual oil and suspended solid removal using natural adsorbents chitosan, bentonite and activated carbon: a comparative study. Chem. Eng. J. 108, 179–185. https://doi.org/10.1016/j.cej.2005.01.016.

Ahmad, A.L., Bhatia, S., Ibrahim, N., Sumathi, S., 2005b. Adsorption of residual oil from palm oil mill effluent using rubber powder. Braz. J. Chem. Eng. 22, 371–379. https://doi.org/10.1590/S0104-66322005000300006.

Ahmad, A.L., Sumathi, S., Hameed, B.H., 2006a. Coagulation of residue oil and suspended solid in palm oil mill effluent by chitosan, alum and PAC. Chem. Eng. J. 118, 99–105. https://doi.org/10.1016/j.cej.2006.02.001.

Ahmad, A.L., Chong, M.F., Bhatia, S., Ismail, S., 2006b. Drinking water reclamation from palm oil mill effluent (POME) using membrane technology. Desalination 191, 35–44. https://doi.org/10.1016/j.desal.2005.06.033.

Albuquerque, M.G.E., Concas, S., Bengtsson, S., Reis, M.A.M., 2010a. Mixed culture polyhydrox-yalkanoates production from sugar molasses: the use of a 2-stage CSTR system for culture selection. Bioresour. Technol. 101, 7112–7122. https://doi.org/10.1016/j.biortech.2010.04.019.

Albuquerque, M.G.E., Torres, C.A.V., Reis, M.A.M., 2010b. Polyhydroxyalkanoate (PHA) production by a mixed microbial culture using sugar molasses: effect of the influent substrate concentration on culture selection. Water Res. 44, 3419–3433. https://doi.org/10.1016/j.watres.2010.03.021.

Albuquerque, M.G.E., Martino, V., Pollet, E., Avérous, L., Reis, M.A.M., 2011. Mixed culture polyhydroxyalkanoate (PHA) production from volatile fatty acid (VFA)-rich streams: effect of substrate composition and feeding regime on PHA productivity, composition and properties. J. Biotechnol. 151, 66–76. https://doi.org/10.1016/j.jbiotec.2010.10.070.

Alias, Z., Tan, I.K.P., 2005. Isolation of palm oil-utilising, polyhydroxyalkanoate (PHA)-producing bacteria by an enrichment technique. Bioresour. Technol. 96, 1229–1234. https://doi.org/10.1016/j.biortech.2004.10.012.

Anfruns, A., Martin, M.J., Montes-Morán, M.A., 2011. Removal of odourous VOCs using sludge-based adsorbents. Chem. Eng. J. 166, 1022–1031. https://doi.org/10.1016/j.cej.2010.11.095.

Angenent, L.T., Karim, K., Al-Dahhan, M.H., Wrenn, B.A., Domíguez-Espinosa, R., 2004. Production of bioenergy and biochemicals from industrial and agricultural wastewater. Trends Biotechnol. 22, 477–485. https://doi.org/10.1016/j.tibtech.2004.07.001.

Arthur, R., Glover, K., 2012. Biomethane potential of the POME generated in the palm oil industry in Ghana from 2002 to 2009. Bioresour. Technol. 111, 155–160. https://doi.org/10.1016/j.biortech.2012.02.065.

Atif, A.A.Y., Fakhru'l-Razi, A., Ngan, M.A., Morimoto, M., Iyuke, S.E., Veziroglu, N.T., 2005. Fed batch production of hydrogen from palm oil mill effluent using anaerobic microflora. Int. J. Hydrog. Energy 30, 1393–1397. https://doi.org/10.1016/j.ijhydene.2004.10.002.

Badiei, M., Jahim, J.M., Anuar, N., Sheikh Abdullah, S.R., 2011. Effect of hydraulic retention time on biohydrogen production from palm oil mill effluent in anaerobic sequencing batch reactor. Int. J. Hydrog. Energy 36, 5912–5919. https://doi.org/10.1016/j.ijhydene.2011.02.054.

Beccari, M., Majone, M., Massanisso, P., Ramadori, R.A., 1998. Bulking sludge with high storage response selected under intermittent feeding. Water Res. 32, 3403–3413. https://doi.org/10.1016/S0043-1354(98)00100-6.

Borja, R., Banks, C.J., 1994a. Anaerobic digestion of palm oil mill effluent using an upflow anaerobic sludge blanket reactor. Biomass Bioenergy 6, 381–389. https://doi.org/10.1016/0961-9534(94)E0028-Q.

Borja, R., Banks, C.J., 1994b. Treatment of palm oil mill effluent by up flow anaerobic filtration. J. Chem. Technol. Biotechnol. 61, 103–109. https://doi.org/10.1002/jctb.280610204.

Castilho, L.R., Mitchell, D.A., Freire, D.M.G., 2009. Production of polyhydroxyalkanoates (PHAs) from waste materials and by-products by submerged and solid-state fermentation. Bioresour. Technol. 100, 5996–6009. https://doi.org/10.1016/j.biortech.2009.03.088.

Chaiprapat, S., Laklam, T., 2011. Enhancing digestion efficiency of POME in anaerobic sequencing batch reactor with ozonation pretreatment and cycle time reduction. Bioresour. Technol. 102, 4061–4068. https://doi.org/10.1016/j.biortech.2010.12.033.

Chakraborty, P., Gibbons, W., Muthukumarappan, K., 2009. Conversion of volatile fatty acids into polyhydroxyalkanoate by *Ralstonia eutropha*. J. Appl. Microbiol. 106, 1996–2005. https://doi.org/10.1111/j.1365-2672.2009.04158.x.

Chan, K.S., Chooi, C.F., 1984. Ponding system for palm oil mill effluent treatment. In: Proceedings of the Regional Workshop on Palm Oil Mill Technology & Effluent Treatment, pp. 185–192.

Chakravarty, P., Mhaisalkar, V., Chakrabarti, T., 2010. Study on poly-hydroxyalkanoate (PHA) production in pilot scale continuous mode wastewater treatment system. Bioresour. Technol. 101 (8), 2896–2899.

Chan, Y.J., Chong, M.F., Law, C.L., 2012. An integrated anaerobic–aerobic bioreactor (IAAB) for the treatment of palm oil mill effluent (POME): start-up and steady state performance. Process Biochem. 47, 485–495. https://doi.org/10.1016/j.procbio.2011.12.005.

Chen, H., Meng, H., Nie, Z., Zhang, M., 2013. Polyhydroxyalkanoate production from fermented volatile fatty acids: effect of pH and feeding regimes. Bioresour. Technol. 128, 533–538. https://doi.org/10.1016/j.biortech.2012.10.121.

Chin, K.K., Lee, S.W., Mohammad, H.H., 1996. A study of palm oil mill effluent treatment using a pond system. Water Sci. Technol. 34 (11), 119–123. https://doi.org/10.1016/s0273-1223(96)00828-1.

Choi, W.H., Shin, C.H., Son, S.M., Ghorpade, P.A., Kim, J.J., Park, J.Y., 2013. Anaerobic treatment of palm oil mill effluent using combined high-rate anaerobic reactors. Bioresour. Technol. 141, 138–144. https://doi.org/10.1016/j.biortech.2013.02.055.

Chong, M.L., Rahim, R.A., Shirai, Y., Hassan, M.A., 2009a. Biohydrogen production by *Clostridium butyricum* EB6 from palm oil mill effluent. Int. J. Hydrog. Energy 34, 764–771. https://doi.org/10.1016/j.ijhydene.2008.10.095.

Chong, M.L., Sabaratnam, V., Shirai, Y., Hassan, M.A., 2009b. Biohydrogen production from biomass and industrial wastes by dark fermentation. Int. J. Hydrog. Energy 34, 3277–3287. https://doi.org/10.1016/j.ijhydene.2009.02.010.

Chow, M.C., Ho, C.C., 2000. Surface active properties of palm oil with respect to the processing of palm oil. J. Oil Palm Res. 12, 107–116. repositorio.fedepalma.org/handle/123456789/83951.

Davis, J.B., Reilly, P.J.A., 1980. Palm oil mill effluent-a summary of treatment methods. Oleagineux 35, 323–330.

Din, M.F., Mohanadoss, P., Ujang, Z., Van Loosdrecht, M., Yunus, S.M., Chelliapan, S., 2012. Development of bio-PORecs system for polyhydroxyalkanoates (PHA) production and its storage in mixed cultures of palm oil mill effluent (POME). Bioresour. Technol. 124, 208–216.

Fang, C.O., Thong, S., Boe, K., Angelidaki, I., 2011. Comparison of UASB and EGSB reactors performance, for treatment of raw and deoiled palm oil mill effluent (POME). J. Hazard. Mater. 189, 229–234. https://doi.org/10.1016/j.jhazmat.2011.02.025.

Gerardi, M.H., 2003. The Microbiology of Anaerobic Digesters. Wiley-Interscience, New Jersey, pp. 51–57.

Gobi, K., Mashitah, M.D., Vadivelu, V.M., 2011. Adsorptive removal of methylene blue using novel adsorbent from palm oil mill effluent waste activated sludge: equilibrium, thermodynamics and kinetic studies. Chem. Eng. J. 171, 1246–1252. https://doi.org/10.1016/j.cej.2011.05.036.

Gulnaz, O., Kaya, A., Dincer, S., 2006. The reuse of dried activated sludge for adsorption of reactive dye. J. Hazard. Mater. 134, 190–196. https://doi.org/10.1016/j.jhazmat.2005.10.050.

Habib, M.A.B., Yusuf, F.M., Phang, S.M., Mohamed, S., 1997. Nutritional values of chironomid larvae grown in palm oil mill effluent and algal culture. Aquaculture 158, 95–105. https://doi.org/10.1016/S0044-8486(97)00176-2.

Hassan, M.A., Shirai, Y., Kusubayashi, N., Karim, M.I.A., Nakanishi, K., Hashimoto, K., 1997a. The production of polyhydroxyalkanoate from anaerobically treated palm oil mill effluent by Rhodobacter spheroids. J. Ferment. Bioeng. 83, 485–488.

Hassan, M.A., Shirai, Y., Umeki, H., Karim, M.I.A., Nakanishi, K., Hashimoto, K., 1997b. Acetic acid separation from anaerobically treated palm oil mill effluent for the production of polyhydroxyalkanoate by *Alcaligenes eutrophus*. Biosci. Biotechnol. Biochem. 61, 1465–1468. https://doi.org/10.1271/bbb.61.1465.

Hassan, M.A., Yee, L.N., Yee, P.L., Ariffin, H., Raha, A.R., Shirai, Y., 2013. Sustainable production of polyhydroxyalkanoates from renewable oil-palm biomass. Biomass Bioenergy 50, 1–9.

James, R., Sampath, K., Alagurathinam, S., 1996. Effects of lead on respiratory enzyme activity, glycogen and blood sugar levels of the teleost Oreochromis mossambicus (Peters) during accumulation and depuration. Asian Fish. Sci. 9, 87–100.

Jamil, Z., Mohamad Annuar, M.S., Ibrahim, S., Vikineswary, S., 2009. Optimization of phototrophic hydrogen production by Rhodopseudomonas palustris PBUM001 via statistical experimental design. Int. J. Hydrog. Energy 34, 7502–7512. https://doi.org/10.1016/j.ijhydene.2009.05.116.

Johnson, K., Kleerebezem, R., Van Loosdrecht, M.C.M., 2010. Influence of ammonium on the accumulation of polyhydroxybutyrate (PHB) in aerobic open mixed cultures. J. Biotechnol. 147, 73–79. https://doi.org/10.1016/j.jbiotec.2010.02.003.

Khalid, R., Wan Mustafa, W.A., 1992. External benefits of environmental regulation: resource recovery and the utilisation of effluents. Environmentalist 12, 277–285. https://doi.org/10.1007/BF01267698.

Leaño, E.P., Babel, S., 2012. The influence of enzyme and surfactant on biohydrogen production and electricity generation using palm oil mill effluent. J. Clean. Prod. 31, 91–99. https://doi.org/10.1016/j.jclepro.2012.02.026.

Li, W.H., Yue, Q.Y., Gao, B.Y., Wang, X.J., Qi, Y.F., Zhao, Y.Q., 2011. Preparation of sludge-based activated carbon made from paper mill sewage sludge by steam activation for dye wastewater treatment. Desalination 278, 179–185. https://doi.org/10.1016/j.desal.2011.05.020.

Liu, Z., Wang, Y., He, N., Huang, J., Zhu, K., Shao, W., 2011. Optimization of polyhydroxybutyrate (PHB) production by excess activated sludge and microbial community analysis. J. Hazard. Mater. 185, 8–16. https://doi.org/10.1016/j.jhazmat.2010.08.003.

Ma, A.N., 2000. Environmental management for the oil palm industry. Palm Oil Dev. 30, 1–10.

Ma, A.N., Singh, G., Lim, K.H., Leng, T., David, L.K., 1999. Oil Palm and the Environment: A Malaysian Perspective. Malaysia Oil Palm Growers' Council, Kuala Lumpur, pp. 113–126.

Mamimin, C., Thongdumyu, P., Hniman, A., Prasertsan, P., Imai, T., Thong, S.O., 2012. Simultaneous thermophilic hydrogen production and phenol removal from palm oil mill effluent by Thermoanaerobacterium-rich sludge. Int. J. Hydrog. Energy. https://doi.org/10.1016/j.ijhydene.2012.04.062.

Manish, S., Banerjee, R., 2008. Comparison of biohydrogen production processes. Int. J. Hydrog. Energy 33, 279–286. https://doi.org/10.1016/j.ijhydene.2007.07.026.

Metcalf Eddy, M.I., 2003. Wastewater Engineering Treatment and Reuse, fourth ed. McGraw Hill.

Mohammadi, M., Hassan, M.A., Phang, L.Y., Shirai, Y., Man, H.C., Ariffin, H., 2012. Intracellular polyhydroxyalkanoates recovery by cleaner halogen-free methods towards zero emission in the palm oil mill. J. Clean. Prod. 37, 353–360. https://doi.org/10.1016/j.jclepro.2012.07.038.

Moita, R., Lemos, P.C., 2012. Biopolymers production from mixed cultures and pyrolysis by-products. J. Biotechnol. 157, 578–583. https://doi.org/10.1016/j.jbiotec.2011.09.021.

Morgan-Sagastume, M.F., Karlsson, A., Johansson, P., Pratt, S., Boon, N., Lant, P., 2010. Production of polyhydroxyalkanoates in open, mixed cultures from a waste sludge stream containing high levels of soluble organics, nitrogen and phosphorus. Water Res. 44, 5196–5211. https://doi.org/10.1016/j.watres.2010.06.043.

MPOB, 2021. Malaysia palm oil board industry website. http://www.mpob.gov.my. (Accessed 10 Oct 2020).

Muhrizal, S., Shamsuddin, J., Fauziah, I., Husni, M.A.H., 2006. Changes in iron-poor acid sulfate soil upon submergence. Geoderma 131, 110–122. https://doi.org/10.1016/j.geoderma.2005.03.006.

Mumtaz, T., Yahaya, N.A., Abd-Aziz, S., Rahman, N.A.A., Yee, P.L., Shirai, Y., 2010. Turning waste to wealth-biodegradable plastics polyhydroxyalkanoates from palm oil mill effluent – a Malaysian perspective. J. Clean. Prod. 18, 1393–1402. https://doi.org/10.1016/j.jclepro.2010.05.016.

Nahrul, H.Z., Astimar, A.A., Juferi, I., Ropandi, M., Mohd Ali, H., Ezyana, K.B., Suriani, A.A., 2017. Microwave assisted pre-carbonisation of palm kernel shell produced charcoal with high heating value and low gas emission. J. Clean. Prod. 142 (4), 2945–2949. https://doi.org/10.1016/j.jclepro.2016.10.176.

Nahrul, H.Z., Astimar, A.A., Juferi, I., Nor Faizah, J., Ropandi, M., Mohamad Faizail, I., Mohd Ali, H., Suriani, A.A., 2018. Reduction of POME final discharge residual using activated bio absorbant from oil palm kernel shell. J. Clean. Prod. 182, 830–837.

Najafpour, G.D., Zinatizadeh, A.A.L., Mohamed, A.R., Hasnain Isa, M., H. Nasrollahzadeh., 2006. High-rate anaerobic digestion of palm oil mill effluent in an upflow anaerobic sludge-fixed film bioreactor. Process Biochem. 41, 370–379. https://doi.org/10.1016/j.procbio.2005.06.031.

Pijuan, M., Werner, U., Yuan, Z., 2009. Effect of long term anaerobic and intermittent anaerobic/aerobic starvation on aerobic granules. Water Res. 43, 3622–3632. https://doi.org/10.1016/j.watres.2009.05.007.

Pijuan, M., Werner, U., Yuan, Z., 2011. Reducing the startup time of aerobic granular sludge reactors through seeding floccular sludge with crushed aerobicgranules. Water Res. 45, 5075–5083. https://doi.org/10.1016/j.watres.2011.07.009.

Rasdi, Z., Mumtaz, T., Abdul, N.A., Rahman Hassan, M.A., 2012. Kinetic analysis of biohydrogen production from anaerobically treated POME in bioreactor under optimized condition. Int. J. Hydrog. Energy 37, 17724–17730. https://doi.org/10.1016/j.ijhydene.2012.08.095.

Rugayah, A.F., Astimar, A.A., Norzita, N., 2014. Preparation and characterisation of activated carbon from palm kernel shell by physical activation with steam. J. Oil Palm Res. 26 (3), 251–264.

Rupani, P.F., Singh, R.P., Ibrahim, M.H., Esa, N., 2010. Review of current palm oil mill effluent (POME) treatment methods: vermicomposting as a sustainable practice. World Appl. Sci. J. 11, 70–81.

Serafim, L.S., Lemos, P.C., Oliveira, R., Reis, M.A.M., 2004. Optimization of polyhydroxybutyrate production by mixed cultures submitted to aerobic dynamic feeding conditions. Biotechnol. Bioeng. 87, 145–160. https://doi.org/10.1002/bit.20085.

Singh, R.P., Hakimi, M.I., Esa, N., 2010. Composting of waste from palm oil mill: a sustainable waste management practice. Rev. Environ. Sci. Biotechnol. https://doi.org/10.1007/s11157-010-9199-2.

Sudesh, K., Abe, H., Doi, Y., 2000. Synthesis, structure and properties of polyhydroxyalkanoates: biological polyesters. Prog. Polym. Sci. 25, 1503–1555. https://doi.org/10.1016/S0079-6700(00)00035-6.

Thong, O.S., Prasertsan, P., Intrasungkha, N., Dhamwichukorn, S., Birkeland, N.K., 2007. Improvement of biohydrogen production and treatment efficiency on palm oil mill effluent with nutrient supplementation at thermophilic condition using an anaerobic sequencing batch reactor. Enzym. Microb. Technol. 41, 583–590. https://doi.org/10.1016/j.enzmictec.2007.05.002.

Thong, O.S., Mamimin, C., Prasertsan, P., 2011. Effect of temperature and initial pH on biohydrogen production from palm oil mill effluent: long-term evaluation and microbial community analysis. Electron. J. Biotechnol. 14, 1–9.

Tsai, J.H., Chiang, H.M., Huang, G.Y., Chiang, H.L., 2008. Adsorption characteristics of acetone, chloroform and acetonitrile on sludge-derived adsorbent, commercial granular activated carbon and activated carbon fibers. J. Hazard. Mater. 154, 1183–1191. https://doi.org/10.1016/j.jhazmat.2007.11.065.

Venkateswar Reddy, M., Venkata Mohan, S., 2012. Influence of aerobic and anoxic microenvironments on polyhydroxyalkanoates (PHA) production from food waste and acidogenic effluents using aerobic consortia. Bioresour. Technol. 103, 313–321. https://doi.org/10.1016/j.biortech.2011.09.040.

Vijayaraghavan, K., Ahmad, D., Aziz, M., 2007. Aerobic treatment of palm oil mill effluent. J. Environ. Manag. 82, 24–31. https://doi.org/10.1016/j.jenvman.2005.11.016.

Waller, J.L., Green, P.G., Loge, F.J., 2012. Mixed-culture polyhydroxyalkanoate production from olive oil mill pomace. Bioresour. Technol. 120, 285–289. https://doi.org/10.1016/j.biortech.2012.06.024.

Wang, D., Zheng, G., Zhou, L., 2012. Isolation and characterization of a nitrobenzene-degrading bacterium *Klebsiella ornithinolytica* NB1 from aerobic granular sludge. Bioresour. Technol. 110, 91–96. https://doi.org/10.1016/j.biortech.2012.01.105.

Wen, Q., Chen, Z., Tian, T., Chen, W., 2010. Effects of phosphorus and nitrogen limitation on PHA production in activated sludge. J. Environ. Sci. 22, 1602–1607. https://doi.org/10.1016/S1001-0742(09)60295-3.

Wu, C.Y., Peng, Y.Z., Wang, S.Y., Ma, Y., 2010. Enhanced biological phosphorus removal by granular sludge: from macro- to micro-scale. Water Res. 44, 807–814. https://doi.org/10.1016/j.watres.2009.10.028.

Wu, C.Y., Peng, Y.Z., Wang, R.D., Zhou, Y.X., 2012. Understanding the granulation process of activated sludge in a biological phosphorus removal sequencing batch reactor. Chemosphere 86, 767–773. https://doi.org/10.1016/j.chemosphere.2011.11.002.

Yokoi, H., Tokushige, T., Hirose, J., Hayashi, S., Takasaki, Y., 1998. H2 production from starch by a mixed culture of *Clostridium butyricum* and *Enterobacter aerogenes*. Biotechnol. Lett. 20, 143–147. https://doi.org/10.1023/A:1005372323248.

Yossan, S., Thong, S.O., Prasertsan, P., 2012. Effect of initial pH, nutrients and temperature on hydrogen production from palm oil mill effluent using thermotolerant consortia and corresponding microbial communities. Int. J. Hydrog. Energy. https://doi.org/10.1016/j.ijhydene.2012.03.151.

Zhang, Y.J., Yan, L., Qiao, X.L., Chi, L., Niu, X.J., Mei, Z.J., 2008. Integration of biological method and membrane technology in treating palm oil mill effluent. J. Environ. Sci. 20, 558–564. https://doi.org/10.1016/S1001-0742(08)62094-X.

Zou, J., Dai, Y., Wang, X., Ren, Z., Tian, C., Pan, K., 2013. Structure and adsorption properties of sewage sludge-derived carbon with removal of inorganic impurities and high porosity. Bioresour. Technol. 142, 209–217. https://doi.org/10.1016/j.biortech.2013.04.064.

PART

3

Composite panel products from oil palm biomass and its applications

12

Classification and application of composite panel products from oil palm biomass

R.A. Ilyas[a,b], S.M. Sapuan[c,d], M.S. Ibrahim[e], M.H. Wondi[f], M.N.F. Norrrahim[g], M.M. Harussani[d], H.A. Aisyah[c], M.A. Jenol[h], Z. Nahrul Hayawin[i], M.S.N. Atikah[j], R. Ibrahim[k], S.O.A. SaifulAzry[c], C.S. Hassan[l], and N.I.N. Haris[m]

[a]School of Chemical and Energy Engineering, Faculty of Engineering, Universiti Teknologi Malaysia, Johor Bahru, Malaysia, [b]Center for Advanced Composite Materials (CACM), Universiti Teknologi Malaysia, Johor Bahru, Malaysia, [c]Laboratory of Biocomposite Technology, Institute of Tropical Forestry and Forest Products (INTROP), Universiti Putra Malaysia, Serdang, Selangor, Malaysia, [d]Advanced Engineering Materials and Composites Research Centre (AEMC), Department of Mechanical and Manufacturing Engineering, Faculty of Engineering, Universiti Putra Malaysia, Serdang, Selangor, Malaysia, [e]Integrated Ganoderma Management, Plant Pathology and Biosecurity Unit, Biology and Sustainability Research (BSR) Division, MPOB, Bandar Baru Bangi, Kajang, Malaysia, [f]Faculty of Plantation and Agrotechnology, Universiti Teknologi MARA, Mukah, Sarawak, Malaysia, [g]Research Centre for Chemical Defence, National Defence University of Malaysia, Kuala Lumpur, Malaysia, [h]Department of Bioprocess Technology, Faculty of Biotechnology and Biomolecular Sciences, Universiti Putra Malaysia, Serdang, Selangor, Malaysia, [i]Engineering and Processing Division, Malaysian Palm Oil Board (MPOB), Kajang, Selangor, Malaysia, [j]Department of Chemical and Environmental Engineering, Universiti Putra Malaysia, Serdang, Selangor, Malaysia, [k]Innovation & Commercialization Division, Forest Research Institute Malaysia, Kepong, Selangor, Malaysia, [l]Mechanical Engineering Department, UCSI University, Kuala Lumpur, Malaysia, [m]Institute of Sustainable and Renewable Energy, Universiti Malaysia Sarawak, Kota Samarahan, Sarawak, Malaysia

Oil Palm Biomass for Composite Panels. https://doi.org/10.1016/B978-0-12-823852-3.00012-X
Copyright © 2022 Elsevier Inc. All rights reserved.

12.1 Introduction

The utilization of natural fiber as a reinforcement in polymer has garnered attention during the last few decades (Asyraf et al., 2020b). This is because of its low weight, density and cost, availability, sustainable materials, and less pollution during production, resulting in minimal health hazards and an eco-friendly nature (Aisyah et al., 2019; Asyraf et al., 2020c,a; Ayu et al., 2020; Nurazzi et al., 2019b; Sari et al., 2020; Syafiq et al., 2020).

The advantages of natural plant fibers have encouraged scientists and engineers to use them to reinforce polymer composites to help minimize the surplus of natural fibers as well as reduce forest source utilization (Abral et al., 2020b, 2019b; Halimatul et al., 2019a; Nadlene et al., 2016). Plant natural fibers are utilized as reinforcements in polymer composites materials to improve the structural and mechanical properties of the composite. To date, natural fibers, such as water hyacinth (Syafri et al., 2019b), sugar palm (Atiqah et al., 2019; Halimatul et al., 2019b; Hazrol et al., 2020; Maisara et al., 2019; Norizan et al., 2020; Nurazzi et al., 2020, 2019a; Rozilah et al., 2020; Sapuan et al., 2020), sugarcane (Asrofi et al., 2020a,b; Ridhwan et al., 2019a), sisal (Yorseng et al., 2020), ramie (Syafri et al., 2019a), oil palm (Ayu et al., 2020), kenaf (Aisyah et al., 2019; Mazani et al., 2019), jute (Islam et al., 2019), hemp (Arulmurugan et al., 2019), ginger (Abral et al., 2019a, 2020a), flax (Akonda et al., 2020), corn (Ibrahim et al., 2020b; Sari et al., 2020), cogon (Ridhwan et al., 2019b, 2020), *Tamarindus indica* nut powder (Kumar et al., 2020), and cassava (Ibrahim et al., 2020a) have been used as reinforcing material in polymer composites. This is because of their benefits, such as ease of processing, excellent mechanical properties, and low-cost manufacturing compared to synthetic fibers (Ilyas and Sapuan, 2020; Nadlene et al., 2018; Nazrin et al., 2020). Plant natural fibers are mainly composed of cellulose, hemicellulose, lignin, pectin, and other waxy substances. Natural fiber-reinforced polymer composites have attracted attention for use in a variety of advanced applications, ranging from plastic packaging to the biomedical industry (Atikah et al., 2019; Sanyang et al., 2018).

The oil palm is a perennial crop that starts yielding palm fruits for oil about 3 years after planting and has a continual productive life span of 25–30 years. Generally, after 25–30 years of oil production, the oil palm tree is discarded for replantation or replacement with new oil palm seedlings. The abundance of oil palm residue is connected to the large-scale production of oil palm products in Malaysia. A large amount of lignocellulosic biomass is generated from oil palm trees, such as the trunk and stem, during the replanting stage. Oil palm trunk (OPT) is comprised of parenchyma and vascular bundle, and according to Abdul Khalil et al. (2012), the amount of OPT residue was estimated to be 3 Mt/year.

Until now, the OPT has not been widely used due to its inconsistent physical and mechanical properties and is normally left to burn or decay in the plantation area. Besides that, some oil palm biomass is burned at the oil palm mills to yield oil palm ash as a partial replacement for cement in concrete, mortar, and other cementitious materials. Thus, to overcome these problems, several research works were carried out to increase the added value of this biomass by turning it into valuable products such as biofuels, furniture, pulp and paper, particleboards, polymer composites, food packaging, medium-density fiber (MDF), and plywood (Abdul Khalil et al., 2012). OPT also is used for the production of paper (Lee et al., 2016), particleboard, and furniture (Ratanawilai et al., 2006; Ratnasingam and Ioras, 2011; Suhaily et al., 2012).

To date, several experiments have been performed using EFB oil palm biomass for particulate boards and fiberboard (Abdul Karim et al., 2020; Oikarinen et al., 1998; Sawawi et al., 2020; Wahab et al., 2017), pulp (Hashim et al., 2020; Kamaludin et al., 2012, 2011; Kim et al., 2016; Kose et al., 2014; Lee and Ryu, 2016; Liu et al., 2019; Martín-Sampedro et al., 2012), packaging (Ayu et al., 2020), and polymer composites (Hanan et al., 2020b; Muhammad Amir et al., 2020; Mustapha et al., 2020; Nor Amira Izzati et al., 2020). Table 12.1 shows the different applications and products of oil palm biomass from different parts of oil palm biomass fibers.

Table 12.1 Various applications and products of oil palm biomass from different parts of oil palm biomass fibers.

Oil palm biomass	Products	References
Oil palm EFB fibers	Plywood	Abdul Khalil et al. (2010), Hoong et al. (2012)
	MDF	Anis, Ibrahim et al. (2014), Azman et al. (2015), Ibrahim et al. (2016), Izani et al. (2012), Jamaludin et al. (2007), Norul Izani et al. (2011, 2013), Ramli et al. (2002)
	Polymer biocomposite	Arif et al. (2017), Saputra et al. (2016), Valášek and Ambarita (2018)
	Hybrid composite	Abdul Karim et al. (2020), Hanan et al. (2020a)
	Particleboards	Ahmad et al. (2011)
	Biofuel	Amin et al. (2012), Asadieraghi and Daud (2015), Mabrouki et al. (2015), Solikhah et al. (2018)
	Packaging	Ayu et al. (2020)
	Pulp and paper	Hashim et al. (2020), Kamaludin et al. (2012), Kamaludin et al. (2011), Kim et al. (2016), Kose et al. (2014), Lee and Ryu (2016), Liu et al. (2019), Martín-Sampedro et al. (2012)
	Wastewater treatment	Amosa (2015), Chan et al. (2018), Septevani et al. (2020)

Continued

220 Chapter 12 Classification and application of composite panel products from oil palm biomass

Table 12.1 Various applications and products of oil palm biomass from different parts of oil palm biomass fibers—cont'd

Oil palm biomass	Products	References
Oil palm frond fibers	Pulp and Paper	Hamid (2008), Nasrullah and Mazlan (2013), Wang et al. (2012a, b)
	Nutrient	Juliantoni et al. (2018)
Oil palm trunk fibers	Fiberboard	Sihabut and Laemsak (2008)
	Biodegradable film	Haliza et al. (2006)
Palm Kernel shell	Animal feed	Dahlan (2000), Ishida and Abu Hassan (1997)
	Downdraft gasifier	Moni et al. (2018), Moni and Sulaiman (2013), Saleh et al. (2020)
	Biofuel	Omar et al. (2018)
	Butanol	Mahmud and Rosentrater (2020)
	Silica	Osman and Sapawe (2020)
	Furfural	Mohamad et al. (2020)
	Carbon fiber	Khalid et al. (2020)
	Wastewater treatment	Teow et al. (2020)
	Cellulose nanocrystal	Nordin et al. (2017)
	Lignin	Shah et al. (2017)
	Composite Board	Rasat et al. (2011)
	Polymer composite	Khalil et al. (2007)
	Particleboard	Wahida and Najmuldeen (2015)
	Biocoal	Nudri et al. (2020)
	Microencapsulation	Maulidna et al. (2020)
	Briquette	Kpalo et al. (2020)
	Veneer panel	Saari et al. (2020)
	Biochar	Razali and Kamarulzaman (2020)
	Bioplastic	Syamani et al. (2020)
	Particleboard	Lee et al. (2020), Amini et al. (2020)
	Methane	Nutongkaew et al. (2020)
	Paper production	Phruksaphithak and Wangprayot (2020)
	Syrup	Sulaiman et al. (2020)
	Cellulose film	Phruksaphithak et al. (2020)
	Bioethanol	Sungpichai and Srinophakun (2020)
	Fuel	Umar et al. (2020), Shrivastava et al. (2020)
	Geopolymer Concrete	Malkawi et al. (2020)
	Plywood	Nuryawan et al. (2020)
	Furniture	Siti Suhaily et al. (2019)
	Fluid loss control agent	Sauki et al. (2020)

Table 12.1 Various applications and products of oil palm biomass from different parts of oil palm biomass fibers—cont'd

Oil palm biomass	Products	References
	Concrete	Alengaram et al. (2013), Buari et al. (2020), Chin et al. (2020), Fanijo et al. (2020), Fapohunda et al. (2020), Oluwasola et al. (2020), Ting et al. (2020), Uchechukwu and Austin (2020)
	Water treatment	Baby and Hussein (2019), Okoroigwe et al. (2014)
	Activated carbon	Abechi et al. (2013), Affam (2020)
	Biochar solid fuel	Amin et al. (2020), Bazargan et al. (2014)
	Polymer composite	Dagwa et al. (2012), Oladele et al. (2020)
	Hydrogen production	Acevedo-Páez et al. (2020)
	Silica nanoparticle	Imoisili et al. (2020)
	Super capacitor	Ayinla et al. (2020)
	Briquette	Mohd Fuad et al. (2020)
	Fuel	Nayaggy and Putra (2019)
	Bio oil	Qarizada et al. (2019)
	Composite board	Zamri et al. (2019)
	Brake pad	Ravikumar and Pridhar (2019)
	Lignin	Chang et al. (2016)
Mesocarp fibers	Cellulose nanocrystal	Chieng et al. (2017)
	Composite	Ahamad Nordin et al. (2016), Campos et al. (2018), Campos et al. (2017), Laftah and Majid (2019), Nordin et al. (2015)
	Cement	Fatina Md Sali and Deraman (2019)
	Biogas production	Saidu et al. (2020)
	Foam	Kormin et al. (2018)
	Sound absorber panel	Latif et al. (2020)
Palm oil mills effluent	Organic fertilizer	Khairuddin et al. (2016)
	Biogas	Aziz et al. (2020), Chin et al. (2013), Choong et al. (2018), Mohtar et al. (2017), Norfadilah et al. (2016), Shakib and Rashid (2019), Uddin et al. (2019)
	Biodegradable plastics	Mumtaz et al. (2010)
	Hydrogen and methane production	Krishnan et al. (2016)
	Biodiesel	Suwanno et al. (2017)
	biosynthesis of silver nanoparticles (AuNps) from gold precursor	Gan et al. (2012)
	Biohydrogen	Vijayaraghavan and Ahmad (2006)

12.2 Oil palm biomass for molded particleboard

According to Hassan and Sukaimi (1993), OPT and oil palm fronds (OPF) have been widely used as raw materials to manufacture molded particleboard. Particleboard mass production using 18% resin content and density of 700% kgm^{-3} had adequate strength, hence, it passed the British Standard BS 5669:1979 Type 1. The procedure of manufacturing molded particleboard is represented in Fig. 12.1.

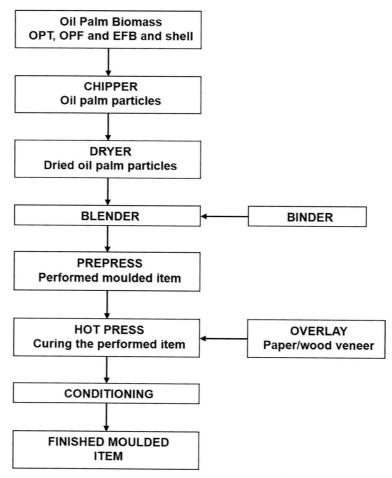

Fig. 12.1 The procedure flow for oil palm molded particleboard manufacturing. Source: Mokhtar, A., Hassan, K., Aziz, A.A., May, C.Y., 2012. Oil palm biomass for various wood-based products. In: Palm Oil. AOCS Press, pp. 625–652.

12.3 Oil palm plywood

Oil palm plywood is widely used for nonstructural materials, such as cabinets and packaging materials, due to the weaker and lower durability of the OPT veneers. Plywood is composed of layers of veneer glued crossways to each other and pressed into a harder sheet. The OPTs were processed under several stages: felling and cutting, debarking, peeling, drying, gluing, hot-pressing, and trimming.

Felling and cross-cutting are applied at the plantation (Fig. 12.2), where the OPT logs are transported and collected into a storage yard. The logs are grouped into the bottom and upper portion for an easier peeling process (Fig. 12.3). Next, the outer bark of the trunks is removed using a rotary cutter prior to veneer production via the peeling

Fig. 12.2 Felling and cutting of oil palm tree during replanting. Extracted with permission from Mokhtar, A., Hassan, K., Aziz, A.A., May, C.Y., 2012. Oil palm biomass for various wood-based products. In: Palm Oil. AOCS Press, pp. 625–652.

Fig. 12.3 OPT grouped into upper and bottom portions were stored in the storage yard. Extracted with permission from Mokhtar, A., Hassan, K., Aziz, A.A., May, C.Y., 2012. Oil palm biomass for various wood-based products. In: Palm Oil. AOCS Press, pp. 625–652.

Fig. 12.4 Debarking of OPT using a rotary lathe machine. Extracted with permission from Mokhtar, A., Hassan, K., Aziz, A.A., May, C.Y., 2012. Oil palm biomass for various wood-based products. In: Palm Oil. AOCS Press, pp. 625–652.

process, as shown in Fig. 12.4. A spindleless rotary lathe machine was used to peel the OPT logs into veneers. During the peeling process, the controlled parameter of the lathe knife angle needs to be altered to provide uniformity of veneer thickness of the peeled log. The thickness of the veneer differs between the outer and inner portions of the OPT. The peeled veneer was cut to 0.9 or 1.2 m (about 4 ft) for easy handling (Fig. 12.5).

A conventional mechanical dryer was used to dry up the veneers (Fig. 12.6A). The temperature applied was in the range of 130–150°C. Steam was also supplied to control the moisture content to remain within 3%–5%. Next, the veneers were exposed to room temperature prior to the gluing stage (Fig. 12.6B). A glue spreader was used

Fig. 12.5 OPT without bark and ready to be peeled into veneers. Extracted with permission from Mokhtar, A., Hassan, K., Aziz, A.A., May, C.Y., 2012. Oil palm biomass for various wood-based products. In: Palm Oil. AOCS Press, pp. 625–652.

Fig. 12.6 (A) Drying of OPT veneers; (B) gluing of OPT veneers. Extracted with permission from Mokhtar, A., Hassan, K., Aziz, A.A., May, C.Y., 2012. Oil palm biomass for various wood-based products. In: Palm Oil. AOCS Press, pp. 625–652.

in this phase, and the glued veneers were pressed using cold- and hot-pressing machines to produce oil palm plywood with an operating temperature ranging from 120°C to 150°C. The applied pressure was 6–15 kg cm^{-3} (Fig. 12.7A).

The produced plywood is taken into the finishing process, including edge trimming using sandpaper grit P-60 (Fig. 12.7B). This process is essential to smooth the plywood surface. According to Mokhtar et al. (2012), about 47.4% from 1 m^3 of OPT were utilized for plywood production.

Fig. 12.7 (A) Pressing of OPT plywood; (B) finished products of OPT plywood. Extracted with permission from Mokhtar, A., Hassan, K., Aziz, A.A., May, C.Y., 2012. Oil palm biomass for various wood-based products. In: Palm Oil. AOCS Press, pp. 625–652.

12.4 Oil palm biomass for medium-density fiberboard (MDF)

Commonly, MDF is made of rubberwood, worked as its raw materials. MDF is usually utilized as a nonstructural fiber-based panel composed of wood fibers bonded with resin under specified heat and pressure. Before the refining process, the OPTs were chipped into smaller fiber pieces. The oil palm fibers were refined using a defibrator or refiner. Next, through the glue blending process, the fibers were mixed with resin and wax as a bonding agent to coat the surface of the fiber (Oikarinen et al., 1998). The resonated fibers are dried and afterward shaped into mats, which are eventually hot-pressed to form a homogeneous board. The oil palm biomass-based MDF production method flow is presented in Fig. 12.8.

12.5 Oil palm lumber as a substitution for wood timber

OPT lumber has been broadly used as a substitute for rubberwood timber either in the production of furniture, interior millwork, or kitchen woodenware (Mokhtar et al., 2008). Based on Mokhtar et al. (2012), OPT could be transformed into lumber via air drying and kiln drying. OPT is a strongly anisotropic product, as its strength and stiffness properties are perpendicular to grain lumber and often shrinks and swells with differing moisture contents.

Oil palm lumber production is much simpler than oil palm plywood. Firstly, the OPT logs are collected during replanting and cut into shorter logs of about 10 or 18 ft. before the sawing process. Next, sawing machines (Fig. 12.9) are used to carry out the operation that requires no machining property control. This is owing to the rough machined surfaces of OPT lumbers, which are mainly due to the raised vascular bundles or tearing of the bundles. Finally, the lumbers were brought to the drying stage via air drying or kiln drying. Different drying methods affected the quality of the lumber boards due to the drying defects that occurred, such as raised grain, warping, and collapse, as shown in Fig. 12.10.

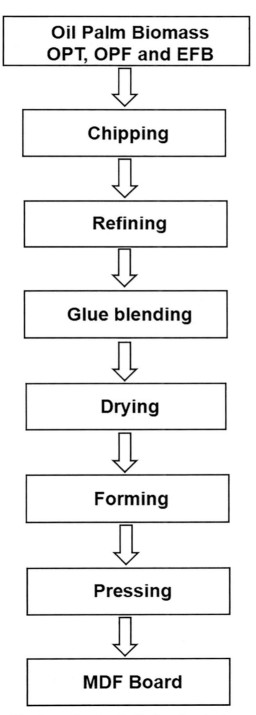

Fig. 12.8 Method flow for medium-density fiberboard production.

Fig. 12.9 Sawing of oil palm trunk into oil palm lumbers. Extracted with permission from Mokhtar, A., Hassan, K., Aziz, A.A., May, C.Y., 2012. Oil palm biomass for various wood-based products. In: Palm Oil. AOCS Press, pp. 625–652.

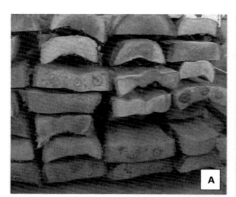 

Fig. 12.10 Drying defects on sawn OPT lumbers: (A) collapse, and (B) cracking effects. Extracted with permission from Mokhtar, A., Hassan, K., Aziz, A.A., May, C.Y., 2012. Oil palm biomass for various wood-based products. In: Palm Oil. AOCS Press, pp. 625–652.

12.6 Conclusions

In term of waste generation, oil palm biomass can be used as a valuable by-product by transforming these biomasses into valuable products such as biofuels, pellets, food packaging, biochar, filler in composites materials, furniture, pulp and paper, particleboards, polymer composites, MDF, plywood, brake pads, foam, and cement concrete.

References

Abdul Karim, M.H., Mohd Shah, M.K., Jundam, M.F., Abdullah, S., 2020. Investigation of tensile properties of the eco-board of hybrid composite that consist of oil palm empty fruit bunch (OPEFB) fiber added with rice husk. IOP Conf. Ser. Mater. Sci. Eng. 834. https://doi.org/10.1088/1757-899X/834/1/012012, 012012.

Abdul Khalil, H.P.S., Nurul Fazita, M.R., Bhat, A.H., Jawaid, M., Nik Fuad, N.A., 2010. Development and material properties of new hybrid plywood from oil palm biomass. Mater. Des. https://doi.org/10.1016/j.matdes.2009.05.040.

Abdul Khalil, H.P.S., Jawaid, M., Hassan, A., Paridah, M.T., Zaidon, A., 2012. Oil palm biomass fibres and recent advancement in oil palm biomass fibres based hybrid biocomposites. In: Composites and Their Applications. InTech, pp. 187–220, https://doi.org/10.5772/48235.

Abechi, S.E., Gimba, C.E., Uzairu, A., Dallatu, Y.A., 2013. Preparation and characterization of activated carbon from palm kernel shell by chemical activation. Res. J. Chem. Sci. 3, 54–61.

Abral, H., Ariksa, J., Mahardika, M., Handayani, D., Aminah, I., Sandrawati, N., Sapuan, S.M., Ilyas, R.A., 2019a. Highly transparent and antimicrobial PVA based bionanocomposites reinforced by ginger nanofiber. Polym. Test. https://doi.org/10.1016/j.polymertesting.2019.106186, 106186.

Abral, H., Basri, A., Muhammad, F., Fernando, Y., Hafizulhaq, F., Mahardika, M., Sugiarti, E., Sapuan, S.M., Ilyas, R.A., Stephane, I., 2019b. A simple method for improving the properties of the sago starch films prepared by using ultrasonication treatment. Food Hydrocoll. 93, 276–283. https://doi.org/10.1016/j.foodhyd.2019.02.012.

Abral, H., Ariksa, J., Mahardika, M., Handayani, D., Aminah, I., Sandrawati, N., Pratama, A.B., Fajri, N., Sapuan, S.M., Ilyas, R.A., 2020a. Transparent and antimicrobial cellulose film from ginger nanofiber. Food Hydrocoll. 98. https://doi.org/10.1016/j.foodhyd.2019.105266, 105266.

Abral, H., Atmajaya, A., Mahardika, M., Hafizulhaq, F., Kadriadi Handayani, D., Sapuan, S.M., Ilyas, R.A., 2020b. Effect of ultrasonication duration of polyvinyl alcohol (PVA) gel on characterizations of PVA film. J. Mater. Res. Technol. 9 (2), 2477–2486. https://doi.org/10.1016/j.jmrt.2019.12.078.

Acevedo-Páez, J.C., Durán, J.M., Posso, F., Arenas, E., 2020. Hydrogen production from palm kernel shell: kinetic modeling and simulation. Int. J. Hydrog. Energy. https://doi.org/10.1016/j.ijhydene.2019.10.146.

Affam, A.C., 2020. Conventional steam activation for conversion of oil palm kernel shell biomass into activated carbon via biochar product. Glob. J. Environ. Sci. Manag. https://doi.org/10.22034/gjesm.2020.01.02.

Ahamad Nordin, N.I.A., Ariffin, H., Hassan, M.A., Shirai, Y., Ando, Y., Ibrahim, N.A., Wan Yunus, W.M.Z., 2016. Superheated steam treatment of oil palm mesocarp fiber improved the properties of fiber-polypropylene biocomposite. Bioresources 12 (1). https://doi.org/10.15376/biores.12.1.68-81.

Ahmad, N., Kasim, J., Mahmud, S.Z., Yamani, S.A.K., Mokhtar, A., Yunus, N.Y.M., 2011. Manufacture and properties of oil palm particleboard. In: 2011 3rd International Symposium & Exhibition in Sustainable Energy & Environment (ISESEE), pp. 84–87, https://doi.org/10.1109/ISESEE.2011.5977115.

Aisyah, H.A., Paridah, M.T., Sapuan, S.M., Khalina, A., Berkalp, O.B., Lee, S.H., Lee, C.H., Nurazzi, N.M., Ramli, N., Wahab, M.S., Ilyas, R.A., 2019. Thermal properties of woven Kenaf/carbon fibre-reinforced epoxy hybrid composite panels. Int. J. Polym. Sci. 2019, 1–8. https://doi.org/10.1155/2019/5258621.

Akonda, M.H., Shah, D.U., Gong, R.H., 2020. Natural fibre thermoplastic tapes to enhance reinforcing effects in composite structures. Compos. A: Appl. Sci. Manuf. 131, 1–8. https://doi.org/10.1016/j.compositesa.2020.105822.

Alengaram, U.J., Al Muhit, B.A., MZB, J., 2013. Utilization of oil palm kernel shell as lightweight aggregate in concrete - a review. Constr. Build. Mater. https://doi.org/10.1016/j.conbuildmat.2012.08.026.

Amin, N., Misson, M., Haron, R., 2012. Bio-oils and diesel fuel derived from alkaline treated empty fruit bunch (Efb). Int. J. Biomass Renew. 1 (1), 6–14.

Amin, S., Bachmann, R.T., Yong, S.K., 2020. Oxidised biochar from palm kernel shell for eco-friendly pollution management. Sci. Res. J. https://doi.org/10.24191/srj.v17i2.10001.

Amini, M.H.M., Sulaiman, N.S., Bakri, M.A., Abu Bakar, M.B., Mohamed, M., Masri, M.N., Nik Yusuf, N.A.A., 2020. Particleboard based on glutardialdehyde treated oil palm trunk particles made at different pressing temperatures. Mater. Sci. Forum 1010, 483–488. https://doi.org/10.4028/www.scientific.net/MSF.1010.483.

Amosa, M.K., 2015. Process optimization of Mn and H2S removals from POME using an enhanced empty fruit bunch (EFB)-based adsorbent produced by pyrolysis. Environ. Nanotechnol. Monit. Manag. 4, 93–105. https://doi.org/10.1016/j.enmm.2015.09.002.

Anis Ibrahim, Z., Aziz, A.A., Ramli, R., Tabar Mokhtar Lee, S., Omar, R., Sirajuddin, A., 2014. Production of medium density fibreboard (mdf) from empty fruit bunch (efb) with wood fibres mix. In: Mpob Information Series.

Arif, Z., Adlie, T.A., Amir, F., Thalib, S., Ali, N., Nazaruddin, N., Mustafa, M., 2017. Tensile loading on composite polymeric foam reinforced by empty fruit bunch waste (EFB). In: International Conference on Science, Technology and Modern Society.

Arulmurugan, M., Prabu, K., Rajamurugan, G., Selvakumar, A.S., 2019. Impact of $BaSO_4$ filler on woven aloevera/hemp hybrid composite: dynamic mechanical analysis. Mater. Res. Express 6 (4). https://doi.org/10.1088/2053-1591/aafb88, 045309.

Asadieraghi, M., Daud, W.M.A.W., 2015. In-depth investigation on thermochemical characteristics of palm oil biomasses as potential biofuel sources. J. Anal. Appl. Pyrolysis 115, 379–391. https://doi.org/10.1016/j.jaap.2015.08.017.

Asrofi, M., Sapuan, S.M., Ilyas, R.A., Ramesh, M., 2020a. Characteristic of composite bioplastics from tapioca starch and sugarcane bagasse fiber: effect of time duration of ultrasonication (bath-type). Mater. Today Proc. https://doi.org/10.1016/j.matpr.2020.07.254.

Asrofi, M., Sujito, Syafri, E., Sapuan, S.M., Ilyas, R.A., 2020b. Improvement of biocomposite properties based tapioca starch and sugarcane bagasse cellulose nanofibers. Key Eng. Mater. 849, 96–101. https://doi.org/10.4028/www.scientific.net/KEM.849.96.

Asyraf, M.R.M., Ishak, M.R., Sapuan, S.M., Yidris, N., Ilyas, R.A., 2020a. Woods and composites cantilever beam: a comprehensive review of experimental and numerical creep methodologies. J. Mater. Res. Technol. 9 (3), 6759–6776. https://doi.org/10.1016/j.jmrt.2020.01.013.

Asyraf, M.R.M., Ishak, M.R., Sapuan, S.M., Yidris, N., Ilyas, R.A., Rafidah, M., Razman, M.R., 2020b. Potential application of green composites for cross arm component in transmission tower: a brief review. Int. J. Polym. Sci. 2020, 1–15. https://doi.org/10.1155/2020/8878300.

Asyraf, M.R.M., Rafidah, M., Ishak, M.R., Sapuan, S.M., Ilyas, R.A., Razman, M.R., 2020c. Integration of TRIZ, morphological chart and ANP method for development of FRP composite portable fire extinguisher. Polym. Compos. 1–6. https://doi.org/10.1002/pc.25587.

Atikah, M.S.N., Ilyas, R.A., Sapuan, S.M., Ishak, M.R., Zainudin, E.S., Ibrahim, R., Atiqah, A., Ansari, M.N.M., Jumaidin, R., 2019. Degradation and physical properties of sugar palm starch/sugar palm nanofibrillated cellulose bionanocomposite. Polimery 64 (10), 27–36. https://doi.org/10.14314/polimery.2019.10.5.

Atiqah, A., Jawaid, M., Sapuan, S.M., Ishak, M.R., Ansari, M.N.M., Ilyas, R.A., 2019. Physical and thermal properties of treated sugar palm/glass fibre reinforced thermoplastic polyurethane hybrid composites. J. Mater. Res. Technol. 8 (5), 3726–3732. https://doi.org/10.1016/j.jmrt.2019.06.032.

Ayinla, R.T., John, O.D., Hasnah, M.T.M.Z., Fahad, U., Asfand, Y., 2020. Effect of particle size on the physical properties of activated palm kernel shell for supercapacitor application. Key Eng. Mater. https://doi.org/10.4028/www.scientific.net/KEM.833.129.

Ayu, R.S., Khalina, A., Harmaen, A.S., Zaman, K., Isma, T., Liu, Q., Ilyas, R.A., Lee, C.H., 2020. Characterization study of empty fruit bunch (EFB) fibers reinforcement in poly(butylene) succinate (PBS)/starch/glycerol composite sheet. Polymers 12 (7), 1571. https://doi.org/10.3390/polym12071571.

Aziz, M.M.A., Kassim, K.A., ElSergany, M., Anuar, S., Jorat, M.E., Yaacob, H., Ahsan, A., Imteaz, M.A., Arifuzzaman, 2020. Recent advances on palm oil mill effluent (POME) pretreatment and anaerobic reactor for sustainable biogas production. Renew. Sust. Energ. Rev. https://doi.org/10.1016/j.rser.2019.109603.

Azman, A.M., Badri, K.H., Baharum, A., 2015. Effect of fibre aspect ratio onto the modulus of palm-based medium-density fibreboard. AIP Conf. Proc. 050011. https://doi.org/10.1063/1.4931290.

Baby, R., Hussein, M.Z., 2019. Application of palm kernel shell as bio adsorbent for the treatment of heavy metal contaminated water. J. Adv. Res. Appl. Mech. 60 (1), 10–16.

Bazargan, A., Rough, S.L., McKay, G., 2014. Compaction of palm kernel shell biochars for application as solid fuel. Biomass Bioenergy. https://doi.org/10.1016/j.biombioe.2014.08.015.

Buari, T.A., Olutoge, F.A., Dada, S.A., Ademola, S.A., Ayankunle, R.A., 2020. Sustainability of palm kernel shell ash (PKSA) as SCM in self-consolidating concrete (SCC) design. Int. J. Eng. Res. Technol. https://doi.org/10.17577/ijertv9is040462.

Campos, A., Sena Neto, A.R., Rodrigues, V.B., Luchesi, B.R., Moreira, F.K.V., Correa, A.C., LHC, M., Marconcini, J.M., 2017. Bionanocomposites produced from cassava starch and oil palm mesocarp cellulose nanowhiskers. Carbohydr. Polym. https://doi.org/10.1016/j.carbpol.2017.07.080.

Campos, A., Sena Neto, A.R., Rodrigues, V.B., Luchesi, B.R., Mattoso, L.H.C., Marconcini, J.M., 2018. Effect of raw and chemically treated oil palm mesocarp fibers on thermoplastic cassava starch properties. Ind. Crop. Prod. 124, 149–154. https://doi.org/10.1016/j.indcrop.2018.07.075.

Chan, C.M., Chin, S.X., Chook, S.W., Chia, C.H., Zakaria, S., 2018. Combined mechanical-chemical pre-treatment of oil palm empty fruit bunch (EFB) fibers for adsorption of methylene blue (MB) in aqueous solution. Malays. J. Anal. Sci. 22 (6), 1007–1013. https://doi.org/10.17576/mjas-2018-2206-10.

Chang, G.Z., Xie, J.J., Yang, H.K., Huang, Y.Q., Yin, X.L., Wu, C.Z., 2016. Structure and pyrolysis characteristics of enzymatic/mild acidolysis lignin isolated from palm kernel shell. J. Fuel Chem. Technol. 44, 1185–1194.

Chieng, B.W., Lee, S.H., Ibrahim, N.A., Then, Y.Y., Loo, Y.Y., 2017. Isolation and characterization of cellulose nanocrystals from oil palm mesocarp fiber. Polymers. https://doi.org/10.3390/polym9080355.

Chin, M.J., Poh, P.E., Tey, B.T., Chan, E.S., Chin, K.L., 2013. Biogas from palm oil mill effluent (POME): opportunities and challenges from Malaysia's perspective. Renew. Sust. Energ. Rev. https://doi.org/10.1016/j.rser.2013.06.008.

Chin, C.O., Yang, X., Paul, S.C., Susilawati, Wong, L.S., Kong, S.Y., 2020. Development of thermal energy storage lightweight concrete using paraffin-oil palm kernel shell-activated carbon composite. J. Clean. Prod. https://doi.org/10.1016/j.jclepro.2020.121227.

Choong, Y.Y., Chou, K.W., Norli, I., 2018. Strategies for improving biogas production of palm oil mill effluent (POME) anaerobic digestion: a critical review. Renew. Sust. Energ. Rev. 82, 2993–3006. https://doi.org/10.1016/j.rser.2017.10.036.

Dagwa, I.M., Builders, P.F., Achebo, J., 2012. Characterization of palm kernel shell powder for use in polymer matrix composites. Int. J. Mech. Mechatron. Eng. 12 (4), 88–93.

Dahlan, I., 2000. Oil palm frond, a feed for herbivores. Asian Australas. J. Anim. Sci. 13 (13), 300–303.

Fanijo, E., Babafemi, A.J., Arowojolu, O., 2020. Performance of laterized concrete made with palm kernel shell as replacement for coarse aggregate. Constr. Build. Mater. https://doi.org/10.1016/j.conbuildmat.2020.118829.

Fapohunda, C., Bello, H., Salako, T., Tijani, S., 2020. Strength, micro-structure & durability investigations of lateritic concrete with palm kernel shell (PKS) as partial replacement of coarse aggregates. Eng. Rev. https://doi.org/10.30765/er.40.2.07.

Fatina Md Sali, N., Deraman, R., 2019. The selection of optimum water-cement ratio for production of low thermal conductivity cement sand brick with oil palm mesocarp fibre as admixture. IOP Conf. Ser. Mater. Sci. Eng. 601. https://doi.org/10.1088/1757-899X/601/1/012037, 012037.

Gan, P.P., Ng, S.H., Huang, Y., Li, S.F.Y., 2012. Green synthesis of gold nanoparticles using palm oil mill effluent (POME): a low-cost and eco-friendly viable approach. Bioresour. Technol. https://doi.org/10.1016/j.biortech.2012.01.015.

Halimatul, M.J., Sapuan, S.M., Jawaid, M., Ishak, M.R., Ilyas, R.A., 2019a. Effect of sago starch and plasticizer content on the properties of thermoplastic films: mechanical testing and cyclic soaking-drying. Polimery 64 (6), 32–41. https://doi.org/10.14314/polimery.2019.6.5.

Halimatul, M.J., Sapuan, S.M., Jawaid, M., Ishak, M.R., Ilyas, R.A., 2019b. Water absorption and water solubility properties of sago starch biopolymer composite films filled with sugar palm particles. Polimery 64 (9), 27–35. https://doi.org/10.14314/polimery.2019.9.4.

Haliza, A.H.N., Fazilah, A., Azemi, M.N., 2006. Development of hemicelluloses biodegradable films from oil palm frond (Elais Guineensis). In: International Conference on Green Adn Sustainable Innovation, Chiang Mai, Thailand.

Hamid, A.K.A., 2008. Production of Cellulose Fiber From Oil Palm Frond Using Steam Explosion Method (Thesis). Universiti Malaysia Pahang. http://umpir.ump.edu.my/id/eprint/697.

Hanan, F., Jawaid, M., Paridah, M.T., 2020a. Mechanical performance of oil palm/kenaf fiber-reinforced epoxy-based bilayer hybrid composites. J. Nat. Fibers 17 (2), 155–167. https://doi.org/10.1080/15440478.2018.1477083.

Hanan, F., Jawaid, M., Paridah, M.T., Naveen, J., 2020b. Characterization of hybrid oil palm empty fruit bunch/woven kenaf fabric-reinforced epoxy composites. Polymers 12 (9), 2052. https://doi.org/10.3390/polym12092052.

Hashim, S.N.A.S., Norizan, B.A.-Z., Baharin, K.W., Zakaria, S., Chia, C.H., Potthast, A., Schiehser, S., Bacher, M., Rosenau, T., Jaafar, S.N.S., 2020. In-depth characterization of cellulosic pulps from oil palm empty fruit bunches and kenaf core, dissolution and preparation of cellulose membranes. Cellul. Chem. Technol. 54 (7–8), 643–652. https://doi.org/10.35812/CelluloseChemTechnol.2020.54.63.

Hassan, K., Sukaimi, J., 1993. Industrial moulding of oil palm particles I. Suitability of oil palm trunk and frond for moulded table-tops. PORIM Bull. 27, 1–7.

Hazrol, M.D., Sapuan, S.M., Ilyas, R.A., Othman, M.L., Sherwani, S.F.K., 2020. Electrical properties of sugar palm nanocrystalline cellulose reinforced sugar palm starch nanocomposites. Polimery 65 (5), 363–370. https://doi.org/10.14314/polimery.2020.5.4.

Hoong, Y.B., Loh, Y.F., Nor Hafizah, A.W., Paridah, M.T., Jalaluddin, H., 2012. Development of a new pilot scale production of high grade oil palm plywood: effect of pressing pressure. Mater. Des. 36, 215–219. https://doi.org/10.1016/j.matdes.2011.10.004.

Ibrahim, Z., Ahmad, M., Aziz, A.A., Ramli, R., Jamaludin, M.A., Muhamed, S., Alias, A.H., 2016. Dimensional stability properties of medium density fibreboard (MDF) from treated oil palm empty fruit bunches (EFB) fibres. Open J. Compos. Mater. 6 (4), 91–99. https://doi.org/10.4236/ojcm.2016.64009.

Ibrahim, M.I., Edhirej, A., Sapuan, S.M., Jawaid, M., Ismarrubie, N.Z., Ilyas, R.A., 2020a. Extraction and characterization of Malaysian cassava starch, peel, and bagasse, and selected properties of the composites. In: Jumaidin, R., Sapuan, S.M., Ismail, H. (Eds.), Biofiller-Reinforced Biodegradable Polymer Composites, first ed. CRC Press, pp. 267–283.

Ibrahim, M.I., Sapuan, S.M., Zainudin, E.S., Zuhri, M.Y., Edhirej, A., Ilyas, R.A., 2020b. Characterization of corn fiber-filled cornstarch biopolymer composites. In: Jumaidin, R., Sapuan, S.M., Ismail, H. (Eds.), Biofiller-Reinforced Biodegradable Polymer Composites, first ed. CRC Press, pp. 285–301.

Ilyas, R.A., Sapuan, S.M., 2020. The preparation methods and processing of natural fibre bio-polymer composites. Curr. Org. Synth. 16 (8), 1068–1070. https://doi.org/10.2174/157017941608200120105616.

Imoisili, P.E., Ukoba, K.O., Jen, T.C., 2020. Green technology extraction and characterisation of silica nanoparticles from palm kernel shell ash via sol-gel. J. Mater. Res. Technol. https://doi.org/10.1016/j.jmrt.2019.10.059.

Ishida, M., Abu Hassan, O., 1997. Utilization of oil palm frond as cattle feed. Jpn. Agric. Res. Q. 31 (1), 41–47.

Islam, F., Islam, N., Shahida, S., Karmaker, N., Koly, F.A., Mahmud, J., Keya, K.N., Khan, R.A., 2019. Mechanical and interfacial characterization of jute fabrics reinforced unsaturated polyester resin composites. Nano Hybrids Compos. 25, 22–31. https://doi.org/10.4028/www.scientific.net/NHC.25.22.

Izani, M.A.N., Paridah, M.T., Astimar, A.A., Nor, M.Y.M., Anwar, U.M.K., 2012. Mechanical and dimensional stability properties of medium-density fibreboard produced from treated oil palm empty fruit bunch. J. Appl. Sci. 12 (6), 561–567. https://doi.org/10.3923/jas.2012.561.567.

Jamaludin, M.A., Nordin, K., Ahmad, M., 2007. The bending strength of medium density Fibreboard (MDF) from different ratios of Kenaf and oil palm empty fruit bunches (EFB) admixture for light weight construction. Key Eng. Mater., 77–80. https://doi.org/10.4028/0-87849-427-8.77.

Juliantoni, J., Zain, M., Ryanto, I., Elihasridas, Kasrad, 2018. Peformance of Bali cattles fed complete feed based on oil palm frond that is added with rumen microbes growth factor (Rmgf). Asian J. Microbiol. Biotechnol. Environ. Sci. 14 (344), 1–9.

Kamaludin, N.H., Ghazali, A., Daud, W.R.W., 2011. Evidence for paper strength improvement by inclusion of fines generated from APMP of EFB. J. Sustain. Sci. Manag. 6 (2), 267–274.

Kamaludin, N.H., Ghazali, A., Daud, W.W., 2012. Potential of fines as reinforcing fibres in alkaline peroxide pulp of oil palm empty fruit bunch. Bioresources. https://doi.org/10.15376/biores.7.3.3425-3438.

Khairuddin, M.N., Zakaria, A.J., Md Isa, I., Jol, H., Rahman, W.M.N.W.A., Salleh, M.K.S., 2016. The potential of treated palm oil mill effluent (POME) sludge as an organic fertilizer. AGRIVITA. J. Agric. Sci. 38 (2). https://doi.org/10.17503/agrivita.v38i2.753.

Khalid, K.A., Karunakaran, V., Ahmad, A.A., Pa'ee, K.F., Abd-Talib, N., Yong, T.L.K., 2020. Lignin from oil palm frond under subcritical phenol conditions as a precursor for carbon fiber production. Malays. J. Anal. Sci. https://doi.org/10.1016/j.matpr.2020.01.252.

Khalil, H.P.S.A., Kumar, R.N., Asri, S.M., Fuaad, N.A.N., Ahmad, M.N., 2007. Hybrid thermoplastic pre-preg oil palm frond fibers (OPF) reinforced in polyester composites. Polym.-Plast. Technol. Eng. https://doi.org/10.1080/03602550600948749.

Kim, D.-S., Sung, Y.J., Kim, C.-H., Kim, S.-B., 2016. Changes in the process efficiency and product properties of pulp mold by the application of oil palm EFB. J. Korea Tech. Assoc. Pulp Paper Ind. 48 (1), 67–74. https://doi.org/10.7584/ktappi.2016.48.1.067.

Kormin, S., Rus, A.Z.M., Azahari, M.S.M., 2018. Polyurethane foams made from liquefied oil palm mesocarp fibre and epoxy. Adv. Sci. Lett. 24 (11), 8803–8807. https://doi.org/10.1166/asl.2018.12349.

Kose, R., Kimura, T., Abdul Aziz, M.K., Okayama, T., 2014. Recycling effects on the properties of pulp Fiber sheets produced from oil palm empty fruit bunch. Sen'i Gakkaishi 70 (11), 259–264. https://doi.org/10.2115/fiber.70.259.

Kpalo, S.Y., Zainuddin, M.F., Manaf, L.A., Roslan, A.M., 2020. Production and characterization of hybrid briquettes from corncobs and oil palm trunk bark under a low pressure densification technique. Sustainability. https://doi.org/10.3390/su12062468.

Krishnan, S., Singh, L., Sakinah, M., Thakur, S., Wahid, Z.A., Alkasrawi, M., 2016. Process enhancement of hydrogen and methane production from palm oil mill effluent using two-stage thermophilic and mesophilic fermentation. Int. J. Hydrog. Energy. https://doi.org/10.1016/j.ijhydene.2016.05.037.

Kumar, T.S.M., Chandrasekar, M., Senthilkumar, K., Ilyas, R.A., Sapuan, S.M., Hariram, N., Rajulu, A.V., Rajini, N., Siengchin, S., 2020. Characterization, thermal and antimicrobial properties of hybrid cellulose nanocomposite films with in-situ generated copper nanoparticles in *Tamarindus indica* nut powder. J. Polym. Environ., 1–10. https://doi.org/10.1007/s10924-020-01939-w.

Laftah, W.A., Majid, R.A., 2019. Development of bio-composite film based on high density polyethylene and oil palm mesocarp fibre. SN Appl. Sci. 1 (11), 1404. https://doi.org/10.1007/s42452-019-1402-7.

Latif, H.A., Zaman, I., Yahya, M.N., Sambu, M., Meng, Q., 2020. Analysis on sound absorber panel made of oil palm mesocarp fibre using delany-bazley and Johnson-champoux-allard models. Int. J. Nanoelectron. Mater. 13, 393–406.

Lee, T.J., Ryu, J.Y., 2016. Applicability of non-wood fibers from empty fruit bunch and palm frond for packaging paper. J. Korea Tech. Assoc. Pulp Paper Ind. 48 (6), 98. https://doi.org/10.7584/JKTAPPI.2016.12.48.6.9.

Lee, J.Y., Kim, C.H., Kim, E.H., Park, T.U., Jo, H.M., 2016. Study on the use of oil palm trunk in paper industry. J. Korea Tech. Assoc. Pulp Paper Ind. 48 (6), 241. https://doi.org/10.7584/JKTAPPI.2016.12.48.6.241.

Lee, S.H., Zaidon, A., Rasdianah, D., Lum, W.C., Aisyah, H.A., 2020. Alteration in colour and fungal resistance of thermally treated oil palm trunk and rubberwood particleboard using palm oil. J. Oil Palm Res. https://doi.org/10.21894/jopr.2020.0009.

Liu, K., Lyu, P., Ping, Y., Hu, Z., Mo, L., Li, J., 2019. Chemical-free thermomechanical pulping of empty fruit bunch and sugarcane bagasse. Bioresources. https://doi.org/10.15376/biores.14.4.8627-8639.

Mabrouki, J., Abbassi, M.A., Guedri, K., Omri, A., Jeguirim, M., 2015. Simulation of biofuel production via fast pyrolysis of palm oil residues. Fuel 159, 819–827. https://doi.org/10.1016/j.fuel.2015.07.043.

Mahmud, N., Rosentrater, K.A., 2020. Techno-economic analysis (TEA) of different pretreatment and product separation technologies for cellulosic butanol production from oil palm frond. Energies. https://doi.org/10.3390/en13010181.

Maisara, A.M.N., Ilyas, R.A., Sapuan, S.M., Huzaifah, M.R.M., Nurazzi, N.M., Saifulazry, S.O.A., 2019. Effect of fibre length and sea water treatment on mechanical properties of sugar palm fibre reinforced unsaturated polyester composites. Int. J. Recent Technol. Eng. 8 (2S4), 510–514. https://doi.org/10.35940/ijrte.b1100.0782s419.

Malkawi, A.B., Habib, M., Aladwan, J., Alzubi, Y., 2020. Engineering properties of fibre reinforced lightweight geopolymer concrete using palm oil biowastes. Aust. J. Civ. Eng. https://doi.org/10.1080/14488353.2020.1721954.

Martín-Sampedro, R., Rodríguez, A., Ferrer, A., García-Fuentevilla, L.L., Eugenio, M.E., 2012. Biobleaching of pulp from oil palm empty fruit bunches with laccase and xylanase. Bioresour. Technol. 110, 371–378. https://doi.org/10.1016/j.biortech.2012.01.111.

Maulidna, Wirjosentono, B., Tamrin, Marpaung, L., 2020. Microencapsulation of ginger-based essential oil (*Zingiber cassumunar* roxb) with chitosan and oil palm trunk waste fiber prepared by spray-drying method. Case Stud. Therm. Eng. https://doi.org/10.1016/j.csite.2020.100606.

Mazani, N., Sapuan, S.M., Sanyang, M.L., Atiqah, A., Ilyas, R.A., 2019. Design and fabrication of a shoe shelf from Kenaf Fiber reinforced unsaturated polyester composites. In: Lignocellulose for Future Bioeconomy. Elsevier, pp. 315–332, https://doi.org/10.1016/B978-0-12-816354-2.00017-7. issue 2000.

Mohamad, N., Abd-Talib, N., Kelly Yong, T.-L., 2020. Furfural production from oil palm frond (OPF) under subcritical ethanol conditions. Mater. Today Proc. https://doi.org/10.1016/j.matpr.2020.01.256.

Mohd Fuad, M.A.H., Razali, M.M., Izal, Z.N.M., Faizal, H.M., Ahmad, N., Rahman, M.R.A., Rahman, M.M., 2020. Torrefaction of briquettes made of palm kernel shell with mixture of starch and water as binder. J. Adv. Res. Fluid Mech. Therm. Sci. https://doi.org/10.37934/ARFMTS.70.2.2136.

Mohtar, A., Ho, W.S., Hashim, H., Lim, J.S., Muis, Z.A., Liew, P.Y., 2017. Palm oil mill efflu- ent (pome) biogas off-site utilization Malaysia specification and legislation. Chem. Eng. Trans. https://doi.org/10.3303/CET1756107.

Mokhtar, A., Hassan, K., Aziz, A.A., Wahid, M.B., 2008. Treatment of Oil Palm Lumber. Malaysian Palm Oil Board (MPOB), Information Series, MPOB TT, p. 379.

Mokhtar, A., Hassan, K., Aziz, A.A., May, C.Y., 2012. Oil palm biomass for various wood-based products. In: Lai, O.-M., Tan, C.-P., Akoh, C.C. (Eds.), Palm Oil: Production, Processing, Characterization, and Uses, first ed. Elsevier, pp. 625–652, https://doi.org/10.1016/B978-0-9818936-9-3.50024-1.

Moni, M.N.Z., Sulaiman, S.A., 2013. Downdraft gasification of oil palm frond: effects of temperature and operation time. Asian J. Sci. Res. https://doi.org/10.3923/ajsr.2013.197.206.

Moni, M.N.Z., Sulaiman, S.A., Baheta, A.T., 2018. Downdraft co-gasification of oil palm frond with other oil palm residues: effects of blending ratio. In: MATEC Web of Conferences., https://doi.org/10.1051/matecconf/201822506018.

Muhammad Amir, S.M., Hameed Sultan, M.T., Md Shah, A.U., Jawaid, M., Safri, S.N.A., Mohd, S., Mohd Salleh, K.A., 2020. Low velocity impact and compression after im- pact properties on gamma irradiated Kevlar/oil palm empty fruit bunch hybrid composites. Coatings 10 (7), 646. https://doi.org/10.3390/coatings10070646.

Mumtaz, T., Yahaya, N.A., Abd-Aziz, S., Abdul Rahman, N., Yee, P.L., Shirai, Y., Hassan, M.A., 2010. Turning waste to wealth-biodegradable plastics polyhydroxyalkanoates from palm oil mill effluent-a Malaysian perspective. J. Clean. Prod. https://doi.org/10.1016/j.jclepro.2010.05.016.

Mustapha, S.N.H., Norizan, C.W.N.F.C.W., Roslan, R., Mustapha, R., 2020. Effect of Kenaf/empty fruit bunch (EFB) hybridization and weight fractions in palm oil blend polyester composite. J. Nat. Fibers, 1–14. https://doi.org/10.1080/15440478.2020.1788686.

Nadlene, R., Sapuan, S.M., Jawaid, M., Ishak, M.R., Yusriah, L., 2016. A review on roselle fiber and its composites. J. Nat. Fibers 13 (1), 10–41. https://doi.org/10.1080/15440478.2014.984052.

Nadlene, R., Sapuan, S.M., Jawaid, M., Ishak, M.R., Yusriah, L., 2018. The effects of chemical treatment on the structural and thermal, physical, and mechanical and morphological properties of roselle fiber-reinforced vinyl ester composites. Polym. Compos. 39 (1), 274–287. https://doi.org/10.1002/pc.23927.

Nasrullah, R.C.L., Mazlan, I., 2013. The effect of mixing virgin pulp from oil palm frond Acetosolv pulp with secondary pulp (old newspaper). In: Hiroshi, O. (Ed.), Proceedings of International Symposium on Resource Efficiency in Pulp and Paper Technology (Reptech 2012). Center for Pulp and Paper, pp. 1–284.

Nayaggy, M., Putra, Z.A., 2019. Process simulation on fast pyrolysis of palm kernel shell for production of fuel. Indian J. Sci. Technol. https://doi.org/10.17509/ijost.v4i1.15803.

Nazrin, A., Sapuan, S.M., Zuhri, M.Y.M., Ilyas, R.A., Syafiq, R., Sherwani, S.F.K., 2020. Nanocellulose reinforced thermoplastic starch (TPS), polylactic acid (PLA), and polybutylene succinate (PBS) for food packaging applications. Front. Chem. 8 (213), 1–12. https://doi.org/10.3389/fchem.2020.00213.

Nor Amira Izzati, A., John, W.C., Nurul Fazita, M.R., Najieha, N., Azniwati, A.A., Abdul Khalil, H.P.S., 2020. Effect of empty fruit bunches microcrystalline cellulose (MCC) on the thermal, mechanical and morphological properties of biodegradable poly (lactic acid) (PLA) and polybutylene adipate terephthalate (PBAT) composites. Mater. Res. Express 7 (1). https://doi.org/10.1088/2053-1591/ab6889, 015336.

Nordin, N.I.A.A., Ariffin, H., Hassan, M.A., Ibrahim, N.A., Shirai, Y., Andou, Y., 2015. Effects of milling methods on tensile properties of polypropylene/oil palm meso-carp fibre biocomposite. Pertanika J. Sci. Technol. 23 (2), 325–337.

Nordin, N.A., Sulaiman, O., Hashim, R., Kassim, M.H.M., 2017. Oil palm frond waste for the production of cellulose nanocrystals. J. Phys. Sci. https://doi.org/10.21315/jps2017.28.2.8.

Norfadilah, N., Raheem, A., Harun, R., Ahmadun, F., 2016. Bio-hydrogen production from palm oil mill effluent (POME): a preliminary study. Int. J. Hydrog. Energy. https://doi.org/10.1016/j.ijhydene.2016.04.096.

Norizan, M.N., Abdan, K., Ilyas, R.A., Biofibers, S.P., 2020. Effect of fiber orientation and fiber loading on the mechanical and thermal properties of sugar palm yarn fiber reinforced unsaturated polyester resin composites. Polimery 65 (2), 34–43. https://doi.org/10.14314/polimery.2020.2.5.

Norul Izani, M.A., Paridah, M.T., Mohd Nor, M.Y., Anwar, U.M.K., 2011. A comparison of different treatment to remove residual oil in oil palm empty fruit bunch (OPEFB) for MDF performances. In: ICCM International Conferences on Composite Materials.

Norul Izani, M.A., Paridah, M.T., Mohd Nor, M.Y., Anwar, U.M.K., 2013. Properties of medium-density fibreboard (MDF) made from treated empty fruit bunch of oil palm. J. Trop. For. Sci. 25 (2), 175–183.

Nudri, N.A., Bachmann, R.T., Ghani, W.A.W.A.K., Sum, D.N.K., Azni, A.A., 2020. Characterization of oil palm trunk biocoal and its suitability for solid fuel applications. Biomass Convers. Biorefinery. https://doi.org/10.1007/s13399-019-00419-z.

Nurazzi, N.M., Khalina, A., Sapuan, S.M., Ilyas, R.A., 2019a. Mechanical properties of sugar palm yarn/woven glass fiber reinforced unsaturated polyester composites: effect of fiber loadings and alkaline treatment. Polimery 64 (10), 12–22. https://doi.org/10.14314/polimery.2019.10.3.

Nurazzi, N.M., Khalina, A., Sapuan, S.M., Ilyas, R.A., Rafiqah, S.A., Hanafee, Z.M., 2019b. Thermal properties of treated sugar palm yarn/glass fiber reinforced unsaturated polyester hybrid composites. J. Mater. Res. Technol. https://doi.org/10.1016/j.jmrt.2019.11.086.

Nurazzi, N.M., Khalina, A., Sapuan, S.M., Ilyas, R.A., Rafiqa, S.A., Hanafee, Z.M., 2020. Thermal properties of treated sugar palm yarn/glass fiber reinforced unsaturated polyester hybrid composites. J. Mater. Res. Technol. 9 (2), 1606–1618. https://doi.org/10.1016/j.jmrt.2019.11.086.

Nuryawan, A., Abdullah, C.K., Hazwan, C.M., Olaiya, N.G., Yahya, E.B., Risnasari, I., Masruchin, N., Baharudin, M.S., Khalid, H., Khalil, H.P.S.A., 2020. Enhancement of oil palmwaste nanoparticles on the properties and characterization of hybrid plywood biocomposites. Polymers. https://doi.org/10.3390/POLYM12051007.

Nutongkaew, T., Prasertsan, P., O-Thong, S., Chanthong, S., Suyotha, W., 2020. Improved methane production using lignocellulolytic enzymes from Trichoderma koningiopsis TM3 through co-digestion of palm oil mill effluent and oil palm trunk residues. Waste Biomass Valoriz. https://doi.org/10.1007/s12649-019-00838-z.

Oikarinen, H., Hochstrate, M., Weber, M., 1998. Medium Density Fibreboard – Manufacturing, Properties and Further Processing. Kymenlaakson ammattikorkeakoulu. http://www.hochstrate.de/micha/reports.

Okoroigwe, E.C., Saffron, C.M., Kamdem, P.D., 2014. Characterization of palm kernel shell for materials reinforcement and water treatment. J. Chem. Eng. Mater. Sci. 5 (1), 1–6.

Oladele, I.O., Ibrahim, I.O., Adediran, A.A., Akinwekomi, A.D., Adetula, Y.V., Olayanju, T.M.A., 2020. Modified palm kernel shell fiber/particulate cassava peel hybrid reinforced epoxy composites. Results Mater. https://doi.org/10.1016/j.rinma.2019.100053.

Oluwasola, E.A., Afolayan, A., Ameen, I.O., Adeoye, E.O., 2020. Effect of curing methods on the compressive strength of palm kernel Shell aggregate concrete. LAUTECH J. Civ. Environ. Stud. https://doi.org/10.36108/laujoces/0202/50(0120).

Omar, N.N., Abdullah, N., Mustafa, I.S., Sulaiman, F., 2018. Characterisation of oil palm frond for bio-oil production. ASM Sci. J. 11 (1), 9–22.

Osman, N.S., Sapawe, N., 2020. Optimization of silica (SiO_2) synthesis from acid leached oil palm frond ash (OPFA) through sol-gel method. Mater. Today Proc. https://doi.org/10.1016/j.matpr.2020.05.300.

Phruksaphithak, N., Wangprayot, J., 2020. Feasibility of paper production from oil palm trunk using arrowroot flour as a binder. Key Eng. Mater. https://doi.org/10.4028/www.scientific.net/kem.841.64.

Phruksaphithak, N., Goomuang, N., Jaema, N., 2020. Enhance cellulose film production from oil palm trunk under NaOH/urea solution at low temperature. Key Eng. Mater. https://doi.org/10.4028/www.scientific.net/KEM.861.383.

Qarizada, D., Mohammadian, E., Alis, A.B., Yusuf, S.M., Dollah, A., Rahimi, H.A., Nazari, A.S., Azizi, M., 2019. Thermo distillation and characterization of bio oil from fast pyrolysis of palm kernel Shell (PKS). Key Eng. Mater. https://doi.org/10.4028/www.scientific.net/kem.797.359.

Ramli, R., Shaler, S., Jamaludin, M.A., 2002. Properties of medium density fibreboard from oil palm empty fruit bunch fibre. J. Oil Palm Res. 14 (2), 34–40.

Rasat, M.S.M., Wahab, R., Sulaiman, O., Moktar, J., Mohamed, A., Tabet, T.A., Khalid, I., 2011. Properties of composite boards from oil palm frond agricultural waste. Bioresources. https://doi.org/10.15376/biores.6.4.4389-4403.

Ratanawilai, T., Chumthong, T., Kirdkong, S., 2006. An investigation on the mechanical properties of trunks of palm oil trees for the furniture industry. J. Oil Palm Res. 2006, 114–121.

Ratnasingam, J., Ioras, F., 2011. Bending and fatigue strength of mortise and tenon furniture joints made from oil palm lumber. Eur. J. Wood Wood Prod. https://doi.org/10.1007/s00107-010-0501-3.

Ravikumar, K., Pridhar, T., 2019. Evaluation on properties and characterization of asbestos free palm kernel shell fibre (PKSF)/polymer composites for brake pads. Mater. Res. Express. https://doi.org/10.1088/2053-1591/ab502d.

Razali, N., Kamarulzaman, N.Z., 2020. Chemical characterizations of biochar from palm oil trunk for palm oil mill effluent (POME) treatment. Mater. Today Proc. https://doi.org/10.1016/j.matpr.2020.02.219.

Ridhwan, J., Ilyas, R.A., Saiful, M., Hussin, F., Mastura, M.T., 2019a. Water transport and physical properties of sugarcane bagasse fibre reinforced thermoplastic potato starch biocomposite. J. Adv. Res. Fluid Mech. Therm. Sci. 61 (2), 273–281.

Ridhwan, J., Saidi, Z.A.S., Ilyas, R.A., Ahmad, M.N., Wahid, M.K., Yaakob, M.Y., Maidin, N.A., Rahman, M.H.A., Osman, M.H., 2019b. Characteristics of cogon grass fibre reinforced thermoplastic cassava starch biocomposite: water absorption and physical properties. J. Adv. Res. Fluid Mech. Therm. Sci. 62 (1), 43–52.

Ridhwan, J., Khiruddin, M.A.A., Asyul Sutan Saidi, Z., Salit, M.S., Ilyas, R.A., 2020. Effect of cogon grass fibre on the thermal, mechanical and biodegradation properties of thermoplastic cassava starch biocomposite. Int. J. Biol. Macromol. 146, 746–755. https://doi.org/10.1016/j.ijbiomac.2019.11.011.

Rozilah, A., Jaafar, C.N.A., Sapuan, S.M., Zainol, I., Ilyas, R.A., 2020. The effects of silver nanoparticles compositions on the mechanical, physiochemical, antibacterial, and morphology properties of sugar palm starch biocomposites for antibacterial coating. Polymers 12 (11), 2605. https://doi.org/10.3390/polym12112605.

Saari, N., Lamaming, J., Hashim, R., Sulaiman, O., Sato, M., Arai, T., Kosugi, A., Wan Nadhari, W.N.A., 2020. Optimization of binderless compressed veneer panel manufacturing process from oil palm trunk using response surface methodology. J. Clean. Prod. https://doi.org/10.1016/j.jclepro.2020.121757.

Saidu, M., Yuzir, A., Salim, M.R., Richard, A., Afiz, B., 2020. Effect of operating parameter on the anaerobic digestion oil palm mesocarp fibre with cattle manure for biogas production. IOP Conf. Ser. Earth Environ. Sci. 476. https://doi.org/10.1088/1755-1315/476/1/012085, 012085.

Saleh, A.R., Sudarmanta, B., Mujiarto, S., Suharno, K., Widodo, S., 2020. Modeling of oil palm frond gasification process in a multistage downdraft gasifier using aspen plus. J. Phys. Conf. Ser. https://doi.org/10.1088/1742-6596/1517/1/012036.

Sanyang, M.L., Ilyas, R.A., Sapuan, S.M., Jumaidin, R., 2018. Sugar palm starch-based composites for packaging applications. In: Bionanocomposites for Packaging Applications. Springer International Publishing, pp. 125–147, https://doi.org/10.1007/978-3-319-67319-6_7.

Sapuan, S.M., Aulia, H.S., Ilyas, R.A., Atiqah, A., Dele-Afolabi, T.T., Nurazzi, M.N., Supian, A.B.M., Atikah, M.S.N., 2020. Mechanical properties of longitudinal basalt/woven-glass-fiber-reinforced unsaturated polyester-resin hybrid composites. Polymers 12 (10). https://doi.org/10.3390/polym12102211.

Saputra, O.A., Rinawati, L., Setia Rini, K., Saputra, D.A., Pramono, E., 2016. Effect of fiber size on mechanical and water absorption properties of recycled polypropylene/empty fruit bunches (rPP/EFB) bio-composites. Appl. Mech. Mater. 842, 7–13. https://doi.org/10.4028/www.scientific.net/AMM.842.7.

Sari, N.H., Pruncu, C.I., Sapuan, S.M., Ilyas, R.A., Catur, A.D., Suteja, S., Sutaryono, Y.A., Pullen, G., 2020. The effect of water immersion and fibre content on properties of corn husk fibres reinforced thermoset polyester composite. Polym. Test. 91. https://doi.org/10.1016/j.polymertesting.2020.106751, 106751.

Sauki, A., Mohd Azmi, M.S., Jarni, H.H., Wan Bakar, W.Z., Tengku Mohd, T.A., Ab, N., Lah, N.K.I., Megat Khamaruddin, P.N.F., Ibrahim, W.A., 2020. Feasibility study of using oil palm trunk waste fiber as fluid loss control agent and lost circulation material in drilling mud. J. Adv. Res. Fluid Mech. Therm. Sci. https://doi.org/10.37934/arfmts.72.1.920.

Sawawi, M., Mohammad, N.H., Sahari, S.K., Junaidi, E., Razali, T.N., 2020. Effects of chemical treatment on mechanical properties of oil palm empty fruit bunch (EFB) with urea formaldehyde (UF) resin particleboard type. Int. J. Recent Technol. Eng. 8 (5), 1330–1334. https://doi.org/10.35940/ijrte.E6094.018520.

Septevani, A.A., Rifathin, A., Sari, A.A., Sampora, Y., Ariani, G.N., Sudiyarmanto, Sondari, D., 2020. Oil palm empty fruit bunch-based nanocellulose as a superadsorbent for water remediation. Carbohydr. Polym. 229. https://doi.org/10.1016/j.carbpol.2019.115433, 115433.

Shah, A.M., Rahim, A.A., Mohamad Ibrahim, M.N., Hussin, M.H., 2017. Depolymerized oil palm frond (OPF) lignin products as corrosion inhibitors for mild steel in 1 M HCl. Int. J. Electrochem. Sci. https://doi.org/10.20964/2017.10.66.

Shakib, N., Rashid, M., 2019. Biogas production optimization from POME by using anaerobic digestion process. J. Appl. Sci. Process Eng. https://doi.org/10.33736/jaspe.1711.2019.

Shrivastava, P., Khongphakdi, P., Palamanit, A., Kumar, A., Tekasakul, P., 2020. Investigation of physicochemical properties of oil palm biomass for evaluating potential of biofuels production via pyrolysis processes. Biomass Convers. Biorefinery. https://doi.org/10.1007/s13399-019-00596-x.

Sihabut, T., Laemsak, N., 2008. Sound absorption efficiency of fiberboard made from oil palm frond. In: FORTROP II International Conference Bangkok, Thailand. vol. 185.

Siti Suhaily, S., Gopakumar, D.A., Sri Aprilia, N.A., Rizal, S., Paridah, M.T., Abdul Khalil, H.P.S., 2019. Evaluation of screw pulling and flexural strength of bamboo-based oil palm trunk veneer hybrid biocomposites intended for furniture applications. Bioresources. https://doi.org/10.15376/biores.14.4.8376-8390.

Solikhah, M.D., Pratiwi, F.T., Heryana, Y., Wimada, A.R., Karuana, F., Raksodewanto, A., Kismanto, A., 2018. Characterization of bio-oil from fast pyrolysis of palm frond and empty fruit bunch. IOP Conf. Ser. Mater. Sci. Eng. 349. https://doi.org/10.1088/1757-899X/349/1/012035, 012035.

Suhaily, S.S., Jawaid, M., Abdul Khalil, H.P.S., Ibrahim, F., 2012. A review of oil palm biocomposites for furniture design and applications: potential and challenges. Bioresources 7 (3), 4400–4423. https://doi.org/10.15376/biores.7.3.4400-4423.

Sulaiman, S., Jafarzadeh, S., Ariffin, F., 2020. Renewable syrup derived from sap of oil palm trunk as an alternative novel feedstock for confectionery product (toffee). Sugar Tech. https://doi.org/10.1007/s12355-020-00861-8.

Sungpichai, P., Srinophakun, T.R., 2020. Bioethanol production from oil palm trunk with scheduling optimization. Int. J. Adv. Res. Eng. Technol. https://doi.org/10.34218/IJARET.11.6.2020.003.

Suwanno, S., Rakkan, T., Yunu, T., Paichid, N., Kimtun, P., Prasertsan, P., Sangkharak, K., 2017. The production of biodiesel using residual oil from palm oil mill effluent and crude lipase from oil palm fruit as an alternative substrate and catalyst. Fuel. https://doi.org/10.1016/j.fuel.2017.01.049.

Syafiq, R., Sapuan, S.M., Zuhri, M.Y.M., Ilyas, R.A., Nazrin, A., Sherwani, S.F.K., Khalina, A., 2020. Antimicrobial activities of starch-based biopolymers and biocomposites incorporated with plant essential oils: a review. Polymers 12 (10), 2403. https://doi.org/10.3390/polym12102403.

Syafri, E., Kasim, A., Abral, H., Asben, A., 2019a. Cellulose nanofibers isolation and characterization from ramie using a chemical-ultrasonic treatment. J. Nat. Fibers 16 (8), 1145–1155. https://doi.org/10.1080/15440478.2018.1455073.

Syafri, E., Sudirman, Mashadi, Yulianti, E., Deswita, Asrofi, M., Abral, H., Sapuan, S.M., Ilyas, R.A., Fudholi, A., 2019b. Effect of sonication time on the thermal stability, moisture absorption, and biodegradation of water hyacinth (Eichhornia crassipes) nanocellulose-filled bengkuang (*Pachyrhizus erosus*) starch biocomposites. J. Mater. Res. Technol. 8 (6), 6223–6231. https://doi.org/10.1016/j.jmrt.2019.10.016.

Syamani, F.A., Nurjayanti, Pramasari, D.J., Kusumaningrum, W.B., Kusumah, S.S., Masruchin, N., Ermawati, R., Supeni, G., Cahyaningtyas, A.A., 2020. Characteristics of bioplastic made from cassava starch filled with fibers from oil palm trunk at various amount. IOP Conf. Ser. Earth Environ. Sci. https://doi.org/10.1088/1755-1315/439/1/012035.

Teow, Y.H., Tajudin, S.A., Ho, K.C., Mohammad, A.W., 2020. Synthesis and characterization of graphene shell composite from oil palm frond juice for the treatment of dye-containing wastewater. J. Water Process Eng. https://doi.org/10.1016/j.jwpe.2020.101185.

Ting, T.Z.H., Rahman, M.E., Lau, H.H., Ting, M.Z.Y., Pakrashi, V., 2020. Oil palm kernel shell – a potential sustainable construction material. In: Encyclopedia of Renewable and Sustainable Materials. Elsevier, https://doi.org/10.1016/b978-0-12-803581-8.11541-3.

Uchechukwu, E.A., Austin, O., 2020. Artificial neural network application to the compressive strength of palm kernel shell concrete. MOJ Civ. Eng. https://doi.org/10.15406/mojce.2020.06.00164.

Uddin, M.N., Rahman, M.A., Taweekun, J., Techato, K., Mofijur, M., Rasul, M., 2019. Enhancement of biogas generation in up-flow sludge blanket (UASB) bioreactor from palm oil mill effluent (POME). Energy Procedia. https://doi.org/10.1016/j.egypro.2019.02.220.

Umar, H.A., Sulaiman, S.A., Ahmad, R.K., Tamili, S.N., 2020. Characterisation of oil palm trunk and frond as fuel for biomass thermochemical. IOP Conf. Ser. Mater. Sci. Eng. https://doi.org/10.1088/1757-899X/863/1/012011.

Valášek, P., Ambarita, H., 2018. Material usage of oil-palm empty fruit bunch (EFB) in polymer composite systems. Manuf. Technol. 18 (4), 686–691. https://doi.org/10.21062/ujep/161.2018/a/1213-2489/MT/18/4/686.

Vijayaraghavan, K., Ahmad, D., 2006. Biohydrogen generation from palm oil mill effluent using anaerobic contact filter. Int. J. Hydrog. Energy. https://doi.org/10.1016/j.ijhydene.2005.12.002.

Wahab, R., Rasat, M.S.M., Samsi, H.W., Mustafa, M.T., Don, S.M.M., 2017. Assessing the suitability of agro-waste from oil palm empty fruit bunches as quality eco-composite boards. J. Agric. Sci. https://doi.org/10.5539/jas.v9n8p237.

Wahida, A.F., Najmuldeen, G.F., 2015. One layer experimental particleboard from oil palm frond particles and empty fruit bunch fibers. Int. J. Eng. Res. Technol. 4 (1), 199–202.

Wang, X., Hu, J., Liang, Y., Zeng, J., 2012a. Tcf bleaching character of soda-anthraquinone pulp from oil palm frond. Bioresources. https://doi.org/10.15376/biores.7.1.0275-0282.

Wang, X., Hu, J., Liang, Y., Zeng, J., 2012b. TCF bleaching character of soda-anthraquinone pulp from oil palm frond. Bioresources 7 (1), 275–282. https://doi.org/10.15376/BIORES.7.1.0275-0282.

Yorseng, K., Rangappa, S.M., Pulikkalparambil, H., Siengchin, S., Parameswaranpillai, J., 2020. Accelerated weathering studies of kenaf/sisal fiber fabric reinforced fully biobased hybrid bioepoxy composites for semi-structural applications: morphology, thermo-mechanical, water absorption behavior and surface hydrophobicity. Constr. Build. Mater. 235. https://doi.org/10.1016/j.conbuildmat.2019.117464, 117464.

Zamri, F.A., Primus, W.C., Shaari, A.H., Sinin, A.E., 2019. Technical note: effects of polyurethane resin on the physical and mechanical properties of wood fiber/palm kernel shell composite boards. Wood Fiber Sci. https://doi.org/10.22382/wfs-2019-043.

13

Laminated veneer lumber from oil palm trunk

Aizat Ghani[a,b], S.H. Lee[a], Fatimah Atiyah Sabaruddin[a], S.O.A. SaifulAzry[a], and M.T. Paridah[a]

[a]*Institute of Tropical Forestry and Forest Products (INTROP), Universiti Putra Malaysia, Serdang, Selangor, Malaysia, [b]Faculty of Forestry and Environment, Universiti Putra Malaysia, Serdang, Selangor, Malaysia*

13.1 Introduction

Laminated veneer lumber (LVL) was first used in World War II as a propeller in airplane body parts, and it started being used as a construction material in the 1970s (Neuvonen et al., 1998). LVL is one of the wood composites consisting of all-veneer that are laminated together, and the veneer is oriented in the same longitudinal direction and then hot-pressed together to obtain maximum stiffness and strength. LVL is usually mistaken for plywood; the distinguishing difference between LVL and plywood is the orientation of the veneer layers. For LVL, the direction of the plies is in the same direction, while the direction of plies is alternated between each veneer in plywood.

The thickness and width of LVL vary. However, the most common dimension produced is in the range of 4.45 cm in thickness, 121.9 cm in width, and a length of generally 18.29 cm. These veneers are usually being used as an alternative to structural lumber after being ripped into narrower dimensions for flange components in wood composites (I joist) and also as headers and beams (Wilson and Dancer, 2007). Other than that, LVL also can be applied to building construction material like framing members such as girders, lintels and columns, and scaffold planks since LVL has already proven its usefulness and efficiency (Diekmann, 1997).

Unlike a glulam beam, LVL is recommended for nonvisual applications that are usually covered or wrapped to achieve a desirable appearance, especially in an industrial application.

It is believed that LVL has great potential in the furniture industry for both economic and esthetic reasons. These wood composites are closer to solid wood in appearance, texture, and structure than any other alternative wood substitute. Moreover, LVL can be easily manufactured to meet consumers-based request, demand, and interest if necessary, hence LVL can be produced in different sizes to suit architectural and structural needs. Interestingly, all these necessities can be obtained from veneer peeled from a log that was low in quality yielding a significant amount of sawn lumber (Eckelman, 1993). According to Diekmann (1997), LVL has an advantage over solid wood in performance predictability, sizes available, strength, dimensional consistency and stability, treatability, and a higher wood utilization rate up to 13% more than that sawn wood.

13.2 Laminated veneer lumber

13.2.1 Raw materials

The manufacturing of LVL is very much the same as the plywood processing industry, with the main differences occurring during the lay-up, hot-pressing, and further processing stages (Hughes, 2015). As mentioned by Kurt (2010), several important factors have to be considered to determine the suitability of wood in manufacturing LVL, such as species quality, wood durability, physical and mechanical properties, ease of treatment, gluing, and seasoning including wood finishing materials. Interestingly, in LVL manufacturing, any tree species can be used as long as the physical and mechanical properties of the wood are acceptable.

LVL is commonly produced from low-density species such as Southern pine, yellow poplar, Douglas fir, and other softwood species (Hiziroglu, 2009; Bal, 2014). However, in Northern America, Douglas fir and Southern pine are two common species usually selected (Diekmann, 1997). Meanwhile, in Germany, a company named Pollemeir choose Beechwood as its raw material to produce an LVL product called "BaubBuche".

Kurt et al. (2012) reported that LVL can be made from hardwood species like aspen and other soft hardwood species. Plus, it has been suggested that LVL can be manufactured from fast-growing species like radiata pine, Scots pine, eucalyptus, poplar, rubberwood, and *Acacia mangium* species (Harding and Orange, 1998; Çolak et al., 2004; Wang, 1992; Gaunt et al., 2003; Aydın et al., 2004; Uysal and Şeref, 2005; Keskin, 2009; Wong et al., 1996).

13.2.2 General process of LVL manufacturing

Initially, the log will be steamed first for an efficient veneer production followed by a log debarking process. Then, the veneer sheet is formed by rotating the log on a veneer lathe with a thickness of about 2.5–4.8 mm, 0.6–1.2 m in width, and 2.4–18.8 m in length. The produced veneer will be dried using a dryer bypass with a hot platen on the veneer surface to reach a targeted moisture content in the range of 6%–10%. Defects in the veneer such as knots will be trimmed and clipped out before the veneer is graded and sorted through ultrasonic screening that can detect the surface of the veneer.

After the grading and trimming process, a phenol-formaldehyde adhesive is applied to the top surface of the veneer sheet usually by a roller glue spreader. Then the veneer with the resin is laid up with the grain in the same direction. The billet of the glued and laid-up veneers goes through a continuous press and heat around 130°C to begin a consolidation and curing process. After hot-pressing, the consolidated LVL is trimmed and sawn to the required dimensions. Finally, the LVL is sanded and graded to dispatch.

13.3 Laminated veneer lumber from oil palm trunk

The use of oil palm trunk (OPT) as a raw material has been a major focus in the veneer-related industry since 2008 (Sulaiman et al., 2008; Wahab et al., 2008). The use of OPT as a raw material in the wood industry could reduce the environmental burden of wood consumption and lessen the usage of tropical hardwood veneers in plywood production (Hashim et al., 2011). In Malaysia, the application of the outer part of the oil palm veneer (OPV) as a core veneer in the plywood manufacturing industry has also rapidly increased (Masseat et al., 2018). According to Nordin et al. (2004a, b), OPT also has the potential to be converted into LVL and be used as an alternative material in structural and nonstructural applications (Hayashi, 1993; Eckelman, 1993).

Much research have been carried out on the production of LVL from OPT. Different researchers utilize oil palm of different ages, portions, and positions of the OPT; adhesive application; and manufacturing processes. Table 13.1 listed some of the research that have been done on LVL made from OPT.

Wahab et al. (2008) manufactured three-layer LVL from different portions and positions of OPT. The results showed that LVL made from the outer part of the bottom portion of OPT resulted in higher density compared to the inner part of OPT, even from the same portion. These density values of the LVL are different because OPT trunk possesses a wide density range depending on the position of the OPT veneer taken. OPT possesses densities that vary based on position along with the height and cross-sectional zones (Paridah and Anis, 2008).

Table 13.1 Fabrication of oil palm trunk (OPT) laminated veneer lumber from the literature.

Veneer material	Oil palm age (years old)	Dimensions	Type of adhesive	Density (kg/m³)	Source
Bottom OPT, outer part	30	240 cm (length) × 120 cm (width) × 25 mm (thickness)	Urea formaldehyde	596	Wahab et al. (2008)
Bottom OPT, inner part				492	
Middle OPT, outer part				589	
Middle OPT, inner part				441	
Matured OPT	25–30	4.5 × 230 × 230 mm for each veneer—3-ply LVL	Emulsions polymeric isocyanate, EPI (without toluene)	570	Hashim et al. (2011)
			EPI (with toluene)	553	
			Polyvinyl acetate, PVAc	590	
Matured OPT	25–30	4.5 × 230 × 230 mm for each veneer—3-ply LVL	Urea formaldehyde	550	Sulaiman et al. (2009)
			Phenol formaldehyde	320	
			Melamine urea formaldehyde	500	
			Phenol resorcinol formaldehyde	610	
Matured oil palm trunk	25–30	40 cm × 40 cm × thickness	Urea formaldehyde	500	Sulastiningsih et al. (2020)
5 layers of LVL—OPT mixed with jabon veneers as face and back layer				507	
5 LVL layers of OPT, with face, back and center core from jabon veneers				518	
5-ply LVL of OPT, with face and back from mahoni species veneers				560	
5-ply LVL of OPT, with face, back and core center core from mahoni veneers				591	
LVL OPT	30	500 mm (width) × 780 mm (length) × 25 mm (thickness) —9 layers	Urea formaldehyde	Not mentioned in the study	Nordin et al. (2004a, b)
LVL OPT mixed with rubberwood veneer as face and back layer					
LVL OPT, mixed with rubberwood veneer as face, back and core center layer					
Matured OPT	25–30	Veneer size - 40 cm × 40 cm	Low molecular weight phenol formaldehyde (LMwPF)	532	Masseat et al. (2018)

Sulaiman et al. (2009) also conducted research on LVL OPT bonded with different types of adhesives. The 3-ply OPT LVL were bonded with urea-formaldehyde (UF), phenol-formaldehyde (PF), melamine urea-formaldehyde (MUF), and resorcinol phenol-formaldehyde (PRF). Remarkably, different types of adhesives give various density results. LVL made from PRF resin resulted in a higher density at $600\,kg/m^3$ than those made using MUF, UF, and PF adhesives.

On the other hand, Sulastiningsih et al. (2020) fabricated 5-ply OPT LVL mixed with jabon wood (*Anthocephalus cadamba* Miq.) and mahoni wood (*Swietenia macrophylla* King.) as face, core, and back layers. The results revealed that incorporating mahoni veneers increased the density of LVL produced. In addition, OPT LVL that incorporated three mahoni wood veneers exhibited the lowest thickness swelling and water absorption. Apart from that, incorporation of jabon and mahoni veneer significantly improved the bending strength, compression strength, and hardness of the OPT LVL compared to that of the 5-ply LVL made purely from OPT veneers. As suggested by Nordin et al. (2004a, b), combining veneers of other species in the layer of LVL OPT could improve the performance of the LVL.

To improve the dimensional stability and strength of the OPT LVL, the OPT veneers could be treated accordingly. In a study by Masseat et al. (2018), the OPT veneers were treated with 30% and 45% low molecular weight phenolic resin. Five-ply LVL were fabricated with the treated OPT veneers used as face layer. As a result, the physical properties of the LVL made from phenolic resin-treated OPT veneers were better than that of their untreated counterparts. However, the physical properties of the treated OPT LVL are still inferior to that of rubberwood LVL.

13.4 Properties of laminated veneer lumber from oil palm trunk

Bonding quality and gluing properties are vital factors that need to be taken into consideration in the manufacturing of OPT LVL. Zalifah et al. (2013) stated that evaluating the extent of adhesive penetration into a veneer is very important, especially for nonwood and porous materials like OPT itself. Overly adhesive penetration will cause undesired high adhesive consumption and consequently increase production cost. In a study by Zalifah et al. (2013), several factors such as pressing temperature, OPT portion, and veneer parts could affect the effective penetration and average penetration of adhesives. Apart from that, the types of adhesives used also play an important role in affecting the penetration extent. Phenol formaldehyde (PF) resin was found to result in higher adhesive penetration compared to that of urea formaldehyde (UF) resin. The following section discusses the properties of OPT LVL manufactured under different variables.

13.4.1 Physical and mechanical properties of OPT LVL

The properties of LVL are closely linked to its structure and the raw materials used in its fabrication. Table 13.2 lists some of the selected mechanical properties of LVL produced from OPT and LVL from OPT mixed with different wood species.

LVL manufactured from OPT veneer alone shows the lowest mechanical strength among the LVLs. As for bending strength, modulus of rupture (MOR), and modulus of elasticity (MOE), LVL made from OPT veneer alone were $15.43\,N/mm^2$ and $570\,N/mm^2$, which is around 7 times and 16 times lower compared to LVL made from rubberwood. LVL made from rubberwood has the highest bending strength in terms of MOR and MOE. However, the bending strength properties of LVL increased when the OPT was mixed with other veneers from other wood species. The improvement in bending strength was attributed to the densification process on the LVL board as a result of inclusions of higher density wood veneers like rubberwood (Nordin et al., 2004a, b). Improvement of the LVL OPT mixed with other wood species could be seen when the LVL layer veneer was mixed with jabon and mahoni wood species (Sulastiningsih et al., 2020).

Table 13.2 Comparison of some selected properties between LVL from rubberwood, oil palm trunk, and other wood species.

LVL types	MOR (N/ mm²)	MOE (N/ mm²)	Shear strength (N/mm²)	Hardness (N/mm²)	Compression (N/mm²)	References
LVL OPT	15.43	570	1.36	–	–	Wahab et al. (2008)
LVL rubberwood	86	9210	10	–	10	Kamala et al. (1999)
OPT mixed with jabon veneer	39	4240	–	26	31	Sulastiningsih et al. (2020)
OPT mixed with mahoni veneers	37	5260	–	22	26	Sulastiningsih et al. (2020)
OPT LVL mixed with rubberwood veneer	59	7613	–	–	26	Nordin et al. (2004a, b)

Different adhesives applied to the LVL give different results on OPT LVL properties. Sulaiman et al. (2009) compared four types of amide adhesives, namely UF, PF, MUF, and PRF with two different level spread rates at 250 and $500 g/cm^2$. This study was aimed to assess their effect on the shear strength of LVL OPT panels. Based on the findings, LVL made from OPT using PRF adhesive with $500 g/cm^2$ spread rate has a higher shear strength of 3.60 MPa than the $250 g/cm^2$ spread rate. PRF-bonded LVL also displayed superior performance than those bonded with other adhesives. Hashim et al. (2011) also compared the effect of PVAc spread rate on OPT LVL, and $500 g/m^2$ spread rate portrayed higher tensile shear (4.99 MPa) with higher wood failure of about 60% than $250 g/m^2$ spread rate.

13.4.2 Dimensional stability and durability of OPT LVL

Siti-Noorbaini et al. (2013) studied the dimensional stability of LVL from OPT bonded with different cold set adhesives and found that the percentage of thickness swelling varies among the adhesives applied compared with LVL made from rubberwood. The study showed that at $500 g/m^2$ spread level, cold setting adhesive (EPI-copolymer vinyl acetate, VAc, and EPI-copolymer styrene-butadiene rubber, SBR) used to bond OPT LVL displayed a nearly similar percentage of thickness swelling through the relative humidity test but remains highest among those compared to OPT LVL bonded with PVAc adhesive. According to Conner (2001), PVAc adhesive is less resistant to humidity and moisture as it consequently tends to creep under a sustained load. Besides, the LVL made from OPT displayed a higher percentage of thickness swelling compared to LVL made from rubberwood.

Furthermore, when the same OPT and rubberwood LVL panels were tested for soil burial test for 8 weeks, there were plenty of organisms found at the site around the samples like termites, ants, worms, bugs, and also traces of their trails. As expected, LVL panel made from OPT was found to be easily deteriorated and degraded compared to those LVL from rubberwood. High amount of starch inside the OPT helps to promote growth of biodeterioration agents like fungi and insects. From the scanning electron microscope (SEM) analysis, both OPT and rubberwood LVL were still intact; however, it was observed that there was fungi wrapped around and entwined among the fibers, starch granules, and parenchyma inside the LVL exposed samples.

During the soil burial test, for the first 4 weeks, it was observed that the OPT LVL bonded with EPI-copolymer SBR exhibited higher degradation than those OPT LVL bonded with EPI-copolymer VAc and PVAc adhesive. OPT LVL bonded with all types of adhesives displayed inferior biological durability than rubberwood LVL (Siti-Noorbaini et al., 2013). Then, in the following 4 weeks until the test was completed, the weight loss percentages of LVL made from rubberwood increased significantly compared to LVL OPT. This is because rubberwood also contains a high amount of starch. However, the presence of laticifers or latex vessel in rubberwood would give the organism ample time to delay the biodeterioration agents to attack the starch content (Gomez, 1982).

Hashim et al. (2011) made LVL from OPT and rubberwood and bonded with EPI (with and without toluene) and PVAc adhesive. They found that LVL made from OPT have a greater thickness swelling and water absorption compared to that of rubberwood LVL regardless of type of adhesive applied. This phenomenon was because oil palm is more hygroscopic than rubberwood. As mentioned in the results, values for both thickness swelling and water absorption for each type of adhesive increased as the immersion time increased. Interestingly, the presence of toluene in the EPI adhesive was found to greatly decrease the thickness swelling of the LVL compared to EPI without toluene.

Moreover, thickness swelling of LVL was significantly affected by layer composition, and the percentage of thickness swelling decreased with the addition of mahoni wood veneers in the OPT LVL structure as well as the percentages of wood absorption (Sulastiningsih et al., 2020). Also, the highest water absorption and thickness swelling of OPT LVL was found in the LVL made with the top portion and inner zone of the OPT and, at the end of the study, researchers suggested that impregnation with PF resin method on the OPT veneer can improve the dimensional stability of the LVL OPT panels.

A recent study by Masseat et al. (2018) reported that OPT veneer treated with 45% and 30% low molecular weight phenol-formaldehyde at 15% initial moisture content resulted in 0.76% and 1.53% thickness swelling, respectively, and these results were better than untreated OPT LVL panels at 2.99%. As for water absorption, the untreated LVL OPT gives a value of 63.03% in which the water absorption value of LVL made from veneer treated with 45% and 30% LmwPF concentration was 24.40% and 28.66%, respectively. From the results obtained, it can be concluded that treated OPT veneer would give better dimensional stability than untreated LVL OPT veneer.

13.5 Conclusions

In conclusion, LVL is known as a high value, reconstituted, engineered wood product that is currently experiencing increased demand due to its versatility properties compared to plywood. It also has been proven that the utilization of OPT as a raw material to fabricate LVL other than wood species is suitable, and it could reduce the environmental burden of wood consumption and minimize air pollution by implementing zero burning practices in the oil palm plantation. It was suggested that the LVL OPT is seemly suitable for structural and nonstructural purposes. Hence, the quality of LVL produced is really crucial and important. The quality and performance of the panel is based on the physical and mechanical properties of the OPT, so it is important to consider several factors such as the type of adhesive and number of LVL layers to produce LVL with acceptable properties and strength. Furthermore, the strength of the LVL OPT could be improved by treating the veneer through resin impregnation. As a result, the dimensional stability and durability of the LVL OPT could be improved. Apart from that, inclusion of other wood species veneer like rubberwood, mahoni, and jabon could also significantly improve the strength of the LVL.

References

Aydın, İ., Çolak, S., Çolakoğlu, G., Salih, E., 2004. A comparative study on some physical and mechanical properties of laminated veneer lumber (LVL) produced from beech (*Fagus orientalis* Lipsky) and Eucalyptus (*Eucalyptus camaldulensis* Dehn.) veneers. Holz Roh Werkst. 62 (3), 218–220.

Bal, B.C., 2014. Some physical and mechanical properties of reinforced laminated veneer lumber. Constr. Build. Mater. 68, 120–126.

Çolak, S., Aydin, I., Demirkir, C., Çolakoğlu, G., 2004. Some technological properties of laminated veneer lumber manufactured from pine (*Pinus sylvestris* L.) veneers with melamine added-UF resins. Turk. J. Agric. For. 28 (2), 109–113.

Conner, A.H., 2001. Wood: Adhesives. In: Encyclopedia of Materials: Science and Technology. Elsevier Science, Ltd., Amsterdam, Netherlands; New York, USA.

Diekmann, E.F., 1997. Engineered wood products: a guide for specifiers, designers, and users. For. Prod. J. 47 (9), 10.

Eckelman, C.A., 1993. Potential uses of laminated veneer lumber in furniture. For. Prod. J. 43 (4), 19.

Gaunt, D., Penellum, B., McKenzie, H.M., 2003. Eucalyptus nitens laminated veneer lumber structural properties. N. Z. J. For. Sci. 33 (1), 114–125.

Gomez, J.B., 1982. Anatomy of hevea and its Influence on Latex Production. Malaysian Rubber Research and Development Board, Kuala Lumpur, Malaysia.

Harding, O.V., Orange, R.P., 1998. The effect of juvenile wood and lay-up practices on various properties of radiata pine laminated veneer lumber. For. Prod. J. 48 (7/8), 63.

Hashim, R., Sarmin, S.N., Sulaiman, O., Yusof, L.H.M., 2011. Effects of cold setting adhesives on properties of laminated veneer lumber from oil palm trunks in comparison with rubberwood. Eur. J. Wood Wood Prod. 69 (1), 53–61.

Hayashi, T., 1993. Bending strength distribution of laminated veneer lumber for structural use. Mokuzai Gakkaishi 39, 985–992.

Hiziroglu, S., 2009. Laminated veneer lumber (LVL) as a construction material. Available from: https://extension.okstate.edu/fact-sheets/laminated-veneer-lumber-lvl-as-a-construction-material.html.

Hughes, M., 2015. Plywood and other veneer-based products. In: Ansell, M.P. (Ed.), Wood Composites. Woodhead Publishing, Sawston, United Kingdom, pp. 69–89.

Kamala, B.S., Kumar, P., Rao, R.V., Sharma, S.N., 1999. Performance test of laminated veneer lumber (LVL) from rubber wood for different physical and mechanical properties. Holz Roh Werkst. 57 (2), 114–116.

Keskin, H., 2009. Effects of impregnation solutions on weight loss during combustion of laminated veneer lumber. Gazi Univ. J. Sci. 22 (3), 235–243.

Kurt, R., 2010. Suitability of three hybrid poplar clones for laminated veneer lumber manufacturing using melamine urea-formaldehyde adhesive. Bioresources 5 (3), 1868–1878.

Kurt, R., Meriç, H., Aslan, K., Çil, M., 2012. Laminated veneer lumber (LVL) manufacturing using three hybrid poplar clones. Turk. J. Agric. For. 36 (2), 237–245.

Masseat, K., Bakar, E.S., Kamal, I., Husain, H., Tahir, P.M., 2018. The physical properties of treated oil palm veneer used as face layer for laminated veneer lumber. IOP Conf. Ser. Mater. Sci. Eng. 368, 012025.

Neuvonen, E., Salminen, M., Heiskanen, J., Hochstrate, M., Weber, M., 1998. Laminated Veneer Lumber: Overview of the Forest Product, Manufacturing, and Marketing Situation. Kymenlaakson Ammattikorkeakoulu, Department of Forest Products Marketing, Wood-Based Panels Technology. Available from: http://www.hochstrate.de/micha/reports/replvl.html.

Nordin, K., Hashim, W.S., Ahmad, M., Jamaludin, M.A., 2004a. Improved properties of oil palm trunk (OPT) laminated veneer lumber (LVL) through the inclusion of rubberwood veneers. Sci. Lett. 1 (1), 51–56.

Nordin, K., Jamaludin, M.A., Ahmad, M., Samsi, H.W., Salleh, A.H., Jalaludin, Z., 2004b. Minimizing the environmental burden of oil palm trunk residues through the development of laminated veneer lumber products. Manag. Environ. Qual. 15 (5), 484–490.

Paridah, M.T., Anis, M., 2008. Process optimization in the manufacturing of plywood from oil palm trunk. In: Proceedings of 7th National Seminar on the Utilization of Oil Palm Tree, Kuala Lumpur, pp. 12–24.

Siti-Noorbaini, S., Wan-Mohd-Nazri, W.A.R., Jamaludin, K., Othman, S., Rokiah, H., Hazandy, A.H., 2013. Study on dimensional stability properties of laminated veneer lumber from oil palm trunk bonded with different cold set adhesives. J. Appl. Sci. 13 (7), 994–1003.

Sulaiman, O., Hashim, R., Wahab, R., Samsi, H.W., Mohamed, A.H., 2008. Evaluation on some finishing properties of oil palm plywood. Holz Roh Werkst. 66 (1), 5–10.

Sulaiman, O., Salim, N., Hashim, R., Yusof, L.H.M., Razak, W., Yunus, N.Y.M., Hashim, W.S., Azmy, M.H., 2009. Evaluation on the suitability of some adhesives for laminated veneer lumber from oil palm trunks. Mater. Des. 30 (9), 3572–3580.

Sulastiningsih, I.M., Trisatya, D.R., Balfas, J., 2020. Some properties of laminated veneer lumber manufactured from oil palm trunk. IOP Conf. Ser. Mater. Sci. Eng. 935, 012019.

Uysal, B., Şeref, K., 2005. Dimensional stability of laminated veneer lumbers manufactured by using different adhesives after the steam test. Gazi Univ. J. Sci. 18 (4), 681–691.

Wahab, R., Samsi, H., Mohamed, A., Sulaiman, O., 2008. Utilization potential of 30 year-old oil palm trunks laminated veneer lumbers for non-structural purposes. Int. J. Sustain. Dev. 1 (3), 109–113.

Wang, Q., 1992. Utilization of laminated-veneer-lumber from Sabah plantation thinnings as beam flanges. III. Production of composite beam and its properties. Mokuzai Gakkaishi 38, 914–922.

Wilson, J.B., Dancer, E.R., 2007. Gate-to-gate life-cycle inventory of laminated veneer lumber production. Wood Fiber Sci. 37, 114–127.

Wong, E.D., Razali, A.K., Kawai, S., 1996. Properties of rubberwood LVL reinforced with acacia veneers. Wood Res. 83, 8–16.

Zalifah, M.S., Hamid, S.A., Izran, K., Mansur, A., Nazip, S.M., 2013. Adhesive penetration in laminated oil palm trunk veneer. J. Trop. For. Sci. 25, 467–474.

14

Enhancement of manufacturing process and quality for oil palm trunk plywood

Y.F. Loh, Y.B. Hoong, A.B. Norjihan, B. Mohd Radzi, M. Mohd Fazli, and H. Mohd Azuar

Fibre and Biocomposite Centre, Malaysian Timber Industry Board, Banting, Selangor, Malaysia

14.1 Introduction

The Malaysian timber industry is an important income generator for Malaysia's economy. In 2020, the export revenue of timber and timber products was valued an excess of RM 22 billion. The export value of plywood in the year 2020 was RM 2.84 billion or 1.5 million m^3. Industry sources estimated that the domestic market requirement is about 900,000 m^3 mainly from domestic production plus some imports for the year. The supply of raw materials is essential to further develop this thriving industry toward continuous growth. With natural forests being kept safe via sustainable forest management practices and with heightened awareness in conservation and green practices, the timber industry is looking at alternative raw material.

The challenges faced by the plywood industry in Malaysia can be summarized as follows:

i. The timber processing industry has increasingly experienced shortages in wood supply.
ii. Supply is declining as natural forests are being depleted and efforts for sustainable forest management increase.
iii. Rubberwood, the prevalent species in Southeast Asia, recently experienced shortages due to:
   - Climatic (prolonged wet periods)
   - Unavailable in more desirable large blocks
iv. Oil palm plantations are increasingly replacing those of rubber.
v. Market competition from China, Indonesia, East Europe, and other resources.

Oil Palm Biomass for Composite Panels. https://doi.org/10.1016/B978-0-12-823852-3.00019-2
Copyright © 2022 Elsevier Inc. All rights reserved.

There are 5.8 million hectares of oil palm plantations in Malaysia. After 25 years, oil palm trees will begin yielding less fruit and need to give way for the replanting of new young trees. This presents a boon to the timber industry as the felled oil palm trunk (OPT) is a renewable source of alternative raw material. It is expected that close to 11 million OPT logs will be felled to make way for replanting every year. This alternative material has come at the right time for the timber industry. With tough competition in the global marketplace, rising costs, and tight supply of timber, the timber industry has been looking for a revival.

Oil palm tree is a rich resource of lignocellulosic raw material and can be used for various applications in the wood industry, e.g., plywood, sawn timber, and particle- and fiber-based products. The use of OPT in the production of plywood is what we believe to be the best substitute for hardwood and other logs that would be made from the former. Such a development would also be able to support forest conservation activities in Malaysia.

There are a huge amount of options in utilizing OPT plywood. It varies from construction use for building housing, industrial, and even skyscraper developments. Besides that, there is a massive market of furniture factories that need quality plywood and timber since it's a very stable material, which makes it ideal for use in furniture. The plywood companies have also manufactured several quality plywood for furniture and construction applications such as phenolic film-laminated plywood, furniture grade plywood, fire-retardant plywood, and others to meet the market's demands.

The solution appears to be in OPT, which is able to be processed into plywood. Initiative carried out by Malaysian Timber Industry Board (MTIB) and the timber industry, both in the past and present, has enabled OPT to become a new and important source of raw material for conversion into veneer, plywood, and then to furniture. To date, MTIB has encouraged and promoted a number of activities particularly in the processing of OPT into value-added downstream products for further utilization in furniture industries.

The main destination for Malaysian plywood in the year 2020 are East Asia (mainly Japan, Korea, and Taiwan), U.S., Australia, and Europe. These traditional markets can be utilized to good advantage for oil palm plywood (Table 14.1).

Efforts have been ongoing to develop and promote oil palm plywood to the first-world markets, which tend to appreciate nonforest timber products, as a sustainable product using plantation resources that are renewable.

Table 14.1 Markets for oil palm plywood.

	Product item	Market target
1.	Exterior and interior grades to Malaysian standard	Domestic market
2.	Exterior quality of oil palm plywood	UK, Europe, U.S.
3.	Marine grade oil palm plywood	Domestic and export
4.	Floor base 6 × 3 ft. to JAS	Japan
5.	Concrete panel 6 × 3 ft. to JAS	Japan
6.	Furniture grade oil palm plywood	U.S., domestic

14.2 Plywood from oil palm trunk

Despite its imperfections, OPT is regarded as one of the potential alternative raw materials for the wood-based industry (Anis et al., 2011). However, plywood manufactured from OPT is still being used in nonstructural applications, for example, packaging materials and cabinets. This is because oil palm has inferior strength properties compared to that of tropical wood. It is very difficult for OPT to match the success of rubberwood or other tropical wood in plywood manufacturing. However, Anis et al. (2011) have taken the initiative to produce plywood from OPT with enhanced quality. The processing flow of OPT plywood is shown in Table 14.2.

In the study, Anis et al. (2011) proved that OPT plywood with acceptable strength could be successfully produced from the trunks of the oil palm tree. However, some modifications on the processing parameters are required. Around 1800 pieces of OPT veneers (3.7 mm) could be peeled in 1 h. A longer drying time of 35–45 min is needed to dry the OPT veneers to a final moisture content (MC) of 5%–10%, around two times longer than the drying of wood veneer. As for glue spread rate, the amount of glue needed for bonding OPT veneers is 20%–30% higher than that of conventional wood veneers. In addition, longer pressing temperature and time are also needed. After the finishing stage, it was noted that a recovery rate of 47% was achieved, which was mainly lost during the peeling stage. In other words, for every 1 m^3 of OPT, 0.47 m^3 can be used for plywood manufacturing.

256 Chapter 14 Enhancement of manufacturing process and quality for oil palm trunk plywood

Table 14.2 Production flow of oil palm trunk plywood.

Processing stage	Description
Selection of OPT logs	– OPT are collected during replanting
	– Oil palm trees with good straightness and uniformity throughout the trunk are preferential
Log handling	– Storage and handling are highly similar to that of conventional plywood mills
	– Normally the storage period of OPT in the open area are within 2–3 weeks
	– At the mill, the OPT logs are cut into shorter dimensions for ease of further processing
Peeling of veneers	– First peeling process to remove bark of the OPT logs
	– Second peeling to obtain OPT veneers
Drying of veneers	– 40–45 min are taken to dry the veneer under high temperature with steam
Gluing of veneers	– Resins are applied by glue spreader
	– Urea formaldehyde and phenol formaldehyde are the most commonly used resins where the former is intended for interior grades and latter is for exterior grades
Pressing	– Cold-pressing followed by hot-pressing
	– More or less similar to the conventional plywood lines with the exception of pressing temperature and time
Finishing	– Edge trimming and sanding

Source: Anis, M., Kamarudin, H., Astimar, A.A., Wahid, M.B., 2011. Plywood from oil palm trunks. J. Oil Palm Res. 23, 1159–1165.

14.3 Production process of oil palm trunk plywood

14.3.1 OPT at the plantation

Oil palm trees at the plantation are usually felled after 25 years for the purpose of replanting. The OPT logs are cut to lengths of 9 and 18 ft. as required by the plywood and timber factories (Fig. 14.1). The fresh cut OPT logs are then loaded onto timber trucks for delivery to factories.

14.3.2 Log preparation

Generally, the OPT logs delivered to the factory are processed fresh, i.e., within 14 days from date of receipt. The first process is cross-cutting wherein the OPT logs are cut to the desired length. Then the OPT logs are debarked or rounded up at the rotary lathe to obtain a cylindrical block.

Fig. 14.1 OPT log bucking at a log yard.

14.3.3 Log peeling

The peeling process is undertaken at the rotary lathe and the spindleless lathe, converting the OPT logs into thin sheets of veneer (Fig. 14.2). The wet OPT veneers are then clipped to the required size and then sorted out according to veneer thickness and veneer quality for the next process.

14.3.4 Veneer drying

OPT veneers have a high MC and need to undergo a drying process, a critical process in the production of plywood. There are two types of dryers used by the manufacturer: a roller dryer and platen dryer (Fig. 14.3).

14.3.5 Gluing

A glue spreader is used to apply glue onto the dried OPT veneers. The common glues used in plywood making are urea formaldehyde and phenol formaldehyde.

14.3.6 Cold-pressing

The cold-pressing process, sometimes called prepressing, uses a cold-press and ensures that the glue is uniformly spread onto the surface of the OPT veneer.

Fig. 14.2 Peeling process of OPT.

Fig. 14.3 Roller conventional drying process for OPT veneers.

14.3.7 Hot-pressing

The prebonded or unformed plywood panel is then fed into a hot-press (Fig. 14.4). This process is undertaken at a predetermined pressure, temperature, and duration to cure the thermosetting glue.

Fig. 14.4 Hot-pressing for OPT plywood.

14.3.8 Panel sizing

The formed plywood panel is then sized, or edge trimmed, to the required size.

14.3.9 Panel sanding

The surfaces of the plywood panel are then sanded to give it a smooth finish.

14.3.10 Grading and packing

The finished palm plywood is then inspected and graded according to the buyer's specifications. The goods are packed and ready for delivery (Fig. 14.5).

14.4 Enhancement of oil palm plywood manufacturing process

The area of greatest potential in the optimal utilization of OPT lies in the production of wood-based products. These abundant and readily available fiber materials make excellent alternative sources of lignocellulosic materials suitable in the manufacturing of plywood. Nevertheless, inherent features in OPT, such as high MC, high density variation, and diverse anatomical structure along the stem, make it less attractive for further processing.

Fig. 14.5 OPT plywood ready for marketing.

In plywood, the strength of a bonded joint depends on how well the adhesive penetrates and forms an anchor with the wood cells. The crucial problems in utilizing OPT in plywood manufacture is its extremely high MC (i.e., 200%–400%) that requires a much longer drying time compared to wood. Depending on how efficient the boiler system is in the mill, a normal drying time for wood veneers is approximately 25 min (on a 15-m roller veneer dryer) while a typical OPT veneer takes at least 1 h. Even though several companies have initiated commercial production of OPT plywood, the long drying time, high density variation (180–460 kg/m^3), and low quality are the three main challenges that need to be solved for full commercialization of this product.

Besides conventional dryers, the predrying process and the auto platen dryer have been developed to ensure higher efficiency to dry OPT veneers and reduce the drying time to save the production cost. Predrying veneer using the heating chamber with platen dryer or roller veneer dryer, on the other hand, is to enhance the quality of the OPT veneers and the resulting plywood at a comparable cost. It is anticipated that the loosely bound parenchyma tissues will significantly compact and increase its density. As a result, the density gradient within the veneer is reduced, and thus a stronger and more stable plywood can be produced. The improvement in palm plywood manufacture related to veneer segregation (density), enhanced drying process, and automation feeders have been carried out to overcome the crucial problems facing the manufacturers as mentioned.

The chemical treatment of veneer with phenolic or equivalent resin polymer or other chemicals, on the other hand, is to enhance the quality of the OPT veneers and the resulting plywood. It is anticipated that the loosely bound parenchyma tissues will significantly absorb the chemicals and, once cured, increase its density. As a result, the density gradient within the veneer is reduced, and thus a stronger and more stable plywood can be produced. The works done on the enhancement of OPT plywood via phenolic resin treatment are shown in Table 14.3.

Table 14.3 List of studies on oil palm plywood enhancement.

Method—variables	Findings	References
Veneer pretreatment—soaking in phenol formaldehyde resin for 20 s	– plywood made from outer layer veneers: improves the MOR by 115%, MOE by 70%, shear strength (dry test) by 134%, and shear strength (wet test) by 197% – plywood made from inner layer veneers: improves the MOR by 60%, MOE by 43%, shear strength (dry test) by 90%, and shear strength (wet test) by 125%	Loh et al. (2010)
Veneer pretreatment—soaking in phenol formaldehyde resin for 20 s	– 44% increment in resistance against termite attack – 69% increment in resistance against white-rot fungi	Loh et al. (2011a)
Veneer from outer and inner layer of oil palm trunk—100% inner, 100 outer, and mixed	– plywood made from outer oil palm stem had better mechanical properties	Loh et al. (2011b)
Veneer pretreatment—soaking in phenol formaldehyde resin for 20 s	– surface characteristic and density of OPS veneers was improved significantly	Loh et al. (2011c)
Veneer pretreatment—soaking for 20 s in phenol formaldehyde resin of three different molecular weights (i.e., 600 (low), 2000 (medium), and 5000 (commercial))	– veneers treated with low molecular weight PF (LMwPF) resin had significantly higher volume percent gain and weight percent gain	Ab Wahab et al. (2012)
Impregnation—resin solid content of 15%, 23%, 32%, and 40%	– higher mechanical properties in plywood made with oil palm veneer treated using higher resin content – formaldehyde emission increased along with increasing solid content	Hoong et al. (2013)

Continued

Table 14.3 List of studies on oil palm plywood enhancement—cont'd

Method—variables	Findings	References
Impregnation—hot-pressing time of 14, 16, 18, and 20 min	– mechanical strength improved along with increasing hot-press time	Hoong and Paridah (2013)
Impregnation—press pressure of 20 bar, 30 bar, 2 stage 20 bar + 30 bar, and 2 stage 20 bar + 50 bar	– two stage pressure resulted in better mechanical properties of the produced plywood	Hoong et al. (2012)

The OPT plywood manufacturing process is the same with the conventional plywood process, from log yard, peeling, drying, gluing, pressing, finishing, etc. Based on the natural characteristics of oil palm wood, such as rough surface, high MC, and density variation as mentioned earlier, the industries need to modify or enhance the process. Researchers in Malaysia have done some work previously like plywood quality enhancement with resin treatment, process/machinery upgrade for dewatering, peeler with segregation, predrying chamber, platen/roller press on the drying process and later on densification, and chemical treatment to suit the furniture requirements.

14.4.1 Enhanced process flow of oil palm plywood

MTIB conducted factory trials with Profina Plywood Sdn. Bhd. located in Kluang, Johor on better processing and plywood improvement with additions and modifications as shown in Fig. 14.6.

14.4.1.1 Green veneer segregator

A device was used to detect green OPT veneer by weight to segregate or separate the OPT veneers into two categories with separate veneer outfeed conveyors (Fig. 14.7). This innovation will have a major positive effect on the efficient handling of OPT veneers, increasing the drying efficiency because of similar density groups, and will have a better distribution of density in the oil palm plywood in later processes.

Chapter 14 Enhancement of manufacturing process and quality for oil palm trunk plywood **263**

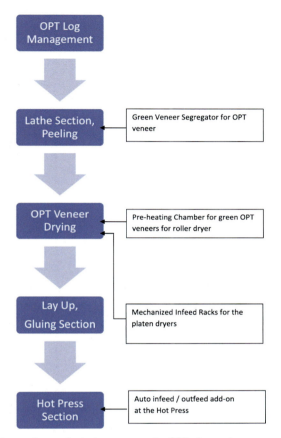

Fig. 14.6 Enhanced manufacturing process for OPT plywood.

Fig. 14.7 Peeler with two separators in the OPT plywood factory.

14.4.1.2 Preheating chamber

A preheating chamber heats up the green OPT veneers to about 100 °C, at which temperature the veneers will start to release moisture and, when infeed into the roller dryer, will dry at a much faster rate.

14.4.1.3 Mechanized infeed rack for platen dryer

Infeed racks are used for the efficient infeed of green OPT veneers into the platen dryers during the drying process. This method of handling better preserves the integrity and quality of the green OPT veneers and also improves the productivity of drying at the platen dryer section.

14.4.1.4 Auto infeed and outfeed for hot-press

A mechanized stacking rack is used for the infeed of the unformed panels and for the outfeed of the hot-pressed formed OPT plywood panels. This add-on to the hot-press can reduce the loading and unloading time of the operation, eliminate the incidence of precured panels, and increase cycle/hour because of shorter feeding time. The workforces were reduced when using the automated mechanism.

14.5 Conclusions

This new processing system can be used as a standard/model that will reduce the drying time of the processing and increase production capacity and quality, thus developing the OPT plywood industry in Malaysia.

The benefits or effects achieved are as follows:

i. Increase in production capacity
ii. The plywood industry has increased production capacity at least 40%
iii. Increase the use of OPT (biomass) in plywood production
iv. With the increase in plywood production capacity, the use of oil palm trunk also increased
v. Produce more competitive plywood in terms of quality
vi. Addition or modification of the machine can also improve the quality of veneer and plywood as follows:
   (a) *Peeler with separator*
      - The veneer obtained has less defect splitting with the automatic separator
   (b) *Drying system*
      - Drying veneers in platen dryer using auto feeders can reduce fraction and breakage during manual handling; preheating chamber with platen dryer can further compress the veneer, and this will increase the density/strength of the veneer

(c) *Hot-press with infeed and outfeed automation system*
- Hot-press with automation improves the quality of the plywood where the delamination problem can be minimized thus enhancing worker safety.

MTIB then proceeded to assemble several types of furniture using the improved plywood manufactured, and initial trials found that they were sufficient to produce good quality furniture. Workability was described as good by the furniture makers. The furniture produced were console tables, cabinets, drawers, etc. There are still avenues that can be explored in the utilization of oil palm trunk as a source of raw material for the plywood industry in Malaysia. It is hoped that with the increased potential of oil palm wood, the uptake of this material will continue to increase in the wood industry. Several problems can be mitigated such as the lack of raw material and also the disposal of oil palm replanting residues.

On the home front, the furniture sector offers many possibilities for palm plywood. In 2020, timber-based furniture achieved earnings in export revenue of about RM 10.6 billion, contributing 48% of the total in revenue for Malaysian timber products. The OPT has shown exciting potential for the making of palm plywood and has become an alternative raw material for the wood-based industry. Malaysian furniture manufacturers produce a wide range of furniture suited to all indoor and outdoor spaces such as office, home, and garden. With the furniture industry being highly export-oriented, it provides a promising future for Malaysia to produce high quality palm plywood for the domestic and export markets. Given also that consumers in developed countries are demanding that wood-based and furniture products in the market be eco-friendly, palm plywood and palm furniture have the potential to be the next wave.

References

Ab Wahab, N.H., Paridah, M.T., Hoong, Y.B., Ashaari, Z., Mohd Yunus, N.Y., MKA, U., Shahri, M.H., 2012. Adhesion characteristics of phenol formaldehyde pre-preg oil palm stem veneers. Bioresources 7 (4), 4545–4562.

Anis, M., Kamarudin, H., Astimar, A.A., Wahid, M.B., 2011. Plywood from oil palm trunks. J. Oil Palm Res. 23, 1159–1165.

Hoong, Y.B., Loh, Y.F., Chuah, L.A., Juliwar, I., Pizzi, A., Paridah, M.T., Jalaluddin, H., 2013. Development a new method for pilot scale production of high grade oil palm plywood: effect of resin content on the mechanical properties, bonding quality and formaldehyde emission of palm plywood. Mater. Des. 52, 828–834.

Hoong, Y.B., Loh, Y.F., Nor Hafizah, A.W., Paridah, M.T., Jalaluddin, H., 2012. Development of a new pilot scale production of high grade oil palm plywood: effect of pressing pressure. Mater. Des. 36, 215–219.

Hoong, Y.B., Paridah, M.T., 2013. Development a new method for pilot scale production of high grade oil palm plywood: effect of hot-pressing time. Mater. Des. 45, 142–147.

Loh, Y.F., Paridah, M.T., Hoong, Y.B., Bakar, E.S., Hamdan, H., Anis, M., 2010. Properties enhancement of oil palm plywood through veneer pretreatment with low molecular weight phenol-formaldehyde resin. J. Adhes. Sci. Technol. 24 (8–10), 1729–1738.

Loh, Y.F., Paridah, M.T., Hoong, Y.B., Bakar, E.S., Anis, M., Hamdan, H., 2011a. Resistance of phenolic-treated oil palm stem plywood against subterranean termites and white rot decay. Int. Biodeterior. Biodegradation 65, 14–17.

Loh, Y.F., Paridah, M.T., Hoong, Y.B., 2011b. Density distribution of oil palm stem veneer and its influence on plywood mechanical properties. J. Appl. Sci. 11 (5), 826–831.

Loh, Y.F., Paridah, M.T., Hoong, Y.B., Yoong, A.C.C., 2011c. Effects of treatment with low molecular weight phenol formaldehyde resin on the surface characteristics of oil palm (Elaeis quineensis) stem veneer. Mater. Des. 32 (4), 2277–2283.

15

Oriented strand board from oil palm biomass

N.I. Ibrahim[a], S.O.A. SaifulAzry[a], M.T.H. Sultan[a,b,c], A.O. Fajobi[b], S.H. Lee[a], and R.A. Ilyas[d,e]

[a]*Laboratory of Biocomposite Technology, Institute of Tropical Forestry and Forest Products (INTROP), Universiti Putra Malaysia, Serdang, Selangor, Malaysia,* [b]*Department of Aerospace Engineering, Faculty of Engineering, Universiti Putra Malaysia, Serdang, Selangor, Malaysia,* [c]*Aerospace Malaysia Innovation Centre (944751-A), Prime Minister's Department, MIGHT Partnership Hub, Cyberjaya, Selangor, Malaysia,* [d]*School of Chemical and Energy Engineering, Faculty of Engineering, Universiti Teknologi Malaysia, Johor Bahru, Malaysia,* [e]*Center for Advanced Composite Materials (CACM), Universiti Teknologi Malaysia, Johor Bahru, Malaysia*

15.1 Introduction

A million hectares of land in Malaysia is currently occupied with oil palm plantations that generate huge amounts of biomass. With more than 19.5 million tons of crude palm oil produced in 2018, the total area planted stood at 5.8 million hectares. The total biomass generated is estimated at approximately 84.74 million tons (Abas et al., 2011) including about 8.2 million tons of oil palm trunks (OPT; Abdul Khalil et al., 2011). All tree components have easy accessibility and renewable capacity, which are extremely reliable. By doing that, the residues of the oil palm can be squeezed into one natural fiber biomass, particularly in the bark of the trunk section. The use of natural biomass is environmentally friendly, sustainable, and biodegradable due to its continuous development. That is why natural fiber biomass is included in the concept of a green applications that are safe and easy to handle.

Usually, an OPT is cut into pieces and burned to prevent events involving insects and diseases, which has led to environmental problems for many years. Questions have been raised about the safety of prolonged use to solve this situation; scientists and engineers must research the OPT as a possible natural source to replace costly, nonrenewable resources. The most widely studied wood-based panel product is plywood. Nonetheless, the oil palm veneer (OPV) has been used for plywood production in Malaysia for many years and was the

Oil Palm Biomass for Composite Panels. https://doi.org/10.1016/B978-0-12-823852-3.00006-4
Copyright © 2022 Elsevier Inc. All rights reserved.

primary veneer for plywood plants (Masseat et al., 2018). From a previous study, according to the Japanese Standard Process, JAS 233:2003, plywood processing from OPT has similar strength properties and is also ideal for the use of nonstructural materials such as packing materials in cabinets (Mokhtar et al., 2018). Consequently, this chapter illustrates the possibilities and new material that is oil palm wood in the development of oriented strands boards (OSB) because this product tends to share closely related plywood elements and properties. It could be possible to create a structure-oriented strand board. OSBs, due to their excellent properties and in particular their competitive price, are being increasingly sought after in the construction materials market. These boards are made of wood chips and adhesives as reinforcement in a composite material (Lunguleasa et al., 2020). The European standards in the field of OSB categorized them as light load-bearing boards and heavy load-bearing boards by field of application; indoor boards are OSB/1 and OSB/2, and outdoor boards are OSB/3 and OSB/4 (ISO16894, 2009); (EN300, 2006). The OSB is effectively derived from a renewable resource in manufacturing. In addition to its highly efficient wood frame construction systems, the natural nature of the wood panel makes it a top energy-saving option. So, the goods and their manufacturing processes have been studied in this chapter.

The mechanical properties of wood panels from raw oil palm (OPT) waste are significantly increased by correct methods and techniques. It also absorbs low production costs that lead to economic growth and is very comprehensive in different industries. Besides, most technologies have been quickly adopted and adapted in industrial countries, some of which have been used, tested, and perfected for more than 25 years, such as medium density fibreboard (MDF). But, there is still no further growth in most countries around the world, such as the OSB, and the market is still expanding (Enters, 1997). Malaysia is the first OSB producer in Southeast Asia since 2015; however, according to MTC (2017), it is also the first in the world to grow tropical hardwood OSBs (e.g., dark red meranti, balau, rubberwood, and others) but not yet materials from oil palm residues.

OSB is a wood-engineered product consisting of thin strands of wood, 8–15 cm in length, 10–30 mm in width, and 0.7–1 mm in thickness (Akrami and Fruehwald, 2009). The density in wood composites is often considered to be a key indicator of board properties (Avramidis and Smith, 1989), as emphasized in this chapter, rather than OSB strand sizes and its material. Some factors, including the types of adhesives, affect the grading the physical and mechanical characteristics of the wood-based panel product. It took approximately three decades and a number of researchers to investigate and evaluate the effects of strand size, raw materials that could be used, the structure of three layers, the different methods of strand alignment, and the effects of alignment on OSB strength to develop product specifications and code that makes OSB panels a primary product for use (Zerbe et al., 2015). Most of the research agrees that the strength of boards is mainly influenced by the mechanical properties of individual strands (Wan-Mohd-Nazri et al., 2011). Nevertheless, the use of OPT as an OSB board is limited in studies, as only 30% covers research conducted in the previous year.

A wide range of research has been carried out to turn the readily available renewable biomass into value-added goods with various applications to avoid problems arising from inappropriate disposal of palm oil biomass (Khalil et al., 2012). Oil palm biomass has also been classified as both a solid and liquid waste, with a total of 83 million tons of solid waste and 60 million tons of liquid waste produced in 2012. By 2020, it is expected that this number will have risen by 85–110 million tons and 70–110 million tons for dry solid and liquid waste, respectively (Kong et al., 2014). During replantation, the OPT produced per hectare of soil was estimated to be 75 tons, resulting in a total of 7.3 million tons of waste. Of this number, 50% was expected to remain in the soil nutrient planting for maintenance. Based on 2014 data, a total of 3.6 million tons of OPT, which could generate more revenue for the Malaysia Government, was available to be transformed into value-added goods (Awalludin et al., 2015).

There have been several studies on the development of plywood from OPT, as described earlier. While the strength of OPT plywood is similar to commercial plywood, due to density differences and OPT instability, its mechanical characteristics give it a poor performance (Feng et al., 2011). As a result, there are significant challenges in translating OPT into a product of acceptable quality and consumer acceptance, as another commercial timber has done (Mokhtar et al., 2011). Maybe this is one of the issues toward using them in the development of OSB. However, the association of southeast asian nations (ASEAN) countries, being the world's largest manufacturers of wood-based furniture, which has grown tremendously over the years with demand for rubberwood furniture on the international market, will benefit greatly from the consideration of OPT-oriented strand board as a new wood material for structural applications that will further boost their economy. Following economic downturns, the current international market for furnishing products remains strong. Malaysia needs to start looking for an alternative to stand out against low-cost countries such as China and Vietnam (Surip et al., 2012). Therefore, this preliminary study was conducted to test the performance of the Oriented Strand Board made from OPT (OSB OPT) with phenol-formaldehyde (PF) resin. Such products' potential is based on being used as substitute materials for structural applications.

15.2 Preparation of OSB panels

In 2019, research was carried out at the Institute of Tropical Forestry and Forest Product, Universiti Putra Malaysia in Serdang, Selangor. OPVs were used as the primary raw material in this study. The veneers were cut into strands with dimensions of $12.5 \times 20 \times 0.3$ cm in length, width, and thickness, using a bandsaw. The strands were dried to reduce the content of moisture. The drying process involves placing the strands in an oven dryer at a temperature of 60°C for 24 h to attain moisture content (MC) of less than 5%. The MC was tested using a moisture analyzer. PF adhesive was used for the manufacture of the OSB. PF was bought from Aica Chemicals (M) Sdn. Bhd. The solid adhesive content (SC) of PF supplied by Aica Chemicals (M) Company was 45.6%, with a viscosity of 66 cps at 30°C, pH 12.0–14.5. The dimensions of the boards produced in this research were 30 cm × 30 cm × 1.3 cm as shown in Fig. 15.1, and target densities were $0.64 \, g/cm^3$ and $0.78 \, g/cm^3$ (Fridiyanti and Massijaya, 2018) with an adhesive level of 7%, 9%, and 12%, respectively.

The manufactured OSB consisted of three layers: face layer, back layer, and core layer. The mats were formed manually with a perpendicular arrangement of the face and back layers to the core layer (Ivakiri et al., 2005; Bufalino et al., 2015; Souza et al., 2018). Upon mixing with the adhesive, the facial–core–back layer weight ratio was set at 25%–50%–25% of the strand's total weight, i.e., 1:2:1 (Bufalino et al., 2015). The formed mat was then pressed under 170°C temperature for 7 min using a peak specific pressure of $25 \, kgf/cm^2$ and a pressure meter gauge value of $50 \, kgf/cm^2$. After hot-pressing, the boards were conditioned at room temperature for at least 1 week to achieve equilibrium MC before physical and mechanical testing; the room temperature ranged from 20°C to 30°C and the relative humidity (RH) was 60% to 65%. Meanwhile, the OSB with different resin levels (7%, 9%, and 12%) was prepared with $0.64 \, g/cm^3$ and $0.78 \, g/cm^3$ densities, following the same procedures as shown in Fig. 15.2.

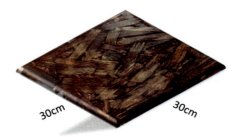

Fig. 15.1 Oriented Strand Board from OPT.

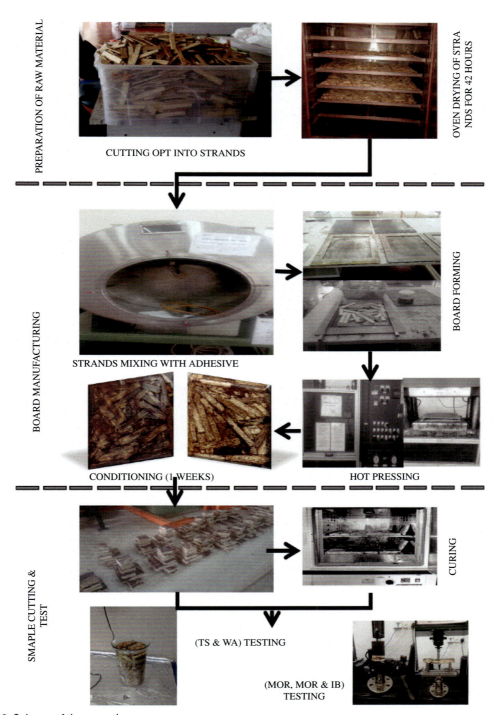

Fig. 15.2 Scheme of the procedures.

In this study, three replicates were produced for every parameter level, and a total of 18 boards (3 resin levels×2 densities×3 replicates) were produced. The boards were trimmed and cut in standard sizes (100×100 mm) before being used for physical tests such as MC. The specimens were dried in an oven dryer for 24 h under temperature 103°C until reaching constant weight board, and then the weight, thickness swelling (TS), and water absorption (WA) (50×50 mm) were measured. The thickness and weight of the specimen were measured before and after soaking in water for 2 and 24 h.

The samples also were cut into smaller specimens for mechanical tests such as modulus of elasticity (MOE) and modulus of rupture (MOR) of the cutting size. Three-points bending was applied over an effective span of 150 mm at a loading speed of 10 mm/minute, and the cutting size for internal bonding (IB) was 50×50 mm. A load was applied to the steel plate at a rate of 5 mm/min. All tests were performed according to the ASTM D143.

15.3 Physical properties

15.3.1 Moisture content

It is important to determine the initial weight of the raw/fresh oil palm strand to ensure that it was adequately dried and met the targeted MC. It is also necessary to calculate the dry oven weight of the oil palm strand using an MC formula. The MC is a physical property representing the board's water content in balance with the RH surrounding it. Rowell and Bart Banks (1987) reported that improvements in the composite board's MC had a major effect on the board's mechanical properties because when the MC decreases below the fiber saturation point, the board's strength will increase. Fig. 15.3 shows the effect of resin level and board density. The sample's MC average ranged between 9.49% and 10.55%, which meets JIS5908, 2003 standard MC that ranges from

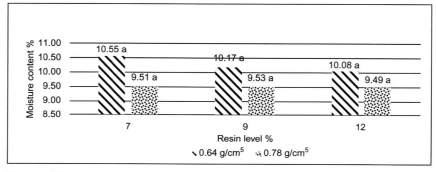

Fig. 15.3 Effect of board density and resin level on moisture content of the OSB.

5% to 13%. This range also was less than 9.30%–11.80% in the previous study (Fridiyanti and Massijaya, 2018). The density and resin level affects the MC of the OSB. The graph shows that, with lower resin levels, the MC of the OSB decreased. It shows that the adhesive will fill the chain's empty space, making it difficult for water to enter and get into OSB. According Hermanto and Massijaya (2018), Hartono et al. et al. (2018), the value of the OPT's MC ranged from 219.9% to 379.4%. This indicates that the OSB humidity quality was much improved.

15.3.2 Thickness swelling (TS) and water absorption (WA)

The TS has been shown to demonstrate OSB sensitivity to water by calculating 2-h (TS2h) and 24-h (TS24h) changes in thickness. Meanwhile, WA is the ability of the board to absorb tested water for the same period when it is soaked in water. The average values of TS2h, TS24h, WA2h, and WA24h are shown in Table 15.1. The average values of TS2h and TS24h soaking ranged from 9.26%–14.23% and 14.81%–18.54%, respectively. This result meets EN300, 2006 standard for grade 0–1 OSB panels, with an average TS24h of 25% and OSB panel grade 0–2 of 20%. Table 15.1 indicates that the greater the density and resin level, the lower the TS and WA percentage. For example, the TS2h immersion of OSB with 7% resin level and density of 0.64 g/cm^3 was higher (14.23%) than OSB with 12% resin level and density of 0.78 g/cm^3, which only has 9.26% TS2h soaking test. Also, in WA2h and WA24h water soaking tests, the results also show the mean value is higher at resin level 7, 0.64 g/cm^3 of density (67.8% and 97.84%),

Table 15.1 Mean values for thickness swelling and water absorption tests.

Resin level (%)	Density (g/cm^3)	Thickness swelling (2 h) %	Thickness swelling (24 h) %	Water absorption (2 h) %	Water absorption (24 h) %
7	0.64	14.23 ± 8.56 a	18.54 ± 05.10 a	67.84 ± 13.09 a	97.84 ± 21.50 a
9	0.64	11.15 ± 6.22 a	16.41 ± 01.07 a	55.52 ± 13.49 a	84.53 ± 20.10 a
12	0.64	10.40 ± 7.83 a	15.25 ± 08.43 a	49.40 ± 15.51 a	73.49 ± 21.10 a
7	0.78	10.26 ± 4.88 a	15.53 ± 05.28 a	46.37 ± 11.57 a	74.48 ± 16.51 a
9	0.78	08.89 ± 5.99 a	14.36 ± 10.90 a	46.14 ± 11.48 a	71.67 ± 15.62 a
12	0.78	09.26 ± 6.41 a	14.81 ± 06.41 a	43.88 ± 01.86 a	66.48 ± 01.99 a

Note: Mean ± standard deviation; means followed with the same letter in the same column are not significantly different at $P < 0.05$.

respectively, than resin level 12% with density 0.78 g/cm^3 (43.88% and 66.48%) respectively. The higher WA was due to the higher consistency between OPT hydrophilic fiber and the higher rate of adhesive distribution (Abdul Khalil et al., 2011).

The research shows that the variation in density and resin ratio distribution in the direction of thickness influence the performance of TS in the OSB. This shows that the inherent swelling features of wood particulate matter contributed to the significant effect on TS. Kawai et al. (1986) stated that the effect was nevertheless reversible. OPT material is strongly hygroscopic and capable of absorbing higher moisture from the surroundings causing TS (Mokhtar et al., 2011; Wahida and Najmuldeen, 2015). Table 15.1 shows that as the resin content and density increase, the OSB WA decreases. The results indicate that the amount of WA percentage at lower density (0.64 g/cm^3) is greater than that at a higher density (0.78 g/cm^3) and decreases at the adhesive rate of 7%, 9%, and 12%, respectively. However, the poor cohesion between the fiber surface and the adhesive contributed to the creation of void structures within the composite, which allowed the absorption of water (Al-Maharma and Al-Huniti, 2019). The test is necessary to determine the resistance of the OSB to water, especially if it is used for external and other meteorologically sensitive uses (rain and moisture). This study shows that the high density and adhesive composition of the board shows good water resistance.

Nevertheless, the findings also show that all average results for all treatments meet the reference value according to EN 300, 2006 standard. The results show that the thickness of the swelling decreases when the resin level increases at a high level, indicating that OSB has better performance at a high resin level of 12% compared to the others. This indicates that the internal adhesion force between the residue adhesive and wood particles is very effective (Souza et al., 2018).

15.3.3 Mechanical properties

The results show an increase in MOR, MOE, and IB, especially when the density is 0.78 g/cm3. In this study, the average values of MOR for all OSB samples in range was 2.9106 N/mm2 to 4.6337 N/mm2. MOR is the major strength of composite boards, such as OSB, and its capacity to sustain the peak pressure of the load. The value of MOR increased with an increase of density and high utilization of resin level at 0.64 g/cm3 OSB with 7%–12% of resin level and also at density 0.78 g/cm3 of OSB manufactured with 7% until 12% of resin level. The statistical analysis indicated that glue spread and density had significantly affected the MOR of composite boards made from OPT waste. For further analysis, using the Tukey method and 95% confidence level, multiple comparison results in grouping information for MOR

Chapter 15 Oriented strand board from oil palm biomass **275**

strength revealed that there is a significantly different effect between an OSB sample with 7% resin level, 0.64 g/cm3 density and an OSB sample with 12% resin level, 0.78 g/cm3 density. This result indicates that the MOR value increased following increasing of glue spread used in composite board production (Hermanto and Massijaya, 2018). The disparity and lower value of MOR shown in Table 15.2 between OSB samples can also be attributed to a low density profile ratio generated from a smooth and even density profile. However, the OSB from this study has a high peak density and low center-density irregular density profile. High peak density improves in bending stress due to higher stiffness. However, to improve the value of MOR, the orientation and sizes of strands should be considered due to the major properties, MOR, and MOE are widely reported to be connected to the density (Wong, 1998; Kelly, 1977).

The results of this finding are similar to those previously conducted by Ahmad et al. (2011), which revealed that particleboard made from OPT with 700 kg/m3 density are 12% resin content (urea-formaldehyde) had the highest values of MOR and MOE compared to the other rates of resin content. The samples also have greater MOE characteristics as MOE increases the board density and adhesive ratio. It showed that the adhesive used in the production of OSB has better stress resistance. The results shown in Table 15.2 shows the trend clearly. The lowest and highest values of MOE were 433.911 and 531.7800 N/mm^2, respectively. These results indicate that the MOE increased with the increase of density and resin level utilization, especially when the density changed from resin level 7%, 0.64 g/cm3 to

Table 15.2 Average values for modulus of rupture (MOR), modulus of elasticity (MOE), and internal bonding (IB).

Resin level (%)	Density (g/cm^3)	Internal bonding (N/mm^2)	Modulus of elasticity (MOE) (N/mm^2)	Modulus of rupture (MOR) (N/mm^2)
7	0.64	0.12080 ± 0.0861 a	433.9113 ± 105.1 a	2.9106 ± 0.05320 b
9	0.64	0.11100 ± 0.0196 a	485.0483 ± 90.30 a	3.8587 ± 0.00003 ab
12	0.64	0.07710 ± 0.0298 a	459.8240 ± 31.20 a	3.7148 ± 0.12250 ab
7	0.78	0.10960 ± 0.0746 a	471.0618 ± 38.20 a	3.8488 ± 0.25500 ab
9	0.78	0.12740 ± 0.0394 a	481.2969 ± 121.2 a	3.9193 ± 0.28700 ab
12	0.78	0.20190 ± 0.1539 a	531.7800 ± 107.1 a	4.6337 ± 1.21300 a

Note: Mean ± standard deviation; means followed with the same letter in the same column are not significantly different at $P < 0.05$.

9%, 0.64 g/cm3 of density. When the density was at 0.78 g/cm3, in line with the increase in the utilization of the resin level, the MOE then increases. This indicates that the relationship between the strength properties and density was related to damage to the cell walls due to densification.

The IB strength is the stress force perpendicular to the panel's plane and is an important force property of composite panels. The IB strength mean values and the pattern of resin level and density varieties are shown in Table 15.2. The highest and lowest values were 0.20190 and $0.07710/mm^2$, respectively, and for OSB, the highest value was 12% resin with $0.78 g/cm^3$ and the lowest IB level was 12% resin with $0.64 g/cm^3$. Even though the lowest values is OSB with 12% resin with $0.64 g/cm^3$ of density, the trend clearly shows that IB increased with the increase of density and resin level. It means that the IB is strengthened with increasing density, and these results are compatible with the values. This study proved that the improvement in board strength and IB was significantly caused by increased pressing temperature, duration, pressure, and adhesive ratio (Nemli, 2003). However, the findings show that all the average results for all treatments are still not meeting the reference value according to EN300: (2006) standard. Sutigno (2000) concluded that extractives found in the cell cavity could minimize IB strength because they could deter adhesives from interacting with components in the cell wall. Nevertheless, this result was attributed to the fact that, in this test, the IB strength to the panel was toward the opposite direction during manufacturing, thus causing the material to be well compressed and glued.

15.3.4 Morphological analysis

Fig. 15.4A and B depict the SEM images of oil palm wood strand and the oil palm OSB panels, respectively, whereas, Figs. 15.5A and B reflect the SEM images of broken surfaces of OSB panels in IB and bending tests, respectively. The SEM images clearly show the surface morphology and microstructure of the OSB panels, indicating the factors that determine the rupture during mechanical testing and revealing its fragile nature, with several stream-like cracks.

Oil palm parenchyma tissues appear to be bowl-shaped, as shown in Fig. 15.4A, consisting of a parallel arrangement of microfibrils with significant amounts of elements on the surface of the fiber (Rosli and Ghazali, 2016). Based on the SEM image taken from the sample surface, it is clear that resin (PF) applied to the particle surface resulted in full contact between the particles during pressing and cured as a function of temperature. Fig. 15.4B indicates the PF on the lumen of the particles, thereby increasing the physical characteristics of OPT biocomposites. Parenchyma also has a high content of starch where

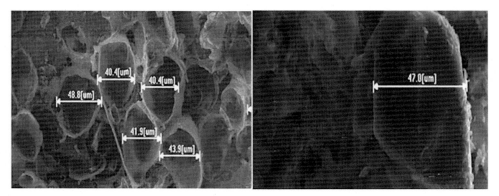

Fig. 15.4 SEM images of oil palm wood strand and OSB panel. (A) Oil palm wood strand and (B) OSB panel.

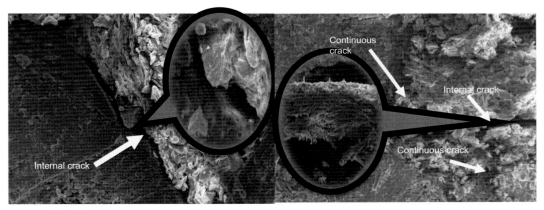

Fig. 15.5 SEM image of OSB samples after mechanical testing. (A) After the internal bonding test and (B) after bending test.

it creates beneficial effects on mechanical properties in line with the variation in size and stiffness of OPT when aiding in the manufacture of binderless panels.

Due to IB test damage, there are some initial internal cracks. As seen in Fig. 15.5A, wood fibers were loosely contacted and the fractured surface of the OSB panel contained some voids. Also, there was even a broken fiber that was pulled apart during the IB test in the OSB panel. The torn fiber was snapped instead of pulled out completely, indicating strong adhesion among the wood fibers, which prevented the recovery of wood fibers and deformation (Ji et al., 2018). It can, therefore, be assumed that the wood fiber adhesion in the OSB panel was high and provided excellent mechanical efficiency and OSB panel water resistance. However, according to Dai et al. (2008), the bonding of two flat-strands reflects more or less the class of veneer-based products from the point of view of material structure, in which the constituent elements are continuous and uniform. So, due to their discontinuous and unpredictable nature, the OSB yield has a substantially lower bonding strength at densification than plywood and laminated veneer lumber. Meanwhile, the specimen in the SEM image has cracks on the bottom side as shown in Fig. 15.5B. In crack morphology following the failure of the bending test, there was an initiative that developed a crack at the edge of the OSB panel. The crack deviated from the initial direction and propagated the load-bearing capability of the OSB panel through the relatively weak interfacial areas between the fiber bundle and the parenchyma tissues. The continuous crack spread rapidly along with the interfacial areas when it started, resulting in drastic interfacial delamination and subsequent fiber bundle failure. With further loading, internal cracks first formed in the parenchyma tissues, while the fiber bundles remained intact spanning the crack wake, functioning as bridges, and inhibiting crack openings by interfacial delamination until their ultimate failure.

Similar to the other composites, the higher strength of the OSB panel composites can be explained by the optimal OSB surface wettability (Flores-Hernández et al., 2014) seen in Figs. 15.4 and 15.5. In the case of natural fibers, it has already been found that unstable interfacial adhesion between fiber and filler frequently leads to fiber pull-out and fiber delamination becoming more likely. As shown in Fig. 15.5, after mechanical analysis, the fracture surface SEM picture of the OSB panel shows numerous prevailing large voids linked to the higher pull-out, weak wetting, and delamination causing mechanism of failure. Some of the cured resins were retained on the fiber surfaces, suggesting insufficient resin penetration. The lack of interfiber bonding was responsible for low IB and bending in all panels containing oil palm fiber.

15.4 Conclusions

OPT retained the original oil palm branch fiber tissue and demonstrated high mechanical properties. As expected, the board's properties increased as board density and adhesive ratio increased. However, in this study for the mechanical and physical properties of the board, there was not a significant difference in the adhesive ratio of 7%, 9%, and 12% for the density of the board. The 7% adhesive panel, however, displays the lowest value performance in all densities for both mechanical and physical properties. In summary, for both physical and mechanical properties measured, boards with a 0.78 g/cm3 density and adhesive ratio of 12% showed high value performance. On the other hand, OSB's method of manufacturing in this study still needs to be improvised by using OPT materials to meet the EN 300 standard requirement for grade 0–1 OSB panel, especially for the mechanical properties. Nevertheless, according to the EN 300 standard for grade 0–1 OSB panels, the physical properties resulting from this study met the standard. The suggested settings would make the goods follow the National Structural Materials Standard requirements. So, due to its good mechanical performance and abundance of supply, OPT can be a promising substitute for the wood resource underwood raw material shortage condition.

Acknowledgments

The authors would like to thank the Institute of Tropical Forestry and Forest Product, Universiti Putra Malaysia, Serdang, Selangor, for financial support as well as research facilities and facilitators that enabled this study to take place.

References

Abas, R., Kamarudin, M.F., Nordin, A., Simeh, M.A., 2011. A study on the Malaysian oil palm biomass sector – supply and perception of palm oil millers. Oil Palm Ind. Econ. J. 11 (1), 28–41.

Abdul Khalil, H.P.S., Nurul Fazita, M.R., Jawaid, M., Bhat, A.H., Abdullah, C.K., 2011. Empty fruit bunches as a reinforcement in laminated bio-composites. J. Compos. Mater. 45 (2), 219–236. https://doi.org/10.1177/0021998310373520.

Ahmad, N., Kasim, J., Mahmud, S.Z., Karim Yamani, S.A., Mokhtar, A., Mohd Yunus, N.Y., 2011. Manufacture and properties of oil palm particleboard. In: 3rd ISESEE 2011 – International Symposium and Exhibition in Sustainable Energy and Environment, No. June, pp. 84–87, https://doi.org/10.1109/ISESEE.2011.5977115.

Akrami, A., Fruehwald, A., 2009. Palms—an alternative raw material for structural application. Centre of Wood Science, University of Hamburg, pp. 1–9.

Al-Maharma, A., Al-Huniti, N., 2019. Critical review of the parameters affecting the effectiveness of moisture absorption treatments used for natural composites. J. Compos. Sci. 3 (1), 27. https://doi.org/10.3390/jcs3010027.

Avramidis, S., Smith, L., 1989. The effect of resin content and face-to-core ratio on some properties of oriented strand board. Holzforschung 43 (2), 131–133. https://doi.org/10.1515/hfsg.1989.43.2.131.

Awalludin, M.F., Sulaiman, O., Hashim, R., Aidawati, W.N., 2015. An overview of the oil palm industry in Malaysia and its waste utilization through thermochemical conversion, specifically via liquefaction. Renew. Sust. Energ. Rev. 50, 1469–1484. https://doi.org/10.1016/j.rser.2015.05.085.

Bufalino, L., Corrêa, A.A.R., De Sá, V.A., Mendes, L.M., Almeida, N.A., Pizzol, V.D., 2015. Alternative compositions of oriented strand boards (OSB) made with commercial woods produced in Brazil. Maderas Cienc. Tecnol. 17 (1), 105–116. https://doi.org/10.4067/S0718-221X2015005000011.

Dai, C., Yu, C., Jin, J., 2008. Theoretical modeling of bonding characteristics and performance of wood composites. Part IV. Internal bond strength. Wood Fiber Sci. 40 (2), 146–160.

EN300: 2006, 2006. Oriented Strand Boards (OSB) – Definitions, Classification and Specifications Platten. issued,.

Enters, T., 1997. "Technology Scenarios In The Asia-Pacific Forestry SECTOR Study." Asia-Pacific Forestry Sector Outlook Study Working Paper Series No: 25.

Feng, L.Y., Tahir, P.M., Hoong, Y.B., 2011. Density distribution of oil palm stem veneer and its influence on plywood mechanical properties. J. Appl. Sci. 11 (5). https://doi.org/10.3923/jas.2011.824.831.

Flores-Hernández, M.A., Reyes González, I., Lomelí-Ramírez, M.G., Fuentes-Talavera, F.J., Silva-Guzmán, J.A., Cerpa-Gallegos, M.A., García-Enríquez, S., 2014. Physical and mechanical properties of wood plastic composites polystyrene-white oak wood flour. J. Compos. Mater. 48 (2), 209–217. https://doi.org/10.1177/0021998312470149.

Fridiyanti, I., Massijaya, M.Y., 2018. Physical and mechanical properties of parallel strand lumber made from hot pre-pressed long Strand oil palm trunk waste. IOP Conf. Ser. Earth Environ. Sci. 141 (1). https://doi.org/10.1088/1755-1315/141/1/012007.

Hartono, R., Iswanto, A.H., Sucipto, T., Lubis, K.M., 2018. Effect of particle pre-treatment on physical and mechanical properties of particleboard made from oil palm trunk. Bioresources 10 (4), 3136–3156. https://doi.org/10.1088/1755-1315/166/1/012006.

Hermanto, I., Massijaya, M.Y., 2018. Performance of composite boards from long strand oil palm trunk bonded by isocyanate and urea formaldehyde adhesives. IOP Conf. Ser. Earth Environ. Sci. 141 (1). https://doi.org/10.1088/1755-1315/141/1/012012.

ISO16894, 2009. ISO 16894:2009 Wood-based panels — Oriented strand board (OSB) — Definitions, classification and specifications buy this standard this standard was last reviewed and confirmed in 2016.

Ivakiri, S., Marin Mendes, L., Karman Saldanha, L., 2005. Produção de Chapas de Partículas Orientadas 'OSB' de Eucalyptus Grandis Com Diferentes Teores de Resina, Parafina e Composição Em Camadas. Ciênc. Florest. 13 (1), 89. https://doi.org/10.5902/198050981726.

Ji, X., Dong, Y., Thang Nguyen, T., Chen, X., Guo, M., 2018. Environment-friendly wood fibre composite with high bonding strength and water resistance. R. Soc. Open Sci. 5 (4). https://doi.org/10.1098/rsos.172002.

JIS5908: 2003, 2003. Japanese Industrial Standard – JIS A 5908: 2003 Particleboards. issued.

Kawai, S., Sasaki, H., Nakaji, M., Makiyama, S., Morita, S., 1986. Physical properties of low-density particleboard. Wood Res. 72, 27–36.

Kelly, M.W., 1977. Critical Literature Review of Relationships between Processing Parameters and Physical Properties of Particleboard.

Khalil, H.P.S.A., Jawaid, M., Hassan, A., Paridah, M.T., Zaidon, A., 2012. Oil Palm Biomass Fibres and Recent Advancement in Oil Palm Biomass Fibres Based Hybrid Biocomposites.

Kong, S.-H., Loh, S.-K., Thomas, R., Abdul, S., Salimon, J., 2014. Biochar from oil palm biomass: a review of its potential and challenges. Renew. Sust. Energ. Rev. 39, 729–739. https://doi.org/10.1016/j.rser.2014.07.107.

Lunguleasa, A., Eliza Dumitrascu, A., Doina Ciobanu, V., 2020. Comparative studies on two types of OSB boards obtained from mixed resinous and fast-growing hard wood. Appl. Sci. 10 (19), 1–15. https://doi.org/10.3390/APP10196634.

Masseat, K., Bakar, E.S., Kamal, I., Husain, H., Tahir, P.M., 2018. The physical properties of treated oil palm veneer used as face layer for laminated veneer lumber. IOP Conf. Ser. Mater. Sci. Eng. 368 (1). https://doi.org/10.1088/1757-899X/368/1/012025.

Mokhtar, A., Hassan, K., Abdul Aziz, A., Basri Wahid, M., 2011. Plywood from oil palm trunks. J. Oil Palm Res. 23 (December), 1159–1165.

Mokhtar, A., Shamim Ahmad, M., Hassan, K., Abdul Hamid, F., Ibrahim, Z., Abdul Aziz, A., 2018. Effect of hot pressing temperature and varying veneer density on the properties of oil palm sandwich board. J. Adv. Res. Fluid Mech. Therm. Sci. 48 (1), 65–79.

MTC, 2017. MTC celebrates silver Jubilee NATIP export target revised to RM25-30 billion best booth award for MTC at Archidex Malaysian Wood Awards 2017 Winners 2nd International Conference on Wood Architecture.

Nemli, G., 2003. Effects of some manufacturing factors on the properties of particleboard manufactured from Alder (Alnus Glutinosa Subsp. Barbata). Turk. J. Agric. For. 27 (2), 99–104. https://doi.org/10.3906/tar-0206-5.

Rosli, F., Ghazali, C.M.R., Al Bakri Abdullah, M.M., Hussin, K., 2016. A review: characteristics of oil palm trunk (OPT) and quality improvement of palm trunk plywood by resin impregnation. Bioresources 11 (2), 5565–5580. https://doi.org/10.15376/biores.11.2.Rosli.

Rowell, R.M., Bart Banks, W., 1987. Tensile strength and toughness of acetylated pine and lime flakes. Br. Polym. J. 19 (5), 479–482. https://doi.org/10.1002/pi.4980190509.

Souza, A.M., Nascimento, M.F., Almeida, D.H., Lopes Silva, D.A., Almeida, T.H., Christoforo, A.L., Lahr, F.A.R., 2018. Wood-based composite made of wood waste and epoxy based ink-waste as adhesive: a cleaner production alternative. J. Clean. Prod. 193 (March 2019), 549–562. https://doi.org/10.1016/j.jclepro.2018.05.087.

Surip, S.S., Jawaid, M., Ibrahim, F., 2012. Design and Applications: Potential and Challenges, No. May 2014.

Sutigno, P., 2000. Effect of aqueous extraction of wood-wool on the properties of wood-wool cement board manufactured from teak (*Tectona Grandis*). In: Proceedings of a Workshop Held at Rydges Hotel, Canberra, Australia, 10 December 2000. 25, pp. 556–568. https://www.cabdirect.org/cabdirect/abstract/19382700446.

Wahida, A.F., Najmuldeen, G.F., 2015. One layer experimental particleboard from oil palm frond particles and empty fruit bunch fibers. Int. J. Eng. Res. Technol. 4 (1), 199–202.

Wan-Mohd-Nazri, W.A.R., Jamaludin, K., Rahim, S., Nor Yuziah, M.Y., Abdul-Hamid, H., 2011. Strand properties of Leucaena *Leucocephala* (lam.) de wit wood. Afr. J. Agric. Res. 6 (22), 5181–5191. https://doi.org/10.5897/AJAR11.514.

Wong, E.D., 1998. Effects of mat moisture content and press closing speed on the formation of density profile and properties of particleboard. J. Wood Sci. 44 (4), 287–295. https://doi.org/10.1007/BF00581309.

Zerbe, J.I., Cai, Z., Harpole, G.B., 2015. An evolutionary history of oriented Strandboard (OSB). In: USDA Forest Service, Forest Products Laboratory, General Technical Report, FPL-GTR-236, 2015; 10 P. 236 (February), pp. 1–10. https://www.fs.usda.gov/treesearch/pubs/49709.

16

Particleboard from oil palm biomass

S.O.A. SaifulAzry[a], S.H. Lee[a], and Wei Chen Lum[b]

[a]Institute of Tropical Forestry and Forest Products (INTROP), Universiti Putra Malaysia, Serdang, Selangor, Malaysia, [b]Institute for Infrastructure Engineering and Sustainable Management (IIESM), Universiti Teknologi MARA, Shah Alam, Selangor, Malaysia

16.1 Introduction

Particleboard is a composite wood engineered for many domestic and industrial usages. Manufacturing of particleboard requires chips, sawdust, or leftover particles from wood. Particleboard is one of the important major timber products in Malaysia. According to IMARC Group (2021), in the year of 2020, the global market of particleboard achieved a total value of $21 billion US. The group forecasts a compound annual growth rate (CAGR) of 4.4% during the next 6 years. In Malaysia, the total export value for major timber products is RM 22,530.70 million in 2019. In 2020, the total export value decreased slightly to RM 22,022.45 million, a decrement of 2.14% compared to the previous year. The import value of major timber products has increased from RM 5964.47 million to RM 6810.61 million, an increment of 14.53% compared to 2019. As for particleboard, in the year of 2019 and 2020, Malaysia has respectively exported a total value of RM 367.09 million and 266.47 million of particleboard. Meanwhile, imported particleboard was recorded at RM 248.29 million and 348.43 million in the year of 2019 and 2020, respectively (Malaysia Timber Industry Board, 2021).

Rubberwood has been used as the main raw material for particleboard production due to the fact that it is a medium density hardwood with a natural color (Balsiger et al., 2000). Before the emergence of oil palm, rubber trees were widely planted due to their economic importance in latex export. Their abundance in number made them suitable for particleboard production. However, rubber plantation area is declining as oil palm generates higher revenues and requires lower input

---

Oil Palm Biomass for Composite Panels. https://doi.org/10.1016/B978-0-12-823852-3.00013-1
Copyright © 2022 Elsevier Inc. All rights reserved.

than rubber plantations. As rubber plantations are fast declining, the use of rubberwood as a raw material for particleboard starts to face shortages. Since then, oil palm residues have been considered to substitute for rubberwood as a raw material for wood-based industries. This is due to abundant amounts of leftover oil palm residues caused by large land use and expansion of oil palm plantations (Hashim et al., 2011a). Moreover, any material that is lignocellulosic can be used in particleboard production (Lathrop and Naffziger, 1994). The leftover oil palm residues are not fully utilized and considered as wastes (Mohammad Padzil et al., 2020). Fully utilizing these residues can prevent wastage while substituting for rubberwood as raw material.

16.2 Oil palm trunk

Oil palm trunk (OPT) has been known as a lignocellulosic material suitable for the manufacturing of value-added composite panels. Conventionally, during the lumber production process, only the outer part of OPT is used because the central region of the trunk contains a high portion of soft parenchyma tissues (Loh et al., 2011). Therefore, only the outer part of the trunk is used to extract lumber. According to Bakar et al. (2006), the unused inner part could made up to 70% of the trunk and is often regarded as waste. These wastes could be a perfect material for the fabrication of particleboard.

Lee et al. (2015) suggested that application of OPT as a core layer of particleboard could result in higher compaction ratio as OPT has lower density than rubberwood. Theoretically, the lower the density of wood materials, the higher the compaction ratio. High compaction is beneficial to the performance of particleboard (Vital et al., 1974). Nevertheless, it is not the case in the study conducted by Lee et al. (2014), where three-layer particleboard using OPT as core layer and rubberwood as surface layers were fabricated. Based on the findings, the authors reported that particleboards made with rubberwood alone has better performance than the mixed species ones. Despite having higher compaction ratio, particleboard made with OPT as core layer still recorded lower strength properties compared to that of pure rubberwood particleboard (Lee et al., 2015). As OPT has lower density, it therefore causes a high volume per unit weight of OPT to occupy the core layer. At the same time, it also led to greater surface area per unit weight, which inhibited adequate coverage of resin applied onto it. As a result, adequate particle-particle bonding could not be achieved, bringing a negative effect on the strength properties of the boards. However, the authors concluded that OPT showed promising potential as an alternative to rubberwood in the production of particleboard provided that the processing parameters, such as pressing temperature and time, are carefully manipulated (Lee et al., 2014, 2015).

Binderless particleboard from OPT has also been fabricated. Baskaran et al. (2015) optimized the pressing temperature and time for binderless OPT particleboards having different thicknesses using Response Surface Methodology (RSM). Based on the findings, for 5-mm-thick particleboard, pressing temperature of 191°C and pressing time of 23 min are the most optimum parameters. As for 10-mm-thick particleboard, the RSM recommended an optimum parameter of 196°C for 24 min. Meanwhile, pressing temperature of 195°C and pressing time of 30 min is the most ideal pressing parameters for particleboard having 15 mm thickness. Using the recommended parameters, particleboards that comply to the Japanese Industry Standard (JIS) A 5908 type 8 (modulus of rupture [MOR] ≥ 8 MPa) could be attained.

A case study was conducted to investigate the properties of particleboard made with OPT. In the study, particleboards were made with OPT particles using four different melamine urea formaldehyde (MUF) resin levels, namely 6%, 8%, 10%, and 12%. The boards with a target density of $700\,kg/m^3$ were hot-pressed at 180°C with 4 MPa pressure for 270 s. Physical properties of OPT particleboard made from different levels of MUF resin are displayed in Table 16.1. OPT particleboards produced in this study have densities ranging from 571.75 to $611.63\,kg/m^3$. Meanwhile, the moisture content (MC) of the particleboard after 24 h of oven drying ranged from 10.62% to 12.25%. The highest MC recorded was 12.25% for the sample made with 10% resin level, while the lowest MC of 10.62% was recorded in the sample made from 12% MUF.

The result for water absorption (WA) and thickness swelling (TS) of the OPT particleboard can be seen in Table 16.1. The lowest value for WA of 75.7% after 24 h of soaking was recorded in the samples made from 8% MUF. On the other hand, the highest WA value was obtained from samples made from 6% MUF resin. It should be noted that there

Table 16.1 Physical properties of oil palm trunk particleboard made from different levels of MUF resin.

Resin level (%)	Water absorption (%)	Thickness swelling (%)	Density (kg/m³)	Moisture content (%)
6	92.55 ± 4.45[b]	43.93 ± 3.23[c]	605.01 ± 39.94[a]	11.45 ± 0.53[a]
8	75.65 ± 3.81[a]	33.37 ± 1.78[b]	611.63 ± 10.83[a]	10.78 ± 2.77[a]
10	83.07 ± 6.46[a]	30.84 ± 1.83[b]	606.05 ± 22.37[a]	12.25 ± 0.24[a]
12	81.81 ± 9.94[a]	25.51 ± 3.26[a]	571.75 ± 33.12[a]	10.62 ± 1.45[a]

Within the same column, mean values followed by different letters a, b, c are significantly different at $P < 0.05$.

is no significant difference between OPT particleboard made from 8% MUF and higher. As for TS, it decreased along with increasing MUF resin levels. The TS of the OPT particleboard produced in this study ranged from 25.51% to 43.94%.

The MOR for the particleboard varied due to different levels of resin ranges from 10.06 to 16.26 MPa as shown in Table 16.2. Meanwhile the modulus of elasticity (MOE) ranged from 1697.804 to 2257.042 MPa. The highest value for MOR was recorded by the sample made with 10% resin, and the lowest MOR was recorded in the samples made with 6% MUF resin. As for MOE, particleboard made from 8% MUF resulted in a highest value, and the lowest value was recorded in the samples made with 12% MUF. There is no significant difference between every level of MUF used. As for internal bonding (IB), the same trend was recorded as the highest IB was recorded in the samples made from 10% MUF resin. Abdul Khalil et al. (2009) reported that by increasing the amount of resin level during OPT particleboard forming, the MOR was expected to increase accordingly.

The correlation between the mechanical properties and resin levels are shown in Table 16.3. From the table, it can be seen that the physical properties of the OPT particleboard decreased along with increasing MUF resin level. Among the tested attributes, TS and MC are most affected. On the other hand, mechanical properties such as MOR and IB increased along with increasing MUF resin level. However, MOE showed a decreasing trend.

16.3 Oil palm frond

On a oil palm plantation, oil palm fronds are generated by pruning and replanting activities. Oil palm frond is still considered as waste, and the common practice is leave it to rot on the site (Fig. 16.1).

As a lignocellulosic material, oil palm frond could be used effectively in the production of composite panels. Unfortunately, the application

Table 16.2 Physical properties of oil palm trunk particleboard made from different levels of MUF resin.

Resin level (%)	Modulus of rupture	Modulus of elasticity	Internal bonding
6	10.06 ± 2.33^a	1909.81 ± 238.33^a	0.41 ± 0.14^a
8	13.56 ± 0.55^{ab}	2257.04 ± 95.77^a	0.43 ± 0.10^a
10	16.26 ± 2.29^b	2151.92 ± 773.30^a	0.69 ± 0.25^a
12	10.89 ± 3.73^a	1697.8 ± 551.66^a	0.56 ± 0.26^a

Within the same column, mean values followed by different letters a, b are significantly different at $P < 0.05$.

Table 16.3 Pearson correlation for mechanical and physical properties of oil palm trunk particleboard as a function of resin level.

Resin level	WA	TS	MOE	MOR	IB
Pearson correlation	−0.488[a]	−0.910[b]	0.176	−0.169	0.390[a]
Sig. (2-tailed)	0.239	0.038	0.490	0.203	0.194

[a] Significant at $P < 0.05$.
[b] Significant at $P < 0.01$.

Fig. 16.1 Oil palm fronds left on plantation site.

of oil palm frond in the production of particleboard is relatively scarce compared to other biomasses. Table 16.4 summarizes the performance of particleboard made with oil palm frond using different processing parameters and binder.

Table 16.4 Properties of particleboard made from oil palm frond.

Density (kg/m³)	Processing parameters	Resin	MOR (MPa)	MOE (MPa)	IB (MPa)	TS$_{24h}$ (%)	WA$_{24h}$ (%)	Reference
700–1100	Temperature—125°C and 150°C Time—5, 10 and 15 s Thickness—6 and 12 mm	Binderless	–	Max 761	Max 1.85		Max 59	Suzuki et al. (1998)
600	Temperature—140°C, 160°C, 180°C, and 200°C Time—10 min Thickness—8 mm	10%, 15%, and 20% citric acid	Max 5.85	Max 1067	≈ 0.35	–	–	Syamani and Munawar (2013)
750	Temperature—160°C and 180°C Time—180 and 300 s Thickness—6 mm	10% Urea formaldehyde (UF)	9.17–11.19	–	0.29–0.42	23–29	67–88	Wahida and Ghazi (2015)
700	Temperature—160°C Time—20 min Thickness—5 mm	UF	11.85–12.40		4.74–5.23	21.60–29.55	127–137	Iling et al. (2019)

Oil palm frond is regarded as a suitable material for the production of binderless particleboard. A study by Laemsak and Okuma (1996) revealed that oil palm frond has 1.5–3 times higher hemicellulose content than wood. Apart from that, a steam-explosion treatment could be used to remove the lignin substance from the cell wall of oil palm frond (Suzuki et al., 1998). During the hot-pressing process, a good bonding strength could be created between oil palm frond fiber and these chemical substances after treatment with steam-explosion. Lignin and polysaccharides in the oil palm frond are the main chemical components that contribute to the self-binding ability (Suzuki et al., 1998). Laemsak and Okuma (2000) fabricated binderless particleboard from steam-exploded oil palm frond fibers at different board densities ranging from 700 to 1200 kg/m^3. The binderless particleboard produced was dark brown in color and had strange, sweet smells due to modification of chemical components during steam-explosion as well as hot-pressing. The mechanical strength of the binderless particleboard improved along with increasing density. The most superior MOR and IB strength was recorded in those boards with density of 1200 kg/m^3. A similar trend was also observed for WA where the denser the board, the lower the WA. In conclusion, the authors successfully proved the feasibility in producing binderless particleboard using oil palm frond.

Apart from that, some studies also demonstrated the feasibility of oil palm frond as an alternative to wood in the manufacturing of particleboard. Fig. 16.2 shows the particleboard made from oil palm fronds in comparison to that of rubberwood. Wahida and Ghazi (2015) reported in their study that a mixture of 30% empty fruit bunch (EFB) and 70% oil palm frond (pressed at 180°C for 300 s) resulted in particleboards that complied with the minimum mechanical strength

Fig. 16.2 Particleboard made from rubberwood (left) and oil palm fronds (right).

requirements stipulated in American and European standards. Iling et al. (2019) recommended that the particleboard made from oil palm frond could be used for indoor applications.

Hashim et al. (2011a) compared the properties of the phenol formaldehyde-bonded particleboard made from various oil palm biomasses such as bark, leaves, fronds, mid-part trunk, and core-part trunk. The findings revealed that particleboard made from fronds displayed the highest MOR among all oil palm biomasses at density of $800 \, kg/m^3$. Meanwhile, at density of $1000 \, kg/m^3$, slightly lower MOR was recorded compared to that of particleboard made from core-part trunk. Hashim et al. (2011b) revealed that the thick fiber walls and high lumen-to-cell wall ratio of oil palm fronds might be the reason for its high bending strength. However, the internal bond strength of the frond particleboard was very much inferior to core-part, mid-part, and bark particleboards, and was only better than leaves particleboard. High TS and WA were also observed in frond particleboards. Using field emission scanning electron microscopy, the authors found that there are some voids that exist on the frond samples, which indicated weak bonding. Consequently, frond particleboard exhibited inferior internal bond strength and dimensional stability compared to the other oil palm biomasses Hashim et al. (2010).

16.4 Empty fruit bunch

EFB is one of the biomasses left behind to waste. Around 4 million tons of fiber are produced annually from EFB (Khalil et al., 2006). The idea of using EFB as a substitution for the raw material in particleboard manufacturing was deemed viable (Ratnasingam et al., 2007). Currently, EFB is solely used as mulch, but high transport costs cause the economics to be marginal. While palm kernel shell (PKS) and fruit fiber are already sufficient for oil palm mill to burn as fuel, EFB still does not have any clear usage in the wood industry (Ellis and Paszner, 1994). EFB is known to have comparable fiber strength to that of rubberwood fiber. Its high toughness and cellulose content make EFB suitable for composite applications (Sreekala et al., 2004; John et al., 2008). However, previous studies showed that mixing EFB with other hardwood species in the production of particleboard is a more feasible approach (SaifulAzry et al., 2015; Lee et al., 2018). SaifulAzry et al. (2015) reported in their study that the particleboard made from 50% EFB and 50% rubberwood exhibited higher bending strength compared to that of the particleboard made with 100% EFB.

A case study has been carried out to evaluate the properties of particleboard made from EFB. Single-layer particleboards with dimensions of $340 \times 340 \times 12$ mm and target density of $700 \, kg/m^3$ were produced. The resin level of MUF used was 6%, 8%, 10% and 12% respectively based on the oven-dry particle weight. The particleboards were hot-pressed for 270 s at 180°C with 4 MPa pressure. The physical and mechanical properties of particleboards, including MOR, density, IB strength, and TS were evaluated.

Table 16.5 summarizes the Pearson correlation between resin level and physical and mechanical properties. The findings indicate that WA and TS of the EFB decrease with increasing MUF resin levels. Meanwhile, all mechanical properties increased as the resin level increased. TS is the most significantly affected property by resin levels.

Figs. 16.3 and 16.4 demonstrate the WA and TS of the EFB particleboard made with different resin loading. The percentage of WA varies from 81% to 97%, as the highest percentage of WA was recorded in particleboard made with 6% MUF resin content level. This result indicates that increasing the level of resin would also reduce the particleboard's

Table 16.5 Pearson correlation for mechanical and physical properties of empty fruit bunch particleboard as a function of resin level.

Resin level	WA	TS	MOE	MOR	IB
Pearson Correlation	− 0.276	− 0.467[a]	0.164	0.297	0.303
Sig. (2-tailed)	0.239	0.038	0.490	0.203	0.194

[a] Significant at $P \leq 0.05$.

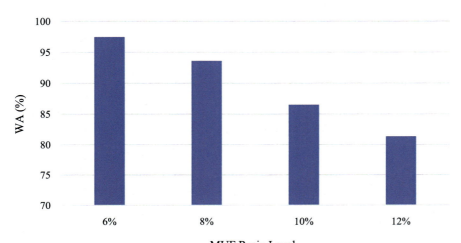

Fig. 16.3 Water absorption (WA) of empty fruit bunch particleboard made with different MUF resin levels.

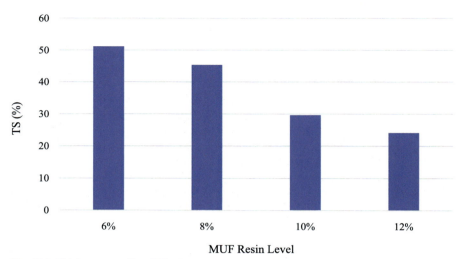

Fig. 16.4 Thickness swelling (TS) of empty fruit bunch particleboard made with different MUF resin levels.

WA. This is because of the existence of the aromatic ring in the melamine molecule, which increased adhesive water resistance (Luo et al., 2019). Another potential reason is that sorption sites within wood cell lumens and cell walls are inhibited by the MUF polymer (Gindl and Gupta, 2002).

As for TS, the highest TS of 51.23% was found in the particleboard made with 6% MUF, while the lowest TS of 24.17% was found in 12% MUF. According to Gatchell et al. (1966), the resin level is the main single variable to control particleboard swelling. The importance of the resin level to improve TS was also demonstrated by Turner (1954). Greater resin additions lead to a board with improved characteristics, especially its thickness stability. Halligan and Schniewind (1974) argued that the thickness of the swelling can be reduced by increasing the resin level. According to Clad (1967), the particleboard TS shows a decreasing pattern when the phenolic resin levels used increased from 4% to 15%. According Ratnasingam et al. (2008), a single fresh EFB comprises 45% cellulose, 32.8% hemicellulose, and 20.5% lignin. EFB's high content in hemicellulose will absorb large amounts of water (Sari et al., 2012). Consequently, the particleboard shows higher than normal TS.

MOE and MOR of EFB particleboard made with different resin levels are shown in Fig. 16.5. The highest MOE and MOR were recorded in the particleboard made from 10% MUF resin. The MOR value of 8.76 MPa surpassed the minimum requirement of bending strength for Type 8 board as stipulated in Japanese Industry Standard (JIS) A 5908. A similar finding was also observed for IB strength (Fig. 16.6), as the highest value was recorded in the particleboard made with 10% MUF resin.

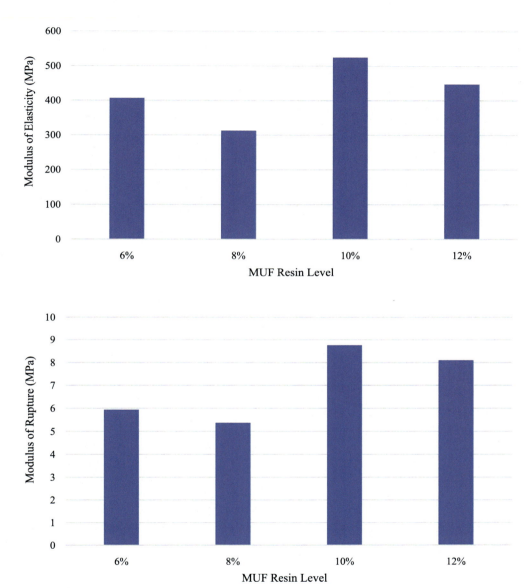

Fig. 16.5 Modulus of elasticity (MOE) and modulus of rupture (MOR) of empty fruit bunch particleboard made with different resin levels.

16.5 Challenges and future prospects

Oil palm biomasses exist abundantly in the vicinity of oil palm plantations as well as processing mills. Effective utilizing of these biomasses could reduce the dependency on natural forests thus reducing deforestation. Moreover, it could assist planters and manufacturers

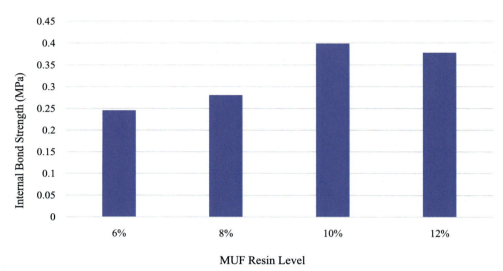

Fig. 16.6 Internal bond strength of empty fruit bunch particleboard made with different resin levels.

in solving the waste disposal problem, which in turn lowers the environmental effects imposed by the oil palm industry (Ooi et al., 2017). Nevertheless, there are several major challenges in utilizing these oil palm biomasses. The first one shall be the confidence of the consumers or end users toward products manufactured from oil palm biomasses. Ratnasingam and Wagner (2009) conducted a survey to evaluate consumer perceptions toward application of EFB particleboard as a furniture material. Most of the respondents agreed that EFB particleboard possesses several advantages such as cheaper in cost, environmentally friendly, and recycled waste fibers. However, the survey also identified some challenges that are faced by EFB particleboard when compared with conventional particleboard including lower attractiveness, inferior machining and finishing characteristics, poorer dimensional stability, ununiform thickness, and rougher surfaces. These are the aspects that must be improved if it is to become widely accepted by consumers. Up until now, the marketing strategy for this material has focused on promoting its environmental friendliness rather than its working properties.

Acknowledgments

This project was funded by Ministry of Higher Education, Malaysia, grant number 5540158, Reference code: FRGS/1/2018/WAB07/UPM//1.

References

Abdul Khalil, H.P.S., Kang, C.W., Khairul, A., Ridzuan, R., Adawi, T.O., 2009. The effect of different laminations on mechanical and physical properties of hybrid composites. J. Reinf. Plast. Compos. 28 (9), 1123–1137.

Bakar, E.S., Fauzi, F., Imam, W., Zaidon, A., 2006. Polygon sawing: an optimum sawing pattern for oil palm stems. J. Biol. Sci. 6 (4), 744–749.

Balsiger, J., Bahdan, J., Whiteman, A., 2000. The Utilization, Processing and Demand for Rubberwood as a Source of Wood Supply. APFC-Working Paper No. APFSOS/WP/50. FAO, Bangkok.

Baskaran, M., Hashim, R., Sulaiman, O., Hiziroglu, S., Sato, M., Sugimoto, T., 2015. Optimization of press temperature and time for binderless particleboard manufactured from oil palm trunk biomass at different thickness levels. Mater. Today Commun. 3, 87–95.

Clad, W., 1967. Phenol-formaldehyde condensates as bonding agents in the production of particleboard. Holz Roh Werkst. 25, 137–147.

Ellis, S., Paszner, L., 1994. Activated self-bonding of wood and agricultural residues. Holzforschung 48 (1), 82–90.

Gatchell, C.J., Heebink, B.G., Hefty, F.V., 1966. Influence of component variables on properties of particleboard for exterior use. For. Prod. J. 16, 46–59.

Gindl, W., Gupta, H.S., 2002. Cell-wall hardness and Young's modulus of melamine-modified spruce wood by nano-indentation. Compos. A: Appl. Sci. Manuf. 33 (8), 1141–1145.

Halligan, A.F., Schniewind, A.P., 1974. Prediction of particleboard mechanical properties at various moisture contents. Wood Sci. Technol. 8 (1), 68–78.

Hashim, R., Saari, N., Sulaiman, O., Sugimoto, T., Hiziroglu, S., Sato, M., Tanaka, R., 2010. Effect of particle geometry on the properties of binderless particleboard manufactured from oil palm trunk. Mater. Des. 31 (9), 4251–4257.

Hashim, R., Nadhari, W.N.A.W., Sulaiman, O., Hiziroglu, S., Sato, M., Kawamura, F., Seng, T.G., Sugimoto, T., Tanaka, R., 2011a. Evaluations of some properties of exterior particleboard made from oil palm biomass. J. Compos. Mater. 45 (16), 1659–1665.

Hashim, R., Nadhari, W.N.A.W., Sulaiman, O., Kawamura, F., Hiziroglu, S., Sato, M., Sugimoto, T., Seng, T.G., Tanaka, R., 2011b. Characterization of raw materials and manufactured binderless particleboard from oil palm biomass. Mater. Des. 32 (1), 246–254.

Iling, E.A., Ali, D.S.H., Osman, M.S., 2019. Effect of pressing pressure on physical and mechanical properties of *Elaeis guineensis* fronds composite board. e-Bangi 16 (3), 1–12.

IMARC Group, 2021. Particle Board Market: Global Industry Trends, Share, Size, Growth, Opportunity and Forecast 2021–2026. Available from: https://www.imarc-group.com/particle-board-market. (Accessed 23 June 2021).

John, M.J., Francis, B., Varughese, K.T., Thomas, S., 2008. Effect of chemical modification on properties of hybrid fiber biocomposites. Compos. Part A 39, 352–363.

Khalil, A.S., Alwani, M.S., Omar, A.M., 2006. Chemical composition, anatomy, lignin distribution, and cell wall structure of Malaysian plant waste fibers. Bioresources 1 (2), 220–232.

Laemsak, N., Okuma, M., 1996. Development of boards made from oil palm frond. I. Manufacturing and fundamental properties of oil palm frond particleboard. In: Proceedings of FORTROP '96 International Conference. November 29–30, Bangkok, Thailand, pp. 71–83.

Laemsak, N., Okuma, M., 2000. Development of boards made from oil palm frond II: properties of binderless boards from steam-exploded fibers of oil palm frond. J. Wood Sci. 46 (4), 322–326.

Lathrop, E.C., Naffziger, T.R., 1994. Evaluation of fibrous agricultural residue for structural building board products-III. A process for the manufacture of high-grade products from wheat straw. Tappi J. 32, 319–330.

Lee, S.H., H'ng, P.S., Lum, W.C., Zaidon, A., Bakar, E.S., Nurliyana, M.Y., Chai, E.W., Chin, K.L., 2014. Mechanical and physical properties of oil palm trunk core particleboard bonded with different UF resins. J. Oil Palm Res. 26, 163–169.

Lee, S.H., Ashaari, Z., Chen, L.W., San, H.P., Tan, L.P., Chow, M.J., Chai, E.W., Chin, K.L., 2015. Properties of particleboard with oil palm trunk as core layer in comparison to three-layer rubberwood particleboard. J. Oil Palm Res. 27 (1), 67–74.

Lee, S.H., Ashaari, Z., Ang, A.F., Halip, J.A., Lum, W.C., Dahali, R., Halis, R., 2018. Effects of two-step post heat-treatment in palm oil on the properties of oil palm trunk particleboard. Ind. Crop. Prod. 116, 249–258.

Loh, Y.F., Paridah, M.T., Hoong, Y.B., Yoong, A.C.C., 2011. Effects of treatment with low molecular weight phenol formaldehyde resin on the surface characteristics of oil palm (*Elaeis guineensis*) stem veneer. J. Mater. Des. 32, 2277–2283.

Luo, J., Zhang, J., Gao, Q., Mao, A., Li, J., 2019. Toughening and enhancing melamine-urea–formaldehyde resin properties via in situ polymerization of dialdehyde starch and microphase separation. Polymers 11 (7), 1167.

Malaysia Timber Industry Board, 2021. Statistics. Available from: https://www.mtib.gov.my/en/information/sources/statistics. (Accessed 23 June 2021).

Mohammad Padzil, F.N.M., Lee, S.H., Ainun, Z.M.A., Lee, C.H., Abdullah, L.C., 2020. Potential of oil palm empty fruit bunch resources in nanocellulose hydrogel production for versatile applications: a review. Materials 13, 1245.

Ooi, Z.X., Teoh, Y.P., Kunasundari, B., Shuit, S.H., 2017. Oil palm frond as a sustainable and promising biomass source in Malaysia: a review. Environ. Prog. Sustain. Energy 36 (6), 1864–1874.

Ratnasingam, J., Wagner, K., 2009. The market potential of oil palm empty-fruit bunches particleboard as a furniture material. J. Appl. Sci. 9, 1974–1979.

Ratnasingam, J., Ioras, F., Macpherson, T.H., 2007. Influence of wood species on the perceived value of wooden furniture: the case of rubberwood. Holz Roh Werkst. 65 (6), 487–489.

Ratnasingam, J., Nyugen, V., Ioras, F., 2008. Evaluation of some finishing properties of oil palm particleboard for furniture application. J. Appl. Sci. 8 (9), 1786–1789.

SaifulAzry, S.O.A., Paridah, M.T., Juliana, A.H., 2015. Optimization of admixture and three-layer particleboard made from oil palm empty fruit bunch and rubberwood clones. In: Agricultural Biomass Based Potential Materials. Springer, Cham, pp. 293–303.

Sari, B., Ayrilmis, N., Nemli, G., Baharoğlu, M., Gümüşkaya, E., Bardak, S., 2012. Effects of chemical composition of wood and resin type on properties of particleboard. J. Lignocellulose 1 (3), 174–184.

Sreekala, M.S., Kumaran, M.G., Thomas, S., 2004. Environmental effects on oil palm fiber reinforced phenol formaldehyde composites: studies on thermal, biological, moisture and high energy radiation effects. Adv. Compos. Mater. 13, 171–197.

Suzuki, S., Shintani, H., Park, S.Y., Saito, K., Laemsak, N., Okuma, M., Iiyama, K., 1998. Preparation of binderless boards from steam exploded pulps of oil palm (*Elaeis guneensis* Jaxq.) fronds and structural characteristics of lignin and wall polysaccharides in steam exploded pulps to be discussed for self-bindings. Holzforschung 52, 417–426.

Syamani, F.A., Munawar, S.S., 2013. Eco-friendly board from oil palm frond and citric acid. Wood Res. J. 4 (2), 72–75.

Turner, H.D., 1954. Effect of particle size and shape on strength and dimensional stability of resin bonded wood-particle panels. For. Prod. J. 4, 210–222.

Vital, B.R., Lehmann, W.F., Boone, R.S., 1974. How species and board densities affect properties of exotic hardwood particleboards. For. Prod. J. 24 (12), 37–45.

Wahida, A.F., Ghazi, F.N., 2015. One layer experimental particleboard from oil palm frond particles and empty fruit bunches fibers. Int. J. Eng. Res. Technol. 4 (1), 199–202.

17

Fiberboard from oil palm biomass

Mansur Ahmad[a], Zawawi Ibrahim[b], Aisyah Humaira Alias[b], Noorshamsiana Abdul Wahab[c], Ridzuan Ramli[c], Fazliana Abdul Hamid[c], and Syaiful Osman[a]

[a]Faculty of Applied Sciences, Universiti Teknologi MARA, Shah Alam, Selangor, Malaysia, [b]Institute of Tropical Forestry and Forest Products (INTROP), Universiti Putra Malaysia, Serdang, Selangor, Malaysia, [c]Malaysian Palm Oil Board (MPOB), Persiaran Institusi, Kajang, Selangor, Malaysia

17.1 Introduction

Malaysia is the world's leading exporter of tropical timber and is also one of the significant manufacturers and exporters of wood-based products. In 2019, Malaysia exported about RM 22.5 billion wood-based products, where furniture products made up the highest export (Anon, 2020). Over the last two decades, the wood-based industry has grown to be one of the country's most significant socio-economic industries. Malaysia's wood-based industry can be divided into several categories: (1) wood and wood products, (2) paper products, and (3) furniture fixtures and medium-density fiberboard (MDF) (Anon, 2009), which is categorized as underwood and wood products. The development of the MDF industry in Malaysia started in 1987 and since then has grown rapidly. The market for MDF has also grown rapidly over the past decade and has significantly contributed to the wood-based sector in Malaysia (Othman and Samad, 2009).

Due to increasing global demand for wood and wood products, government policies open new opportunities to use alternative raw materials, such as oil palm, bamboo, kenaf, and various agricultural residues. At present, wood is the main raw material in particleboard and fiberboard industries worldwide, and Malaysia is dependent on rubberwood (RW) as the main material. The demand for raw materials for MDF is continuously increasing. As a lesser-known and underutilized wood species, agricultural residues have become alternative raw materials to fulfill the demands for lignocellulose-based products. Due to the shortage of RW supply and higher prices, oil palm biomass is a potential source of non-wood materials for this industry. On the other

Oil Palm Biomass for Composite Panels. https://doi.org/10.1016/B978-0-12-823852-3.00003-9
Copyright © 2022 Elsevier Inc. All rights reserved.

hand, the government of Malaysia has promoted the use of other natural fibers for this deficiency and utilized plantation wastes. Oil palm biomass can be a viable replacement for wood that has shown potential due to its high versatility where it has been used for various products.

17.2 Fiberboard and its manufacturing process

Fiberboard is a composite material formed in a refining process by breaking down the mass structure of the material into the constituent fibers at high temperatures and pressure. The processed fibers are then mixed in the binder with resin, normally formaldehyde-based resin, and waxed to form a board by pressing into a fiber-resin structure. In general, fiberboard can be categorized according to its density: low, medium, and high density. Low-density fiberboards (LDF) have a density ranging from 150 to 450 kg/m^3, while the density of medium-density fiberboard (MDF) is 600 to 800 kg/m^3. High-density fiberboard (HDF) or hardboard has a density of 850 to 1200 kg/m^3 and is normally used for floors and the construction industry. In the furniture and construction industry, MDF is more favored than LDF and HDF due to its flexibility in strength, decent surface durability, greater dimensional stability, and lower cost than plywood. MDF has a more uniform board density and has closer edges than other panel composites (Fig. 17.1).

Compared to particleboard, fiberboard is typically made using different processes and for different applications. Particleboard consists of small irregular particles, shavings, and dust from the chipping process. Meanwhile, fiberboard, which is typically MDF, is composed of individual fibers softened by a pressurized steam process. The material is then processed through refiner plates that separate the material

Fig. 17.1 Medium-density fiberboard (MDF).

to produce uniformly sized fibers. In general, fiberboard is considered higher quality than particleboard due to its uniform density, structure, and strength.

A pulping and refining process is typically used in the MDF industry to refine raw materials such as wood or oil palm biomass into fibers. First, raw material is converted into chips using a chipper machine and then screened into the desired sizes. During refining, the chips are exposed to a mechanical operation at elevated temperatures and pressures and are ultimately broken down into individual fibers and fiber particles (Reme and Helle, 1998). Refining is a two-stage method that begins with defibration and ends with the production of fibers. The fiber is then blended with resin, normally urea-formaldehyde (UF) or phenol-formaldehyde (PF). UF resin is usually used for interior applications and PF resin for exterior application mat-forming. The mat is then pressed under pressure and heat to cure the resin at a specified thickness and forms a board. The board is cooled prior to the trimming and machining process. Fig. 17.2 shows the general process flow of MDF production.

Fig. 17.2 General process of medium-density fiberboard (MDF) production. (Source: Mokhtar, A., Hassan, K., Abdul Aziz, A., Yuen May, C. 2012. Oil palm biomass for various wood-based products. In: Palm Oil: Production, Processing, Characterization, and Uses. AOCS Press. https://doi.org/10.1016/B978-0-9818936-9-3.50024-1.)

Pulping is a method where lignocellulosic material is processed into fiber (Karlsson, 2006). Pulping works at breaking down the material's bulk structure into constituent fibers. Mechanical, thermal, chemical, or a mixture of these processes can be used in the pulping phase to separate the wood and other lignocellulosic material into fibers. The idea of mechanical pulping is that lignocellulosic fibers are freed by mechanical means. The advantage of mechanical pulp is its high yield (more than 95%), but it has high energy consumption. The common process used is stone groundwood (SGW) and refiner mechanical pulp (RMP).

On the other hand, thermo-mechanical pulping (TMP) is a variation of RMP, where fiber from the TMP process is produced by using a heat source and a mechanical motion to refine. The TMP consists of steaming the chip materials for a short time before and after the refining process. This chip has a moisture content of approximately 25%–30%, and mechanical force is applied to the chips in a crushing or grinding process, which creates heat and water vapor that softens the lignin, resulting in the separation of the individual fibers (Karlsson, 2006).

Small quantities of chemicals may be added to the TMP process before being defibrated in a refiner. When materials need to be strengthened or mixed with mechanical pulp, chemical pulping is used to provide various characteristics to the component. It is known as chemical mechanical pulp (CMP) or chemical thermomechanical pulp (CTMP). Chemical degradation of lignocellulosic materials occurs in the CTMP process, resulting in low molecular weight compounds, like volatile organic compounds generated by fibers and pulps (Konn et al., 2006). The most successful approach for achieving high yield (between 55% and 90%) and good quality fiber has been a combination of chemical and mechanical pulp (Karlsson, 2006). Chemicals soften the chips during this phase, and the remaining pulping process is mechanically performed in a disc refiner. One of the typical semichemical processes is semichemical neutral sulfite (NSSC). The NSSC method uses sodium sulfite cooking liquor buffered with sodium carbonate to neutralize the organic acids produced from the material during cooking. Table 17.1 compares the parameters between mechanical, thermal, and chemical pulping processes.

TMP is the primary process of MDF processing facilities for the industrial processing of fibers. It is favored because it produces high fiber strength and yield, contributing to better raw material use. In the TMP process, fiber is generated by steam-heating the chips and mechanically separating the fibers in a pressurized refiner. Two basic steps are involved in manufacturing TMP fibers. Fiberization (defibration) is the first step that transforms the original wood structure into single fibers. The goal is the development of pure, long fibers with

Table 17.1 Comparison of pulping processes.

	Refining parameters		
	Mechanical	**Thermal**	**Chemical**
Source	Pulping by mechanical energy	Pulping with heat	Pulping with chemicals and heat
Yield	High (90%–95%)	Intermediate (55%–90%)	Low (40%–55%)
Fiber properties	A high number of fines and fractions	Average fiber properties	Long and pure fibers
Example	Stone groundwood RMP	TMP	Kraft Pulp

minimal debris. The second stage is fibrillation, involving transforming a part of the whole fiber into fibrils and parts of the cell wall (Kano et al., 1982).

The base and center of the TMP phase is grinding, where the extraction and production of fibers are included. Refining means the processing of fibers by applying friction between two rotating discs, as the fibers are compressed and decompressed, leading to geometric changes (Fig. 17.3A and B). A batch of chips is added to the first stage of the refiner (Fig. 17.3C) in refining, and the chips are broken down as they contact the rotor breaking bars. The energy is transmitted to the pulp through compression and frictional forces due to the movement of the disc, the rotating disc, and the surface of the refining segment (Illikainen, 2008).

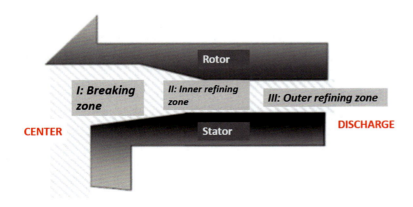

Fig. 17.3 The zones in refiner disc gap. (Source: Illikainen, M. 2008. Mechanisms of Thermomechanical Pulp Refining. Oulu University Press, Oulu.)

TMP extraction strongly affects aggregate fiber properties. Digestion pressure and time during processing in traditional TMP processes are the two key manufacturing parameters that affect fiber properties, thereby affecting almost all MDF panel properties. The pressure and time of digestion are important factors for deciding the isolation and development of fibers. Each adjustment of the pressure control in the refiner changes the temperature (Sikter et al., 2007).

Fiber developments occur by unraveling and peeling the outer layer of fibers, while refinement is used to cut fibers. Karnis (1994) provided a clear example of the process of refining production, as shown in Fig. 17.4. The initial breakdown of chips into shives occurs by particle disintegration, and fiber processing occurs through disintegration and peeling, with peeling being the main mechanism.

During TMP, there are many variables influencing fiber separation. Li et al. (2011) stated that the quality of fibers was determined by two main variables: (1) the quality of chips and (2) the method used in the

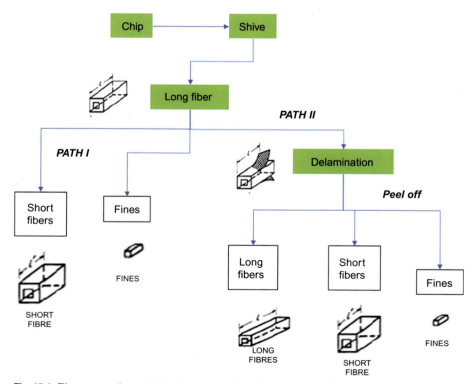

Fig. 17.4 Fiber separation and development in the refining process. (Source: Karnis, A. 1994. The mechanism of fibre development in mechanical pulping. J. Pulp Pap. Sci. 20 (10), J280–J288. https://pascal-francis.inist.fr/vibad/index.php?action=getRecordDetail&idt=4252971.)

refining process. An analysis by Overend et al. (1987) noted how well the final consistency of the refined fibers is determined by the steaming and grinding conditions set during TMP. The mechanism consists of two distinct phases: chip defibration into fibers and fibrillation into fine fibrils of those fibers. Defibration requires broad deformation of plastic where fibers are released from the matrix. Positions such as basic energy and refining strength during defibration are, according to Miles and Karnis (1995), crucial for deciding the final consistency of the generated fiber. Two key factors affect the defibration: the speed of refining and the location where the defibration of the chips occurs. Basic capacity, digestion pressure and temperature, output rate, quality, rotational speed, and plate pattern were among the refining strength parameters (Gorski et al., 2011).

There are two different mechanisms involved during TMP. First, the fibers are softened by the steaming process. They then break up into single or smaller bundles of fiber. Second, these fibers are refined further to have optimal properties, such as fiber length, diameter, and width. The main components of lignocellulosic material are three viscoelastic polymer materials: cellulose, hemicellulose, and lignin, and their softening behavior determines how the material is broken down. The proportion of these polymers differs depending on the form of fiber and the cell wall composition of the fiber, with lignin being the most abundant in the middle lamella.

The raw material is saturated with water throughout the refining process. According to Back and Salmen (1982), cellulose and hemicellulose become softened in water-saturated environments at a temperature of 20°C. Lignin is a viscoelastic polymer that is rigid and glassy at low temperatures (below the glass transition), and where lignin softens is crucial in the fiber separation phase. The hardness of lignin reduces in the transition area of refining, and at high temperatures, it acts like a rubbery substance (Irvine, 1985). If the lignin is too stiff during refining, the fracture occurs throughout the more softened fiber wall, and the refining results in broken fibers and high fine content. If the lignin is too soft, refining results in fibers coated in a layer of lignin, making it more difficult to refine. Refining conditions of 120°C to 135°C were estimated to be essential to make the lignin soft, but the softening temperatures ranging from 100°C to 170°C were also reported to be much larger (Atack, 1982).

Clark (1978) claimed that a decrease in fiber length and debris generation was the main effect of refining. Myers (1983) found that, as the refining becomes more severe, the shortening length of the fiber occurs with more damage. According to Xing et al. (2006), the quality of fiber, which later dominates the output of the final composites, is a key factor in the refining process. This was confirmed by Belini et al. (2008), who discovered that increasing the digestion pressure and heating period reduces fiber length, thus increasing the percentage of broken fiber. A mild condition generated rough fiber, while an extreme condition of refining led to fine material.

In the TMP process, at higher pressure, the defibration occurs deeper inside the fiber structure, through S1 or S2 fiber walls, and results in thinner fiber, resulting in small fiber width and finer fibers (Fig. 17.5). At lower digestion pressures, chip defibration occurs through the middle lamella, releasing coarse, thick fibers with intact lignin layers on the surface from the matrix (Johansson et al., 2011).

There is a great deal of concern about the mechanism that affects fiber surface during the TMP process. Pearson (1983) believed that during the TMP process, the compression and expansion between the bars of the refiner plates with excess steam resulted in higher digestion pressure and longer refining time causing cleavage at the S2 layer. Sundholm (1993) suggested multiple steps to achieve fines, including separating the middle lamella, accompanied by delamination and fibrillation of the fiber wall, as shown in Fig. 17.6. In addition, Karnis (1994) proposed the removal of the main and S1 layers revealing the S2 layer. Gorning (1971) also showed that at high temperatures, lignin softens. Therefore, the middle lamella becomes a region of failure at high temperatures, which makes the neighboring fibers easily cleaved.

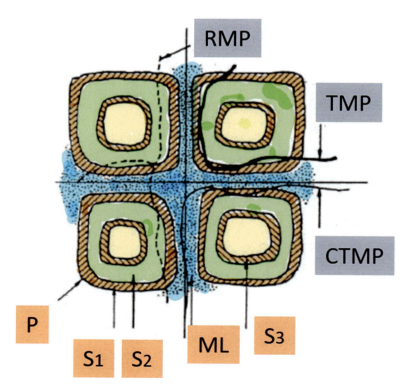

Fig. 17.5 Fracture zones forming upon defibration of chips under different conditions. (Source: Franzen, R.G. 1985. General and selective upgrading of mechanical pulps. Nord. Pulp Pap. Res. J. 1, 13–14.)

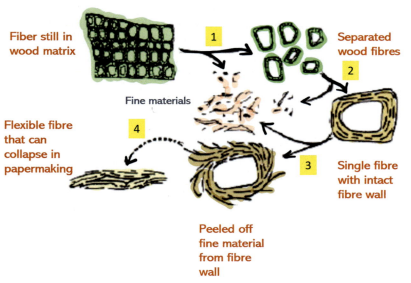

Fig. 17.6 Phases of mechanical pulp refining. (Source: Sundholm, J. 1993. Can we reduce energy consumption in mechanical pulping? In: International Mechanical Pulping Conference, Oslo, 133. https://ci.nii.ac.jp/naid/10011399850/.)

17.3 Fiberboard from oil palm biomass

As a renewable agricultural option for composite materials, oil palm biomass is becoming important. The utilization of oil palm fibers has been developed worldwide. Many researchers have studied the utilization of oil palm fibers and examined the potential of using oil palm frond (OPT), oil palm trunk (OPT), and oil palm empty fruit bunches (EFB) as the main raw material in fiberboard production. Table 17.2 summarizes the production of fiberboard from oil palm fibers.

Table 17.2 Production of fiberboard from oil palm biomass fibers.

Fiber types	Research details	References
Oil palm fronds, trunk, and empty fruit bunches	Production of MDF from OPF, OPT, and EFB fibers, with polyethylene at different percentages, namely 0%, 5%, 12%, and 20%. UF resin was used in this study. The properties of MDF were found to increase when 20% of polyethylene was used.	Mohd Ariff (2005)
Oil palm fronds (OPF)	The binderless fiberboards' mechanical properties (MOR, MOE, IB strength) increased with increasing board density. Refining at steam explosion condition of 25 kgf/cm^2 of steam pressure and 5 min of digestion period displayed the maximum strength of boards.	Laemsak and Okuma (2000)

Continued

306 Chapter 17 Fiberboard from oil palm biomass

Table 17.2 Production of fiberboard from oil palm biomass fibers—cont'd

Fiber types	Research details	References
	Sound-absorbing fiberboards were produced using OPF fibers at different densities and thicknesses. The ideal density of the fiberboard was 0.276 g/cm³· and the sound absorption coefficients of ¾-inch fiberboard were higher than ½-inch fiberboard at low frequencies.	Sihabut and Laemsak (2008)
	Sound-absorbing fiberboards without a cold-press process show better sound absorption capacity	Sihabut and Laemsak (2010)
	Fiberboards produced from OPF cooked for 21 min at a temperature of $162 \pm 2°C$ exhibited the best sound absorption capacity but low fiber yield.	Sihabut and Laemsak (2012)
	Blended OPF fiber with mixed tropical hardwood (MTH) at 20% resulted in better MDF properties and cost-savings.	Ibrahim et al. (2012)
	Fiberboards at lower density, i.e., 1,7 g/cm³, showed better sound absorption at high frequencies.	Bubparenu et al. (2018)
Oil palm trunk (OPT)	Properties of MDF from OPT fibers and MTH sawmill residues were evaluated. All board properties have acceptable strength, with MDF made of tropical hardwood had a higher rupture modulus (MOR) than OPT fibers and mixed fibers.	Onuorah (2005)
	Refining pressure (2, 4, 6, and 8 bar) and preheating time (100, 200, 300, and 400 s) influenced the properties of MDF from OPT fiber. Boards from intermediate refining conditions were found to have better board properties.	Ibrahim et al. (2013)
	The physical and mechanical properties of MDF were affected by density and parenchyma content. MDF at 750 kg/m³ and 10% of parenchyma content are higher in all MDF properties.	Ibrahim et al. (2014)
Oil palm empty fruit bunches (EFB)	Treated EFB fiber with two methods of pretreatment, boiling in water and soaking in 2% of NaOH for 30 min, were used for MDF making.	Ramli et al. (2002)
	The bending strength of the MDF was tested in various ratios of kenaf, oil palm EFB, and EFB-Kenaf admixtures. The results indicated that kenaf and EFB could be mixed to produce MDF for many applications.	Jamaludin et al. (2007)
	Treatment on EFB fiber for MDF making used PF as a matrix. Modification of the EFB fiber treated with the acetic and propionic anhydride enhanced the IB strength of the MDF by increasing the matrix resin surface compatibility and fiber surface hydrophobic properties.	Abdul Khalil et al. (2007)
	Study on the effects of storage time and relative humidity of MDF made from RW and EFB fibers by assessing the mechanical and physical characteristics of the board were at 2-week intervals for 10 weeks.	Abdul Khalil et al. (2008)
	Study on the mechanical and dimensional properties of MDF from treated EFB fiber. The EFB fiber was subjected to the different types of treatment: (1) soaking in 2% of NaOH for 30 min, (2) boiling in hot water for 30 min, (3) a combination of soaking in the NaOH and boiling, and (4) untreated fiber.	Norul Izani et al. (2012)

Table 17.2 Production of fiberboard from oil palm biomass fibers—cont'd

Fiber types	Research details	References
	Production of MDF from mixtures of RW and EFB fibers at three different ratios; 70RW:30EFB, 50RW:50EFB, and 30RW:70EFB. The findings showed that the MDF containing more RW fibers showed higher MOR and MOE. Meanwhile, the MDF with higher content of EFB fibers has higher IB values.	Harmaen et al. (2013)
	EFB fiber was treated with boiling water, soaked in sodium hydroxide (NaOH), or a combination using phenol formaldehyde (PF) at 8%, 10%, and 12% levels. All types of treatments enhanced the dimensional stability of the MDF board. Treated with NaOH and boiling water, the EFB fibers resulted in MDF and decreased bending characteristics.	Norul Izani et al. (2013)
	The effect of different concentrations of sodium hydroxide (NaOH) and acetic acid treatment on the morphological structure of oil palm EFB fibers was investigated.	Ibrahim et al. (2015)
	Study the mechanical properties and fire resistivity of MDF from EFB fibers by adding flame retardant, namely Aluminum hydroxide (ATH), in the PU resin mixture.	Musbah Redwan et al., (2015)
	Better modulus of rupture (MOR), bonding, and swelling was found in MDF from EFB fibers refined at a pressure of 90 bar and pressing time of 7 min.	Zainuddin et al. (2016)

17.4 Fiberboard from oil palm frond (OPF)

OPF was produced during the pruning process and fruit harvesting activities. Traditionally, OPF has been left in the plantation as organic fertilizer and used as animal feed. It has been discovered that oil palm fiber has the highest hemicellulose composition compared to EFB, OPT, coir, softwood, and hardwood fibers (Khalil et al., 2006). OPF fibers were selected as one of the more favored fibers in papermaking due to their shorter and thicker fibers (Wanrosli et al., 2007). Mohd Nor (1997) studied the potential and suitability of fronds as raw material for papermaking and found that 63% yield was obtained from OPF pulping, with tensile strength and tear indices comparable to the softwood pulp indices.

Fiberboard from OPF fibers is widely used for reinforced and laminated materials in a wide variety of applications, including furniture, nonstructural applications, and sound barriers. Binderless OPF boards with a density of $1.2 \, g/cm^3$ made from steam-explosion at a pressure of $25 \, kgf/cm^2$ exhibited good mechanical properties, namely MOR, MOE, and IB. On the other hand, higher IB was found in fiberboard with a thickness of 6 mm compared to 12 mm board. The advantage of using a hot-pressing process of steam-exploded in the fiber processing is that it creates lignin-furfural linkages (Laemsak and Okuma, 2000).

Sihabut and Laemsak (2010) reported that the OPF fiberboard for sound absorption purposes shows a high sound absorption capacity. In the study, they produced fiberboard mats with different finishing processes; (1) self-bond fibers in the mat structure without any pressing, (2) the mat was cold-pressed, and (3) the mat was dried cold-pressed and about 5% of the mat's surface area was then punctured (Fig. 17.7). They concluded that fiberboards made a cold-press process produced low sound absorption capacity due to the tight structure, while the perforated surface fiberboard showed a significantly higher sound absorption at high frequencies than others.

A further study on sound barrier fiberboard from OPF under different refining conditions was studied by the same researchers (Sihabut and Laemsak, 2012). Three cooking times (16,19, and 21 min), refining disc clearance (0.5, 0.6, and 0.7 mm), and density of fiberboard were applied in the manufacturing process. It was found that the longest cooking duration produced fine fibers and better sound absorption capacity. The softness of refined fibers and abundance of fibrillated fibers created an optimum porosity board having a structure that is not too tight or loose. These properties are important for sound barrier applications, where the friction, air viscosity, and vibration of sound movement in the fiber structure determine the sound absorption capacities.

Ibrahim et al. (2012) has shown that 100% OPF, RW, and MTH fibers developed by the MDF pilot scale have excellent properties. The sample MDF boards are 100% OPF and RW and MTH fiber on the MPOB's pilot scale. After the chipping phase, OPF was mixed with RW and MTH and refined at an elevated steam pressure of 5–6 bar for 300 s in an inclined digester where the refined fiber was produced. A mechanical blender was used to spray resin and wax onto the fiber until it

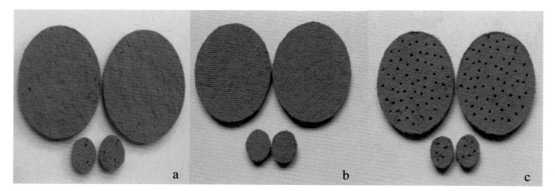

Fig. 17.7 Fiberboards from OPF fibers with various finishing: (A) rough surface (no pressing), (B) screen surface (cold-press), (C) perforated surface. (Source: Sihabut, T., Laemsak, N. 2010. Sound absorption capacity of oil palm frond fiberboard with different finishing 1. Environ. Nat. Resour. 8 (1), 38–43. https://ph02.tci-thaijo.org/index.php/ennrj/article/view/82561.)

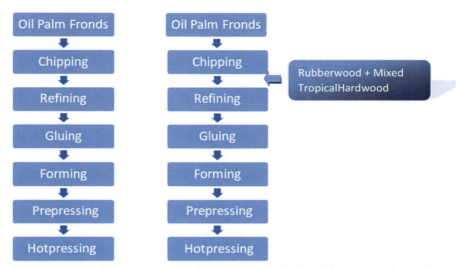

Fig. 17.8 Process flow of MDF production from oil palm fronds, rubberwood, and mixed tropical hardwood (MTH) fibers. (Source: Ibrahim, Z., Aziz, A.A., Mokhtar, A., Ramli, R., Mamat, R. 2012. Production of medium density fibreboard (MDF) from oil palm fronds and its admixture. MPOB Inf. Ser. 601. www.mpob.gov.my.)

formed into a mat. The mat was then consolidated at 2000°C for 300 s in the final phase of hot-pressing. Fig. 17.8 depicts the method flow for making MDF from OPF.

The findings showed that MDF made entirely of OPF and its combination has strong mechanical and swelling properties (Table 17.3). Internal bonding tests on fiberboards made completely of OPF yielded outstanding performance. The overall value of the bending tests and swelling properties met the European standard's specifications (EN 622-5,2006). It suggests that OPF is an appropriate raw material for MDF processing. Blending ratios of OPF from 5% to 20% with RW and MTH were used to determine the properties of partially substituting RW and MTH in panel processing. According to the study, the mechanical properties of the boards improved as the OPF blending ratio was raised. In comparison to the standard, the bending test and internal bonding findings showed higher values, suggesting that the substance is ideal for load-bearing applications. The water intake values for OPF mixture boards are within the acceptable range. MDF from OPF needs a higher volume of a hydrophobic material such as filler and special additive for better dimensional stability.

Bubparenu et al. (2018) fabricated a series of OPF fiberboards with different board densities, namely 0.17, 0.21, 0.26, and 0.30 g/cm^3, and tested them for their sound absorption coefficients (Fig. 17.9). They found that a board with a density of 0.17 g/cm^3 offered an outstanding

Table 17.3 Some physical and mechanical properties of MDF from OPF fiber with fiber ratios.

OPF fiber blending ratio (%)	MOE (N/mm^2)	MOR (N/mm^2)	IB (N/mm^2)	TS (%)
100	2870.33	32.42	1.12	14.26
5	2693.84	32.87	1.17	13.62
10	3514.95	43.89	1.05	13.54
15	3861.49	44.36	1.26	13.76
20	3953.23	45.43	1.28	13.44
EN STD (622–52,006)	>2500	>22	>0.6	< 15

Note: *MOE*, modulus of elasticity; *MOR*, modulus of rapture; *IB*, internal bonding; *TS*, thickness swelling.
Source: Ibrahim, Z., Aziz, A.A., Mokhtar, A., Ramli, R., Mamat, R. 2012. Production of medium density fibreboard (MDF) from oil palm fronds and its admixture. MPOB Inf. Ser. 601. www.mpob.gov.my.

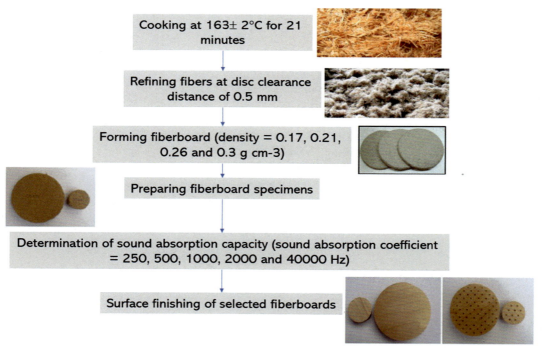

Fig. 17.9 Process flow of OPF fiberboard for sound absorption. (Source: Bubparenu, N., Laemsak, N., Chitaree, R., Sihabut, T. 2018. Effect of density and surface finishing on sound absorption of oil palm frond. Asia Pac. J. Sci. Technol. 23 (4), 1–7. https://doi.org/10.14456/apst.2018.15.)

sound absorption capacity at many frequencies, mainly high frequencies. They also tested the sound absorption capacity of $0.17 \, \text{g/cm}^3$ board covered with wood veneer, but the sound absorption of the materials at high frequencies decreased. However, by using a punched veneer surface area at 5%, the sound absorption of all measured frequency ranges of the material was significantly improved.

17.5 Fiberboard from oil palm trunk (OPT)

Ariff (2005) reported that fiberboard was successfully developed from OPT, OPF, and EFB fibers using polyethylene (PE) as a plasticizer at four different levels, i.e., 0%, 5%, 12%, and 20%. The resin used was UF at 10% of resin level. The results displayed improved performance in the MOE values when incorporated with OPT fibers due to residual parenchyma in the fibers. However, good MOR strength was achieved when incorporating OPF fibers in the fiberboard. They also concluded that the mechanical strength of OPT fiberboard was improved when the PE content was increased to 20%.

Onuorah (2005) studied the effect of dry-formed fiberboard and wet-formed hardboard of OPT fibers, tropical hardwood fibers, and their admixtures on the properties of UF bonded board. They produced two types of hardboard surfaces: (1) two smooth surfaces and (2) one smooth surface. They found that the dry-formed process of MDF of OPF increased the mechanical properties such as MOE, IB, and dimensional stability than the hardboard form OPF, with the MOR value that was found to be higher in the hardboard by 7.58%. However, the MDF and hardboard values were higher in the tropical hardwood boards due to their fiber properties and chemical components. Among hardboard surfaces in OPF hardboard, two smooth surfaces exhibited superior MOR and MOE values than one surface or screen meshed back, contributed by the more cohesive and integral surface to resist force.

A study conducted by Ibrahim et al. (2013) revealed that MDF from OPT fibers could produce certain processing parameters. In this study, the properties of MDF from OPT at four refining pressures (2, 4, 6, and 8 bar) and four different preheating times (100, 200, 300, and 400 s) were investigated. The density of $720 \, \text{kg/m}^3$ and UF resin at 10% was used. They concluded that medium refining conditions, i.e., 6 bar and 300 s, were the best for producing better board properties when compared to low and severe refining conditions. At this refining condition, the OPF fibers produced have effective fiber length, consequently improving the network in the fiber-to-fiber bonding.

The researchers conducted a further study on utilizing OPF fiber in MDF production (Ibrahim et al., 2014). The parameters studied were the density (650 and 750 kg/m³), parenchyma content (10%, 20%, and 30%), and UF content (8%, 9%, and 10%). Some physical and mechanical properties, namely TS, IB, MOR, and MOE, were evaluated according to European Standard EN 622-5:2006. The findings revealed that board density and parenchyma material have a significant effect on MDF properties. The results showed that the properties of the MDF panel at a higher density of 750 kg/m³ are better than that of 650 kg/m³. Compared to panels with higher parenchyma content, boards with a low parenchyma content (10%) (Fig. 17.10) had greater dimensional stability, bonding, and bending strength.

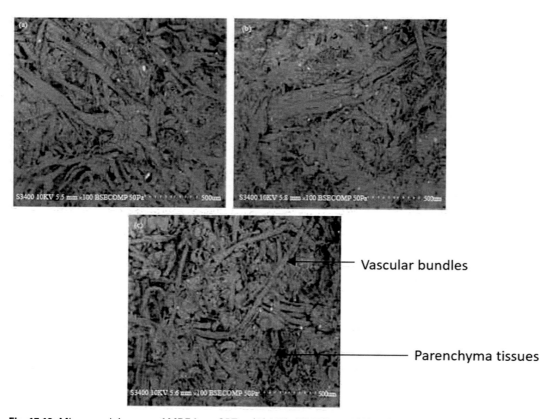

Fig. 17.10 Micrograph images of MDF from OPT at (A) 10%, (B) 20%, and (C) 30% of parenchyma content. (Source: Ibrahim, Z., Aziz, A.A., Ramli, R., Mokhtar, A., Omar, R., Lee, S. 2014. Production of medium density fibreboard (MDF) from oil palm trunk (OPT). J. Appl. Sci. 14 (11), 1174–79. https://doi.org/10.3923/jas.2014.1174.1179.)

It was observed that the resin content affects all panel properties with the exception of MOE. Higher UF resin content resulted in better dimensional stability of the boards, as well as the mechanical properties. As a result, the optimum parameters for MDF made from OPT fibers were 750 kg/m^3, 10% parenchyma content, and a minimum of 8% UF content to produce acceptable MDF properties.

17.6 Fiberboard from oil palm empty fruit bunches (EFB)

The EFB fiber includes about 4.5% of the remaining oil in the form of a coating on the outside surface of EFB fibers (Bakar et al., 2006; Rozman et al., 2001). Several researchers (Ramli et al., 2002; Paridah et al., 2000; Nor Yuziah et al., 1997) argued that the residual oil interferes with the penetration of the binding agent and affects the properties of the final EFB materials such as polymer-based composite, particleboard, and MDF (Ramli et al., 2002; Yusop, 2001; Norul Izani et al., 2012, 2013). The residual oil and silica in the EFB fiber also affected the fiber properties and, thus, the composites. To overcome this problem, fiber pretreatment is required to clean the fiber surface and chemically modify the surface, thus enhancing the final properties of the composites. The treatment can be physical, chemical, mechanical, or a combination of several treatments. Several experiments have been conducted on utilizing EFB for fiber production. For example, some researchers are using acid and alkaline-based chemicals to improve the physical and mechanical properties of EFB fibers (Zuhri Mohamed Yusoff, 2009; Abood et al., 2012).

Ramli et al. (2002) used EFB for the development of MDF and found that MDF can be generated from EFB fiber but required pretreatment. Two pretreatments eliminated the oil: boiling in water and 2% of sodium hydroxide (NaOH). They discovered that the fiber pretreatment eliminated the residual oil and greatly increased the efficiency of the MDF. Pretreatment using NaOH was more effective than boiling in water in extracting the residual oil but resulted in weaker fibers with high bulk density and lowered the properties of the MDF. Meanwhile, water-treated fiberboards were improved in physical and mechanical properties, including MOR, MOE, IB, and NaOH-treated fibers.

The dimensional stability behavior of MDF was analyzed by Abdul Khalil et al. (2007) by treating the EFB fiber with acetic and propionic anhydride and using PF as a matrix. When the acetic and propionic anhydrides were added to the EFB fibers, they reduced the number of hydrophobic groups in the fibers. The blocking of hydroxyl groups also affects the fiber's hydrophilic existence, leading to water being extracted from the substrate. Due to the greater compatibility of the matrix resin surface, the modification of the EFB fiber with acetic and propionic anhydride also increased the IB strength of the MDF. The alteration in the fiber's chemical structure modified the fiber's moisture absorption capacity, thus increasing the IB strength due to better adhesion properties in the fiber-matrix network.

A study was conducted by (Khalil et al. 2010) on utilizing EFB and RW to produce a hybrid MDF with a density of 650 kg/cm^3. The ratio of EFB:RW is 20:80 and 50:50 and mixed with 12% of the commercial UF resin. Both EFB and RW fibers were refined separately under 6–7 bar of refining pressure and 195–200°C of steam pressure to produce refined fibers. It was found that hybrid MDF from EFB:RW, 20:80 have higher flexural strength, flexural modulus, IB strength, and better physical properties. The presence of a large amount of EFB in MDF with the ratio of EFB:RW, 50:50, contributed to the low board properties. Thus, the treatment of EFB fibers is crucial to improve the fibers' properties and finally the board performance.

Harmaen et al. (2013) also evaluated the performance properties of EFB and RW fibers at different fiber ratios; 100RW, 70RW:30EFB, 50RW:50EFB, 30RW:70EFB, and 100EFB as shown in Fig. 17.11. They discovered that the ratio of RW fibers affected the strength of the MDF, with MDF containing more RW fibers being stronger than MDF containing more EFB fibers. In comparison, MDF with higher content of EFB fibers has higher IB values.

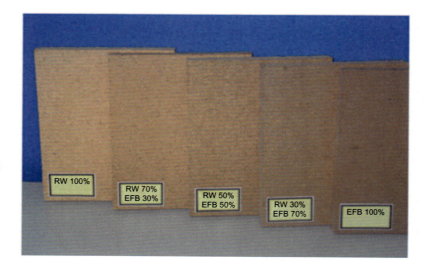

Fig. 17.11 MDF from EFB and RW fibers. (Source: Harmaen, A.S., Jalaluddin, H., Paridah, M.T. 2013. Properties of medium density fibreboard panels made from rubberwood and empty fruit bunches of oil palm biomass. J. Compos. Mater. 47 (22), 2875–83. 10.1177/0021998312459868.)

A similar study was also conducted by the Malaysian Palm Oil Board (MPOB) by using EFB fibers but blended with RW fibers and MTH fibers (Ibrahim et al., 2014). In this study, the ratio of MTH fiber is fixed at 30%, while the EFB:RW fiber ratios were 10:60, 20:50, and 30:40. The shredded fibers were refined at a pressure of 5–6 bars for 300s, mixed with UF resin and emulsion wax, and molded into a mat using a mechanical blender. The mat was then pressed at 200°C for 300s, and the properties of the boards were evaluated. The general manufacturing process of MDF from blended EFB fibers with RW and MTH is illustrated in Fig. 17.12.

Particularly on mechanical properties, the results were very promising. The best results were achieved with a blending ratio of up to 20% EFB fibers. With further EFB fibers loaded, the swelling property deteriorated to the point that it was no longer suitable. Both properties are less than the minimum standards (EN 622-5,2006) for MDF panels made entirely of EFB fibers. Ibrahim et al. (2015) also investigated the effect of fiber treatment on the EFB fiber surface for MDF production. The EFB fibers were treated at different treatments using NaOH and acetic acid at 0.2%, 0.4%, 0.6%, and 0.8% of concentration levels. The study showed that fibers treated with NaOH had markedly removed more oil content than the acetic acid. However, the surface of the NaOH-treated fiber was uneven and had some substances deposited on it, whereas the acetic acid-treated fiber had clear cleavage, a rough surface, and fractures, as shown in Fig. 17.13.

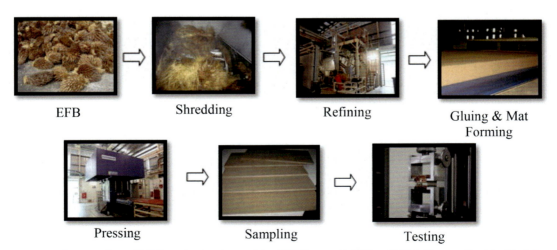

Fig. 17.12 Process flow of MDF production from mixing EFB fibers with RW and MTH. (Source: Ibrahim, Z., Aziz, A.A., Ramli, R., Mokhtar, A., Omar, R., Lee, S. 2014. Production of medium density fibreboard (MDF) from oil palm trunk (OPT). J. Appl. Sci. 14 (11), 1174–79. https://doi.org/10.3923/jas.2014.1174.1179.)

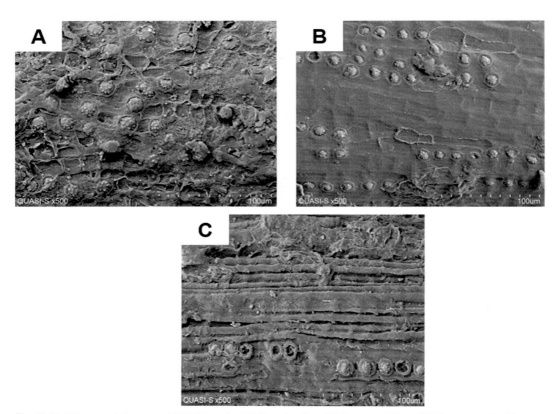

Fig. 17.13 Micrograph images of EFB fibers for MDF production. (A) Untreated fiber; (B) fiber treated with NaOH; and (C) fiber treated with acetic acid. (Source: Ibrahim, Z., Aziz, A.A., Ramli, R., Jusoff, K., Ahmad, M., Ariff Jamaludin, M. 2015. Effect of treatment on the oil content and surface morphology of oil palm (*Elaeis guineensis*) empty fruit bunches (EFB) fibres. Wood Res. 60 (1), 157–66.)

17.7 Conclusions

Oil palm biomass has tremendous potential as a raw material for fiberboard and other composite processing. The database from researchers showed that some of the performance of fiberboard from oil palm fibers has good properties and is comparable with the products made from wood. Nevertheless, specific processing practices and treatments for oil palm fibers were required based on the oil palm biomass properties. The fiberboard from oil palm biomass can be used in many applications such as sound barriers, furniture, wall panels, as well as for nonstructural materials.

References

Abdul Khalil, H.P.S., Issam, A.M., Ahmad Shakri, M.T., Suriani, R., Awang, A.Y., 2007. Conventional agro-composites from chemically modified fibres. Ind. Crop. Prod. 26 (3), 315–323. https://doi.org/10.1016/j.indcrop.2007.03.010.

Abdul Khalil, H.P.S., Nur Firdaus, M.Y., Anis, M., Ridzuan, R., 2008. The effect of storage time and humidity on mechanical and physical properties of medium density fiberboard (MDF) from oil palm empty fruit bunch and rubberwood. Polym.-Plast. Technol. Eng. 47 (10), 1046–1053. https://doi.org/10.1080/03602550802355644.

Abood, F., Kamal, I., Ashaari, Z., Abdul Malek, A.R., Soon, C.W., 2012. Fire performance and properties of medium density fibreboard made from rubberwood (*Hevea brasiliensis*) treated with AG° PI. Lignocellulose J. 1 (1), 13–21. http://clls.sbu.ac.ir/index.php/Lignocellulosic/article/view/1389.

Anon., 2009. NATIP: National Timber Industry Policy. Ministry of Plantation Industries and Commodities Malaysia, p. 27.

Anon., 2020. Malaysia exports RM22.5 Bln worth of wood-based products in 2019. Selangor J. https://selangorjournal.my/2020/09/malaysia-exports-rm22-5-bln-worth-of-wood-based-products-in-2019/.

Ariff, M., 2005. Physical and mechanical properties of fibreboards from oil palm fibres and polyethylene. Sci. Lett. 2 (1), 49–54.

Atack, D., 1982. On the characterization of pressurized refiner mechanical pulps. Sven. Pap. 75 (3), 89–94. https://ci.nii.ac.jp/naid/10011399852/.

Back, E.L., Lennart Salmen, N., 1982. Glass transitions of wood components hold implications for molding and pulping processes. TAPPI J. 65 (7), 107–110. https://ci.nii.ac.jp/naid/80001286115/.

Bakar, A.A., Hassan, A., Yusof, A.F.M., 2006. The effect of oil extraction of the oil palm empty fruit bunch on the processability, impact, and flexural properties of PVC-U composites. Int. J. Polym. Mater. Polym. Biomater. 55 (9), 627–641. https://doi.org/10.1080/00914030500306446.

Belini, U.L., Tomazello, F.M., Oliveira, J.T.S., 2008. Alterations of the anatomical structure of eucalypt wood chips in three refining conditions for MDF panels production. In: Proceedings of the 51st International Convention of Society of Wood Science and Technology, pp. 1–7. http://www.swst.org/meetings/AM08/proceedings/WS-07.pdf.

Bubparenu, N., Laemsak, N., Chitaree, R., Sihabut, T., 2018. Effect of density and surface finishing on sound absorption of oil palm frond. Asia Pac. J. Sci. Technol. 23 (4), 1–7. https://doi.org/10.14456/apst.2018.15.

Clark, J.A., 1978. Pulp Technology and Treatment for Paper. Miller Freeman Publications, p. 752. https://www.cabdirect.org/cabdirect/abstract/19800665281.

Gorning, D.A.I., 1971. Polymer properties of lignin and lignin derivatives. Sarkanen KV Lignins 2017, 695–768. https://ci.nii.ac.jp/naid/10006135610/.

Gorski, D., Mörseburg, K., Axelsson, P., Engstrand, P., 2011. Peroxide-based ATMP refining of spruce: energy efficiency, fibre properties and pulp quality. Nord. Pulp Pap. Res. J. 26 (1), 47–63.

Harmaen, A.S., Jalaluddin, H., Paridah, M.T., 2013. Properties of medium density fibreboard panels made from rubberwood and empty fruit bunches of oil palm biomass. J. Compos. Mater. 47 (22), 2875–2883. https://doi.org/10.1177/0021998312459868.

Ibrahim, Z., Aziz, A.A., Mokhtar, A., Ramli, R., Mamat, R., 2012. Production of medium density fibreboard (MDF) from oil palm fronds and its admixture. MPOB Inf. Ser., 601. www.mpob.gov.my.

Ibrahim, Z., Aziz, A.A., Ramli, R., Jusoff, K., Ahmad, M., Ariff Jamaludin, M., 2015. Effect of treatment on the oil content and surface morphology of oil palm (*Elaeis guineensis*) empty fruit bunches (EFB) fibres. Wood Res. 60 (1), 157–166.

Ibrahim, Z., Aziz, A.A., Ramli, R., Mokhtar, A., Lee, S., 2013. Effect of refining parameters on medium density fibreboard (MDF) properties from oil palm trunk. Open J. Compos. Mater. 3 (4), 127–131. https://doi.org/10.4236/ojcm.2013.34013.

Ibrahim, Z., Aziz, A.A., Ramli, R., Mokhtar, A., Omar, R., Lee, S., 2014. Production of medium density fibreboard (MDF) from oil palm trunk (OPT). J. Appl. Sci. 14 (11), 1174–1179. https://doi.org/10.3923/jas.2014.1174.1179.

Illikainen, M., 2008. Mechanisms of Thermomechanical Pulp Refining. Oulu University Press, Oulu.

Irvine, G.M., 1985. The significance of the glass transition of lignin in thermomechanical pulping. Wood Sci. Technol. 19 (2), 139–149. https://doi.org/10.1007/BF00353074.

Jamaludin, M.A., Kamarulzaman, N., Ahmad, M., 2007. The bending strength of medium density fibreboard (MDF) from different ratios of Kenaf and oil palm empty fruit bunches (EFB) admixture for light weight construction. Key Eng. Mater. 334–335, 77–80. https://doi.org/10.4028/www.scientific.net/kem.334-335.77.

Johansson, L., Hill, J., Gorski, D., Axelsson, P., 2011. Improvement of energy efficiency in TMP refining by selective wood disintegration and targeted application of chemicals. Nord. Pulp Pap. Res. J. 26 (1), 31–46. https://doi.org/10.3183/npprj-2011-26-01-p031-046.

Kano, T., Iwamida, T., Sumi, Y., 1982. Energy consumption in mechanical pulping. Pulp Pap. Can. 83 (6), 80–84.

Karlsson, H., 2006. Fibre Guide – Fibre Analysis and Process Applications in the Pulp and Paper Industry. Lorentzen & Wettre, p. 9.

Karnis, A., 1994. The mechanism of fibre development in mechanical pulping. J. Pulp Pap. Sci. 20 (10), J280–J288. https://pascal-francis.inist.fr/vibad/index.php?action=getRecordDetail&idt=4252971.

Khalil, A., Abdul Khalil, H.P.S., Siti Alwani, M., Mohd Omar, A.K., 2006. Chemical composition, anatomy, lignin distribution, and cell wall structure of Malaysian plant waste fibers. Bioresources 1 (2), 220–232.

Khalil, H.P.S.A., Nur Firdaus, M.Y., Jawaid, M., Anis, M., Ridzuan, R., Mohamed, A.R., 2010. Development and material properties of new hybrid medium density fibreboard from empty fruit bunch and rubberwood. Mater. Des. 31 (9), 4229–4236. https://doi.org/10.1016/j.matdes.2010.04.014.

Konn, J., Pranovich, A., Holmbom, B., 2006. Dissolution of fibre material in alkaline pre-treatment and refining of spruce CTMP. Holzforschung 60 (1), 32–39. https://doi.org/10.1515/HF.2006.007.

Laemsak, N., Okuma, M., 2000. Development of boards made from oil palm frond II: properties of binderless boards from steam-exploded fibers of oil palm frond. J. Wood Sci. 46. https://doi.org/10.1007/BF00766224.

Li, B., Li, H., Zha, Q., Bandekar, R., Alsaggaf, A., Ni, Y., 2011. Effects of wood quality and refining process on TMP pulp and paper quality. Bioresources 6 (3), 3569–3584.

Miles, K.B., Karnis, A., 1995. Wood characteristics and energy consumption in refiner pulps. J. Pulp Pap. Sci. 21 (11). https://pascal-francis.inist.fr/vibad/index.php?action=getRecordDetail&idt=2910831.

Mohd Nor, M.Y., 1997. High yield pulping of oil palm frond fibres in Malaysia. In: Nanjing International Symposium on High Yield Pulping, pp. 21–23.

Musbah Redwan, A., Haji Badri, K., Tarawneh, M.A., 2015. The effect of aluminium hydroxide (ATH) on the mechanical properties and fire resistivity of palm-based fibreboard prepared by pre-polymerization method. Adv. Mater. Res. 1087, 287–292. https://doi.org/10.4028/www.scientific.net/AMR.1087.287.

Myers, G.C., 1983. Relationship of fiber preparation and characteristics to performance of medium-density hardboards. For. Prod. J. 33 (1), 43–51. http://128.104.77.228/documnts/pdf1983/myers83b.pdf.

Nor Yuziah, M.Y., Ramli, M., Jalaludin, H., 1997. The effectiveness of selected adhesives in the fabrication of oil palm particleboard. In: Proceeding of Utilisation of Oil Palm Tree: Oil Palm Residue: Progress Towards Commercialization, pp. 106–110.

Norul Izani, M.A., Paridah, M.T., Astimar, A.A., Mohd Nor, M.Y., Anwar, U.M.K., 2012. Mechanical and dimensional stability properties of medium density fibreboard produced from treated oil palm empty fruit bunch. J. Appl. Sci. 12 (6), 561–567.

Norul Izani, M.A., Paridah, M.T., Mohd Nor, M.Y., Anwar, U.M.K., 2013. Properties of medium-density fibreboard (MDF) made from treated empty fruit bunch of oil palm. J. Trop. For. Sci. 25 (2), 175–183.

Onuorah, E.O., 2005. Properties of Fibreboards made from oil palm (*Elaeis guineensis*) stem and/or mixed tropical hardwood sawmill residues. J. Trop. For. Sci. 17 (4), 497–507. https://www.jstor.org/stable/23616587.

Othman, M.S.H., Samad, A.R.A., 2009. A preliminary study of strategic competitiveness of MDF industry in peninsular Malaysia by using SWOT analysis. Int. J. Bus. Manag. 4, 205–214. https://doi.org/10.5539/ijbm.v4n8p205.

Overend, et al., 1987. Fractionation of lignocellulosics by steam-aqueous pretreatments. Philos. Trans. R. Soc. Lond. Ser. A Math. Phys. Sci. 321 (1561), 523–536. https://doi.org/10.1098/rsta.1987.0029.

Paridah, M.T., Wong, I., Zaidon, A., 2000. Finishing system for oil palm empty fruit bunch medium densiry fibreboard. In: Proceedings of the 5th national Seminar on the Utilization of Oil Palm Tree. Oil Palm Tree Utilization Committee, Malaysia, pp. 127–132.

Pearson, A.J., 1983. Towards a unified theory of mechanical pulping and refining. In: Proceedings of the Technical Association of the Pulp and Paper Industry (USA). TAPPI Press, pp. 131–138. http://agris.fao.org/agris-search/search.do?recordID=US8275384.

Ramli, R., Shaler, S., Jamaludin, M.A., 2002. Properties of medium density fibreboard from oil palm empty. J. Oil Palm Res. 14 (2), 34–40.

Reme, P.A., Helle, T., 1998. Fibre characteristics of some mechanical pulp grades. Nord. Pulp Pap. Res. J. 13 (4), 263–268.

Rozman, H.D., Tay, G.S., Kumar, R.N., Abusamah, A., Ismail, H., Mohd, Z.A., 2001. The effect of oil extraction of the oil palm empty fruit bunch on the mechanical properties of polypropylene-oil palm empty fruit bunch-glass fibre hybrid composites. Polym.-Plast. Technol. Eng. 40 (2), 103–115. https://doi.org/10.1081/PPT-100000058.

Sihabut, T., Laemsak, N., 2008. Sound absorption efficiency of fiberboard made from oil palm frond. In: FORTROP II International Conference, p. 185.

Sihabut, T., Laemsak, N., 2010. Sound absorption capacity of oil palm frond fiberboard with different finishing 1. Environ. Nat. Resour. 8 (1), 38–43. https://ph02.tci-thaijo.org/index.php/ennrj/article/view/82561.

Sihabut, T., Laemsak, N., 2012. Production of oil palm frond fiberboard and its sound absorption characteristics. Agric. Nat. Resour. 46 (1), 120–126. https://li01.tci-thaijo.org/index.php/anres/article/view/242757.

Sikter, D., Karlström, A., Engstrand, P., Czmaidalka, J., 2007. Using the refining zone temperature profile for quality control. In: International Mechanical Pulping Conference 2007, TAPPI. 2, pp. 7–9.

Sundholm, J., 1993. Can we reduce energy consumption in mechanical pulping? In: International Mechanical Pulping Conference, Oslo, p. 133. https://ci.nii.ac.jp/naid/10011399850/.

Wanrosli, W.D., Zainuddin, Z., Law, K.N., Asro, R., 2007. Pulp from oil palm fronds by chemical processes. Ind. Crop. Prod. 25 (1), 89–94. https://doi.org/10.1016/j.indcrop.2006.07.005.

Xing, C., Deng, J., Zhang, S.Y., Riedl, B., Cloutier, A., 2006. Properties of MDF from black spruce tops as affected by thermomechanical refining conditions. Holz Roh Werkst. 64 (6), 507–512. https://doi.org/10.1007/s00107-006-0129-5.

Yusop, A., 2001. Finishing System for Oil Palm Empty Fruit Bunch (EFB) Medium Density Fiberboard (Ph.D. Thesis). Universiti Putra Malaysia.

Zainuddin, Z., Jani, S.M., Aznan, W.M., 2016. The effects of pressure and pressing time on the mechanical and physical properties of oil palm empty fruit bunch medium density fibreboard. J. Trop. Agric. Food Sci. 44 (1), 111–120.

Zuhri Mohamed Yusoff, M., Sapuan, S.M., 2009. Tensile properties of single oil palm empty fruit bunch (OPEFB) fibre. Sains Malays. 38 (4), 525–529.

18

Other types of panels from oil palm biomass

A.H. Juliana[a,b], S.H. Lee[b], S.O.A. SaifulAzry[b], M.T. Paridah[b,c], and N.M.A. Izani[d]

[a]Faculty of Technology Management and Business, Universiti Tun Hussein Onn Malaysia, Parit Raja, Batu Pahat, Johor, Malaysia, [b]Institute of Tropical Forestry and Forest Products (INTROP), Universiti Putra Malaysia, Serdang, Selangor, Malaysia, [c]Faculty of Forestry and Environment, Universiti Putra Malaysia, Serdang, Selangor, Malaysia, [d]Faculty of Agriculture and Food Sciences, Universiti Putra Malaysia, Bintulu Sarawak Campus, Bintulu, Sarawak, Malaysia

18.1 Introduction

Recently, the volume of palm oil production has increased globally to meet the demand for palm oil and its products. Today, Indonesia is the leading palm oil producer in the world with a value of 42.5 million metric tons, followed by Malaysia (18.50 million metric tons), Thailand (2.80 million metric tons), Colombia (1.53 million metric tons), Nigeria (1.02 million metric tons), and others (5.93 million metric tons) (Shahbandeh, 2020). This high production of palm oil indirectly produces a high amount of biomass from oil palm plantations (i.e., oil palm frond (OPF) and oil palm trunk (OPT)) and mills (i.e., empty fruit bunches (EFB), palm kernel shells, mesocarp, and palm oil mill effluent (POME)) (Aljuboori and Rajab, 2013). Therefore, many studies have investigated the properties and potential application of oil palm biomass, especially for composite panels.

According to Stark et al. (2010) and Maloney (1986), there are four classifications of wood-based composites, namely veneer-based (i.e., plywood, laminated veneer lumber (LVL)), laminates (i.e., glue-laminated timbers, laminated wood), composite (i.e., fiberboard, particleboard, flakeboard, oriented strand board, laminated strand lumber), and wood-non-wood composite (i.e., wood fiber-polymer composite, inorganic-bonded composite). Numerous studies have been carried on the properties of wood-based composites. Nevertheless, numerous studies were also reported on the properties

Oil Palm Biomass for Composite Panels. https://doi.org/10.1016/B978-0-12-823852-3.00010-6
Copyright © 2022 Elsevier Inc. All rights reserved.

of oil palm biomass composites (Faizi et al., 2017;Khalid et al., 2015; Suhaily et al., 2012; Hassan et al., 2010). This current chapter emphases the other types of oil palm panels, mainly laminated composites, polymer composites, and cement composites.

In oil palm laminated composite, oil palm biomass is commonly cut into strips or lumber elements and serves as the main component (Srivaro et al., 2019; Khalid et al., 2015). Thus, only wood-like parts of the oil palm biomass like OPT and OPF can be used for laminated composite manufacturing. On top of that, oil palm biomass element serves as a filler or reinforcing agent in polymer and inorganic-bonded composites. Polymer composite has significant advantages, particularly in terms of its low cost, good durability, and low weight (Zaaba and Ismail, 2019; Andrady and Neal, 2009). Thermoplastic composites, particularly wood plastic/polymer composite, have become a mostly commercial composite product in the applications of construction, automotive, and furniture (Oksman Niska and Sain, 2008). The advantages of inorganic-bonded composite are low operation cost, requires no heat for curing, and resistant to deterioration (Stark et al., 2010).

18.2 Laminated-based composites

Laminated composites are unlike veneer-based composite (Stark et al., 2010). Veneer-based composites are commonly made from veneer, namely plywood and LVL, while laminated composites are made from solid wood or non-wood (woody part) excluding veneer. Most laminated composites are dominated by wood species, and fewer studies have reported on oil palm-based materials. Table 18.1 tabulates the previous studies on laminated oil palm biomass composites. As tabulated in Table 18.1, only OPT and OPF in strip form were reported as a potential biomass for laminated-based composites.

18.2.1 Laminated-based composites made from OPT

The trunk part of oil palm, also known as oil palm stem, has the highest volume of oil palm biomass and is the size of timber. This makes it practical to be processed into lumber, scantling, or strip elements. A recent study conducted by Prabuningrum et al. (2020) investigated a process for improving the performance of OPT laminated composite. In their study, the lumber was pretreated with heat and pressure prior to the composite manufacturing. The findings revealed that the densification of the low-density lumber (inner part of OPT) has no significant effect on the performance of the composite. Nonetheless, the modulus of rupture (MOR) and modulus of elasticity (MOE) values of laminated OPT composite with high density lumbers at the face and back layers meet the requirement of the Japanese

Table 18.1 Previous studies on laminated oil palm biomass composites.

Palm oil biomass	Adhesive	Application	Number of layers	Glue spread (g/m^2)	Resin (%)	References
OPT	Isocyanate	n/a	3	300	–	Prabuningrum et al. (2020)
	Polyvinyl acetate (PVAc) - Edge Melamine urea formaldehyde (MUF) - Face	Low-rise building construction	3	250, 500	–	Srivaro et al. (2019)
	Polyvinyl acetate (PVAc)	n/a	2	250, 500	–	Nordin et al. (2015)
	Low molecular weight Phenol formaldehyde (LmwPF)	n/a	3	–	40	Aizat et al. (2014)
	Isocyanate	n/a	2	200, 250, 300	–	Darwis et al. (2014)
	Polyvinyl acetate (PVAc)	n/a	n/a	250, 500	–	Nordin et al. (2013)
OPF	Phenol formaldehyde (PF)	n/a	6, 8, 10	200, 250, 300	–	Khalid et al. (2015)
	Phenol formaldehyde (PF) Urea formaldehyde (UF)	n/a	n/a	–	12, 15	Rasat et al. (2011)

Agricultural Standard (JAS, 234-2003). Additionally, Darwis et al. (2014) also used isocyanate as a binder for fabrication of two layers of OPT glued laminated composite. In their study, glued laminated OPT composite bonded with isocyanate has no adhesive failure. The findings mentioned that the shear strength of glued laminated OPT increased with increasing of isocyanate spread rate. As reported by Zheng (2002), isocyanate and phenolic resin are two frequently used exterior thermosetting adhesives in the wood-based industry.

The evaluation of OPT as raw material for cross-laminated timber (CLT), particularly for building construction, was carried out by Srivaro et al. (2019). In their study, the OPTs were cut into lumber and dried prior to the manufacturing of three-layer CLT. The OPT lumbers were classified in two densities, medium ($0.342\,g/cm^3$) and high ($0.541\,g/cm^3$), and two adhesives used, polyvinyl acetate (PVAc) (edge-to-edge jointing) and melamine urea formaldehyde (face-to-face jointing). Findings revealed that the densities of OPT CLT composites were greater than that of initial OPT lumber. Additionally, the density of CLT composites made from high-density lumber was significantly greater compared to CLT made from pine and spruce; however, those made from medium-density lumber have slightly lower density (Li, 2017; Sikora et al., 2016). As a result, CLT applied with a highly controlled strain level (20.8%) produced a less thick panel but gave better physical and mechanical properties.

Other studies on the performance evaluation of OPT laminated composites were done by Nordin et al. (2013, 2015). In both studies, the OPT lumbers were steamed, dried, compressed, and bonded at two levels of PVAc spread rates. Solid lumber of OPT and uncompressed OPT laminated composite were prepared for control purposes (Nordin et al., 2013). Compression of OPT prior to the composite manufacturing contributes to low surface roughness, high contact angle, and low wettability compared to uncompressed. Although low contact angle offered good wettability, deeply absorbed adhesive may cause starved joints and bonding failure of the composite (Nordin et al., 2013). Laminated composites made from uncompressed OPT have significantly higher thickness swelling (TS) and water absorption (WA) values yet low mechanical properties compared with compressed laminated composites. Interestingly, the WA values of uncompressed laminated composites and compressed laminated composites range from 214% to 218% and 42% to 43%, respectively. This might be due to the pores' anatomical structure of uncompressed OPT, which leads to high water uptake. The study also observed that there are no significant effects on the mechanical properties of solid lumber OPT and uncompressed OPT laminated composites. In conclusion, compression was significantly enhanced, approximately more than 50% in both physical and mechanical performances of laminated OPT (Nordin et al., 2013).

Meanwhile, in another study conducted by Nordin et al. (2015), all laminated composite samples were made from compressed OPT at the glue spread rates of $250\,g/m^2$ and $500\,g/m^2$. In their study, laminated compressed OPT with $500\,g/m^2$ spread rate displays greater durability compared to the composite with lower spread rate. Composite made with high adhesive spread rate also has significantly more durability against bio-organisms.

18.2.2 Laminated-based composites made from OPF

Apart from OPT, there are few studies on OPF laminated composite. In a study done by Khalid et al. (2015), 25-year-old defect-free OPF was cut into thick strips at a thickness of 10 mm and compressed into 3–4 mm prior to the fabrication of laminated board with similar final thickness. The laminated frond board was manufactured using three different numbers of layers and adhesive spread rates. Findings reported that a 10-layer laminated board has better compaction and higher density (0.63 to $0.72\,g/cm^3$) compared to 6-layer (0.42 to $0.50\,g/cm^3$) and 8-layer (0.57 to $0.64\,g/cm^3$). However, 6-layer laminated frond board was observed to have more void, less compaction, and low density. The density of 10-layer frond laminated board was comparable with compressed OPT and greater than solid OPT and rubberwood (Nurjannah, 2013; Sulaiman et al., 2012). As reported by previous studies, the density of OPF, OPT, and rubberwood ranged between 0.6 and $1.2\,g/cm^3$, 0.5 to $1.1\,g/cm^3$, 0.62 to $0.64\,g/cm^3$, respectively (Abdul Khalil et al., 2012, Sulaiman et al., 2012; Ratnasingam and Scholz, 2009).

As to physical and mechanical properties, 10-layer laminated frond board has greater performance compared to those with a smaller number of layers. Composite with the lowest layer and adhesive spread rate possesses the highest values of swelling and water uptake. Meanwhile, 10-layer composite bonded with $250\,g/m^2$ PF spread rate has the greatest MOR and MOE of 35.20 MPa and 2961.27 MPa, respectively (Khalid et al., 2015). Again, the MOR values of 10-layer frond laminated composite are slightly higher compared to steam-compressed OPT (31.36 MPa) and OPT-compressed (28.62 MPa) (Nurjannah, 2013). This might be due to higher density and higher cellulose content of OPF. According to Onoja et al. (2019), the cellulose content of OPF, OPT, and EFB are 40.01%, 34.44%, and 23.70%, respectively. Meanwhile, Rasat et al. (2011) evaluated the physical and mechanical properties of OPF laminated composites at different maturity (mature, intermediate, and young), sections (bottom, middle, and top), and adhesives (phenol formaldehyde (PF) and urea formaldehyde (UF)). From their observations, the composite density increased from the top to the bottom sections. Results stated that the cement composite made from matured OPF from the bottom section possessed the highest density, MOR, MOE, and compression strength parallel to the grain, especially those bonded with phenolic resin.

18.3 Polymer-based composites

Polymer can be classified into two classes: thermoplastic-based and thermoset-based. Thermoplastic polymers used for natural fiber usually melt below the thermal degradation temperature of the wood; 200°C to 220°C (Zaaba and Ismail, 2019; Asim et al., 2017; Stark et al., 2010). Common thermoplastics used in fiber composites are polypropylene (PP), polyethylene (PE), polystyrene (PS), vinyls, low-density polyethylene (LDPE), and high-density polyethylene (HDPE) (Tazi et al., 2016; Stark et al., 2010). Meanwhile, thermoset polymer requires heat for curing and offers better thermal stability and good rigidity such as UF, PF, and epoxy. There are two main functions of wood polymer composites: (i) the wood or natural fiber element acts as a reinforcing filler in a polymer composite and (ii) the polymer matrix acts as a binder to the wood or natural fiber elements. This study will discuss the properties of oil palm biomass as a reinforcing filler in a polymer matrix composite. Table 18.2 tabulates the type of oil palm biomass and matrices used in oil palm biomass polymer-based composites.

18.3.1 Polymer-based composites made from OPT

UF and PF resins are two popular adhesives used by the wood-based products industry. These two adhesives are commonly used in conventional wood-based composites, and a study done by Abdullah et al. (2012) verified that it can be used as a polymer composite. In their study, UF and PF resins were impregnated in the OPT lumber at 25%, 50%, and 75% resin loadings. Result shows that impregnated OPT polymer composites have lower TS and WA values compared to untreated OPT. OPT polymer composites at 75% UF and PF resin loading displayed lower TS and WA values than those polymer composites at 25% and 50% resin level. In addition, Abdul Khalil et al. (2010) studied the properties of oil palm biomass (OPT, OPF, and EFB) fiber-polypropylene composites at five different fiber ratios (10%, 20%, 30%, 40%, and 50%). In general, the tensile strength declined as fiber ratio increased, while the tensile modulus, flexural strength, and flexural modulus values increased as fiber ratio increased. Tensile strength, tensile modulus, and flexural modulus were higher for OPT compared to OPF and EFB. Meanwhile, the flexural strengths are higher for EFB compared to OPT and OPF.

Table 18.2 Types and matrices used in oil palm biomass polymer-based composites.

Palm oil biomass	Matrix	Treatment (concentration)	Time (h)	Application	References
OPT	Phenol formaldehyde & Urea formaldehyde	–	–	n/a	Abdullah et al. (2012)
	Polypropylene	–	–	n/a	Abdul Khalil et al. (2010)
OPF	Starch	Ionic liquids	3	n/a	Mahmood et al. (2016)
	Polypropylene	–	–	n/a	Abdul Khalil et al. (2010)
	Unsaturated polyester	–	–	n/a	Abdul Khalil et al. (2007)
EFB	Polypropylene	–	–	n/a	Lah and Mazni Ismail (2019)
	Polypropylene	–	–	n/a	Othman et al. (2018)
	Epoxy	Untreated NaOH (6%)	– 12	n/a	Valášek et al. (2017)
	Epoxy	–	–	n/a	Tshai et al. (2016)
	Polypropylene Ethylene vinyl acetate Polyamide-6	–	–	n/a	Mohsin et al. (2015)
	Polypropylene	–	–	n/a	Abdul Khalil et al. (2010)

18.3.2   Polymer-based composites made from OPF

In a study done by Mahmood et al. (2016), the thermal and mechanical properties of treated OPF-thermoplastic starch composite were evaluated. OPF particles were pretreated using ionic liquids (ILs), and thermoplastic starch was mixed with the OPF fibers at a ratio of 1:1, then pressed at 170°C for 10 min. In their finding, the thermal stability of the treated fibers and treated OPF-thermoplastic starch composite were remarkably improved. In addition, the treated OPF-thermoplastic starch composite exhibited superior flexural strength compared to the untreated OPF-thermoplastic starch composite. Other studies using a polypropylene and unsaturated polyester also have been conducted (Abdul Khalil et al., 2007, 2010).

18.3.3 Polymer-based composites made from EFB

There are various studies on the properties of polymer-based composites made from EFB and incorporated with other natural and man-made fibers. Nonetheless, current studies only discussed polymer composites made from EFB. In a recent work done by Lah and Mazni Ismail (2019), effects of fiber loadings (5%, 10%, and 15%) with addition of 5% maleic anhydride grafted polyethylene (MAPE) were studied. The results stated that addition of EFB fibers increased the tensile elastic modulus, then reduced the impact strength of EFB-PP composites. Meanwhile, Othman et al. (2018) manufactured EFB-PP at 0%, 5%, 10%, 15%, 20%, 25%, and 30% of EFB and 0% and 3% of maleic anhydride-grafted polypropylene. In their study, increasing EFB increased the MOR but reduced the tensile and elongation at break. However, the addition of a compatibilizer slightly increased the properties of EFB-PP composites.

Valášek et al. (2017) evaluated the properties of both untreated and treated EFB fibers as a reinforcing element in epoxy polymer composite. In their study, the effects of fiber length (3 mm, 6 mm, and 9 mm) and fiber loading (2.5 to 10.0 vol%) on the mechanical properties of EFB-epoxy composite were determined. Findings reported that the alkaline treatment increased the interfacial interaction of the fibers. Alkaline treatment gives less effect to the performance of EFB-epoxy composite. However, the presence of short EFB fibers significantly increased the MOE and decreased the tensile strength up to 47% and 26%, respectively. Meanwhile, Tshai et al. (2016) described the inclusion of untreated EFB fibers with epoxy enhanced the overall mechanical properties of the EFB-epoxy composite. They also suggested that a 27.3% mass fraction of EFB is the greatest amount of fiber loading. In another study conducted by Mohsin et al. (2015), three types of polymers, namely polypropylene (PP), ethylene vinyl acetate (EVA), and polyamide-6 (PA6), were incorporated to rPP/EFB composites. It was found that the rPP/EFB/PP, rPP/EFB/EVA, and rPP/EFB/PA6 composites demonstrated the greatest mechanical properties, better thermal properties, and exhibited significant enhancement in both mechanical and thermal properties, respectively.

18.4 Cement composites

Inorganic-bonded composites are defined as molded products or composites that consist of 10% to 70% (w/w) particles or fibers and 90% to 30% (w/w) inorganic binder (Stark et al., 2010). Inorganic binders include Portland cement, gypsum, and magnesia, whereas those composites composed of Portland cements are more durable compared to gypsum and magnesia. The current study emphases the use of oil palm biomass in cement composites. Natural fibers such as oil palm biomass are used as a reinforcement to enhance the ductility of cement composites (Momoh and Osofero, 2019). As tabulated in Table 18.3, the most common oil palm biomass used in cement composites is EFB.

Table 18.3 Types and treatment used in oil palm biomass cement composites.

Palm oil biomass	Treatment				Application	References
	Medium	Level	Temp.	Time (h)		
OPT	–	–	–	–	n/a	Nurnabilah (2016)
	–	–	–	–	Beam	Ofuyatan and Olutoge (2013)
	–	–	–	–	Lightweight construction materials	Mazlan and Abdul Awal (2012)
	–	–	–	–	n/a	Ahmad et al. (2010)
OPF	–	–	–	–	n/a	Ayrilmis et al. (2017)
	–	–	–	–	n/a	Hermawan et al. (2001)
EFB	NaOH	1%, 10%	–	2 6 24	n/a	Bonnet-Masimbert et al. (2020)
	NaOH	1%	–	24	n/a	Peter et al. (2020)
	Heat	–	103	24		
	Water		Cold,	2	Building	Omoniyi (2019)
	NaOH	2, 4, 6, 8, 10	60, 100	24		
	NaOH	0.4%	–	24	n/a	Akasah et al. (2019)
	NaOH	n/a	n/a	n/a	Partition walls	Abas et al. (2018)
	Water	n/a	n/a	24	Residential housing component	Mohtar et al. (2015)
	NaOH			n/a		
	–	–	–	–	n/a	Purwanto (2016)
	Boiled water	–		2	Residential building	Lertwattanaruk and Suntijitto (2015)
Broom fibers	–	–	–	–	Low-cost earthquake resistant housing	Momoh and Osofero (2019)

18.4.1 Properties of cement composites made from OPT

Trunk used in cement composites is normally processed into strips, particles, strand, and sawdust elements. In a previous study conducted by Nurnabilah (2016), OPT strand was mixed with Portland cement at different densities and different OPT strand sizes, and the physical and mechanical properties were evaluated. The findings reported that the OPT cements composite with the lowest density of $900\,kg/m^3$ has the highest WA and the lowest mechanical properties compared to higher densities composites. The highest MOR and MOE values are owned by the cement composite at a density of $1300\,kg/m^3$ and 50 mm OPT strand size. Studies also reported that addition of OPT fiber loading increased the load carrying capacity of cement composite (Ofuyatan and Olutoge, 2013; Mazlan and Abdul Awal, 2012). Mazlan and Abdul Awal (2012) showed that flexural and compressive strength has been found to increase up to a certain limit of 3% fiber content. However, findings by Ahmad et al. (2010) mentioned that the mechanical properties of cement composite reinforced with OPT are improved by the addition of 1%, beyond which the strength declined steadily. However, the increasing amount of OPT fiber increased the WA of composite. Overall, findings obtained propose OPT fiber as a good potential reinforcing material of cement composite.

18.4.2 Properties of cement composites made from OPF

OPF, known as the main veins of oil palm, is also used in the production of cement composite. Ayrilmis et al. (2017) investigated the performance of OPF cement composites at three levels of particles, namely 10%, 15%, and 20%, and three levels of calcium chloride ($CaCl_2$) (0%, 3%, and 6%). In their findings, the increment of OPF particles decreased the MOR, MOE, and IB of the OPF cement composite. However, the mechanical properties improved with the addition of $CaCl_2$ up to 6%. Therefore, the cement composite made from a combination of 10% OPF particles and 6% $CaCl_2$ shows preferable mechanical and physical properties. Meanwhile, other studies fabricated cement composites at cement:oven dried particle:water ratios of 2.2:1.0:1.32 and 2.7:1.0:1.62, respectively, with 0% to 15% magnesium chloride ($MgCl_2$) (Hermawan et al., 2001). Similar to a study done by Ayrilmis et al. (2017), the addition of $MgCl_2$ improved the hydration reaction of cement and consequently increased the mechanical and physical properties of OPF cement composites. The MOR and MOE properties of OPF cement composite at a cement-particle ratio of 2.7:1.0 were slightly greater than that at 2.2:1.0 (Hermawan et al.,

2001). Therefore, findings propose that the OPF can be considered as a reinforcing element for cement composite with the addition of an accelerator.

18.4.3 Properties of cement composites made from EFB

As tabulated in Table 18.3, EFB fibers require treatment prior to the fabrication of composite. The most common treatments of EFB are alkaline treatments that encourage the hydration process. The latest study done by Bonnet-Masimbert et al. (2020) reported that alkaline treatments significantly improve compatibility and allow more fluid to further enhance the properties of the interface. They concluded that the tensile properties of EFB fibers could be improved by exposing 10% of sodium hydroxide at 6 h soaking time. Besides, the durability of EFB cement composites in accelerated aging condition was investigated by Peter et al. (2020). The accelerated aging assessment was done at 70°C for 10 days, 30 days, and 60 days and 5 cycles, 15 cycles, and 30 cycles for hot water immersion and wet-dry cycle, respectively. The heat-treated cement composite has higher MOR, MOE, and IB values compared to those exposed with alkaline treatment. Pull-out test revealed that treated EFB fibers have a greater interface compared to the untreated fibers due to the rough surfaces.

Apart from alkaline treatment, the hot water treatment at 60°C of EFB fibers significantly improved the properties of the cement composites (Omoniyi, 2019). A significant difference of treated and untreated EFB fibers was also observed in TS and WA values. Low values of composite consist of EFB-treated fibers indicated good stability of treated cement composite. Akasah et al. (2019) investigated the effect of EFB fiber length on the mechanical performance of EFB cement composites. Additionally, cement composite made from a mixture of short EFB fiber produced low mechanical properties.

In most cases, the values of mechanical properties declined with the increasing amount of EFB fibers (Abas et al., 2018; Purwanto, 2016; Lertwattanaruk and Suntijitto, 2015). Lertwattanaruk and Suntijitto (2015) investigated the properties of coconut coir-EFB cement composite. Both fibers were pretreated using boiled water for 2 h. During the manufacturing of cement composite, 5%, 10%, and 15% of fibers (fiber-to-cement weight) were added in a mixture of cement mortar. In their result, increments of fiber loading decreased the density, compressive strength, and flexural strength of the cement composite. Among two fibers, samples consisting of EFB have lower porosity (8.75% to 9.50%), lower WA (3.88% to 4.83%), lower compressive strength (288 to 377 kg/cm^2), and lower flexural strength (78 to 85 kg/cm^2) compared to those with coconut coir fibers. It was found that high fiber loading at 15% of both fibers yielded acceptable physical and mechanical properties. Therefore, findings recommend that fiber treatment of EFB fiber is the utmost vital process in the manufacturing of the cement board.

18.5 Application of other types of panels from oil palm biomass

Nowadays, the demand for raw materials from natural and renewable resources increased due to the depleting of the log supply. Abundance of oil palm residues will be an advantage to the industry by converting the residues into value-added products, specifically laminated, polymer, and cement composites. For example, OPT and OPF can be used as CLT and glued-laminated timber, while OPT, OPF, and EFB can be used as fillers or reinforcements in polymer and cement composites. As shown in Tables 18.1, 18.2, and 18.3, fewer studies reported on the application of laminated, polymer, and cement composites from oil palm biomass. Most studies of laminated oil palm biomass mentioned that the composites would be considered for alternative value-added material for structural applications (Nordin et al., 2013, 2015; Rasat et al., 2011). Only Srivavo et al. (2019) specifically stated the potential application of OPF laminated composite (Table 18.1). In their study, laminated composite made from OPF can be used in low-rise building construction. The large accessibility and low value of oil palm biomass makes it an attractive raw material for value-added applications with some modification.

On the other hand, no particular application was stated for oil palm biomass polymer composite (Table 18.2). However, it can be improved to be commercialized as a wood polymer composite. According to Ngo (2018), wood polymer composite was commercialized for modular house construction. Apart from that, polymer composite is in demand from the automotive sector. In Germany, natural fibers are consumed the most by the automotive industry, and each car manufactured in Germany comprises of 3.6 kg natural fibers (Carus et al., 2015). According to Ofuyatan and Olutoge (2013), OPT cement composite can be used as reinforcement in concrete beams. Even though the properties of oil palm strips as reinforcement element in concrete is lesser than steel, it possibly can be considered for low-stress structure applications. Other potential applications of OPT cement composites are for lightweight construction materials (Mazlan and Abdul Awal, 2012). Cement composite made from oil palm broom has good ductility and energy absorption capacity, thus it is recommended as a cheap potential material for seismic applications (Momoh and Osofero, 2019).

Omoniyi (2019) reported that high sorption properties of untreated EFB cement composite are unsuitable for outdoor applications. However, composite with treated fibers has good dimensional stability and is suitable for interior and exterior building applications. As reported by Abas et al. (2018), the cement board made from EFB is suitable for internal applications, namely partition walls. However, low MOR and MOE values makes the product inappropriate for use in load-bearing applications. The potential of use of natural fibers as composite reinforcement depends mainly on the properties and increasing regulations (Peças et al., 2018). They also stated that the development of information concerning natural fibers will let consumers have higher levels of trust regarding mechanical and chemical properties.

18.6 Conclusions

In conclusion, the woody parts of oil palm biomass such as OPT and OPF are commonly used for laminated-based composite. OPT is the most used for the fabrication of laminated composites when cut into a lumber or strip form before being processed into laminated composites. Thermosetting resins such as isocyanate, melamine UF, and PF were commonly used as a binder of OPT and OPF in the oil palm biomass laminated composite. On the other hand, both thermoplastic and thermoset polymers can be used as matrices in oil palm polymer composite. From observation, EFB fibers are the most used for both polymer- and cement-based composites manufacturing. This study recommends heat or alkaline fiber treatments of EFB fiber prior to the fabrication of the cement composite. OPT and OPF laminated composites have the potential to be utilized as raw material for CLT and glued-laminated timber. Apart from that, oil palm biomass cement composite can be commercialized into low-cost building and construction materials.

Acknowledgment

The authors gratefully thank Universiti Tun Hussein Onn Malaysia (UTHM) and Universiti Putra Malaysia (UPM) for their support and opportunity.

References

Abas, N.H., Jaafar, A., Seman, A.S.M., 2018. Some Basic Properties of Cement Board Produced by Using Oil Palm Empty Fruit Bunch (EFB) Fibre. Vol. 1 Penerbit UTHM.

Abdul Khalil, H.P.S., Jawaid, M., Hassan, A., Paridah, M.T., Zaidon, A., 2012. Oil palm biomass fibres and recent advancement in oil palm biomass fibres based hybrid biocomposites. In: Composites and Their Applications. IntechOpen, pp. 187–220, https://doi.org/10.5772/48235.

Abdul Khalil, H.P.S., Kumar, R.N., Asri, S.M., Nik Fuaad, N.A., Ahmad, M.N., 2007. Hybrid thermoplastic pre-preg oil palm frond fibers (OPF) reinforced in polyester composites. Polym.-Plast. Technol. Eng. 46 (1), 43–50. https://doi.org/10.1080/03602550600948749.

Abdul Khalil, H.P.S., Poh, B.T., Issam, A.M., Jawaid, M., Ridzuan, R., 2010. Recycled polypropylene-oil palm biomass: the effect on mechanical and physical properties. J. Reinf. Plast. Compos. 29 (8), 1117–1130. https://doi.org/10.1177/0731684409103058.

Abdullah, C.K., Jawaid, M., Abdul Khalil, H.P.S., Zaidon, A., Hadiyane, A., 2012. Oil palm trunk polymer composite: morphology, water absorption, and thickness swelling behaviours. Bioresources 7 (3), 2948–2959.

Ahmad, Z., Saman, H.M., Tahir, P.M., 2010. Oil palm trunk fiber as a bio-waste resource for concrete reinforcement. Int. J. Mech. Mater. Eng. 5 (2), 199–207.

Aizat, A.G., Zaidon, A., Nabil, F.L., Bakar, E.S., Rasmina, H., 2014. Effects of diffusion process and compression on polymer loading of laminated compreg oil palm (*Elaeis guineensis*) wood and its relation to properties. J. Biobased Mater. Bioenergy 8 (5), 519–525. https://doi.org/10.1166/jbmb.2014.1470.

Akasah, Z.A., Soh, N.M.Z.N., Dullah, H., Abdul Aziz, A., Aminudin, E., 2019. The influence of oil palm empty fruit bunch fibre geometry on mechanical performance of cement bonded fibre boards. Int. J. Mech. Eng. Robot. Res. 8 (4), 547–552. https://doi.org/10.18178/ijmerr.8.4.547-552.

Aljuboori, Rajab, A.H., 2013. Oil palm biomass residue in Malaysia: availability and sustainability. Int. J. Biomass Renew. 2 (1), 13–18.

Andrady, A.L., Neal, M.A., 2009. Applications and societal benefits of plastics. Philos. Trans. R. Soc. B 364, 1977–1984. https://doi.org/10.1098/rstb.2008.0304.

Asim, M., Jawaid, M., Saba, N., Nasir, M., Sultan, M.T.H., 2017. Processing of hybrid polymer composites—a review. In: Hybrid Polymer Composite Materials. Woodhead Publishing, pp. 1–22.

Ayrilmis, N., Khalil Hosseinihashemi, S., Karimi, M., Kargarfard, A., Soleimani Ashtiani, H., 2017. Technological properties of cement-bonded composite board produced with the main veins of oil palm (*Elaeis guineensis*) particles. Bioresources 12 (2), 3583–3600.

Bonnet-Masimbert, P.-A., Gauvin, F., Brouwers, H.J.H., Amziane, S., 2020. Study of modifications on the chemical and mechanical compatibility between cement matrix and oil palm fibres. Results Eng. 7, 100–150. https://doi.org/10.1016/j.rineng.2020.100150.

Carus, M., Eder, A., Dammer, L., Korte, H., Scholz, L., Essel, R., Breitmayer, E., Barth, M., 2015. Wood-Plastic Composites (WPC) and Natural Fibre Composites (NFC). 16 Nova-Institute, Hürth, Germany.

Darwis, A., Massijaya, M.Y., Nugroho, N., Alamsyah, E.M., Nurrochmat, D.R., 2014. Bond ability of oil palm xylem with isocyanate adhesive. J. Trop. Wood Sci. Technol. 12 (1), 39–47. https://doi.org/10.51850/jitkt.v12i1.81.g78.

Faizi, M.K., Shahriman, A.B., Abdul Majid, M.S., Shamsul, B.M.T., Ng, Y.G., Basah, S.N., Cheng, E.M., 2017. An overview of the oil palm empty fruit bunch (OPEFB) potential as reinforcing fibre in polymer composite for energy absorption applications. In: MATEC Web of Conferences. vol. 90. EDP Sciences, p. 01064.

Hassan, A., Salema, A.A., Ani, F.N., Bakar, A.A., 2010. A review on oil palm empty fruit bunch fiber-reinforced polymer composite materials. Polym. Compos. 31 (12), 2079–2101. https://doi.org/10.1002/pc.21006.

Hermawan, D., Subiyanto, B., Kawai, S., 2001. Manufacture and properties of oil palm frond cement-bonded board. J. Wood Sci. 47 (3), 208–213.

Khalid, I., Sulaiman, O., Hashim, R., Razak, W., Jumhuri, N., Rasat, M.S.M., 2015. Evaluation on layering effects and adhesive rates of laminated compressed composite panels made from oil palm (*Elaeis guineensis*) fronds. Mater. Des. 68, 24–28. https://doi.org/10.1016/j.matdes.2014.12.007.

Lah, N.C., Mazni Ismail, N., 2019. Empty fruit bunch (EFB) fibers as reinforcement in polypropylene. J. Mod. Manuf. Syst. Technol. 2, 84–92.

Lertwattanaruk, P., Suntijitto, A., 2015. Properties of natural fiber cement materials containing coconut coir and oil palm fibers for residential building applications. Constr. Build. Mater. 94, 664–669. https://doi.org/10.1016/j.conbuildmat.2015.07.154.

Li, M., 2017. Evaluating rolling shear strength properties of cross-laminated timber by short-span bending tests and modified planar shear tests. J. Wood Sci. 63 (4), 331–337.

Mahmood, H., Moniruzzaman, M., Yusup, S., Md Akil, H., 2016. Pretreatment of oil palm biomass with ionic liquids: a new approach for fabrication of green composite board. J. Clean. Prod. 126, 677–685. https://doi.org/10.1016/j.jclepro.2016.02.138.

Maloney, T.M., 1986. Terminology and product definitions—a suggested approach to uniformity worldwide. In: Proceedings, 18th International Union of Forest Research Organization World Congress; September. IUFRO World Congress Organizing Committee, Ljubljana, Yugoslavia.

Mazlan, D., Abdul Awal, A.S.M., 2012. Properties of cement based composites containing oil palm stem as fiber reinforcement. Malays. J. Civil Eng. 24 (2), 107–117.

Mohsin, M.E.A., Ibrahim, A.N., Arsad, A., Rahman, M.F.A., Alothman, O.Y., 2015. Effect of polypropylene, ethylene vinyl acetate and polyamide-6 on properties of recycled polypropylene/empty fruit bunch composites. Fibers Polym. 16 (11), 2359–2367.

Mohtar, A., Azman Hassan, N., Nizam, M.F.M., 2015. Utilization of empty fruit bunch (EFB) fiber from palm oil waste for cement production. Meta.

Momoh, E.O., Osofero, A.I., 2019. Behaviour of oil palm broom fibres (OPBF) reinforced concrete. Constr. Build. Mater. 221, 745–761. https://doi.org/10.1016/j.conbuildmat.2019.06.118.

Ngo, T.-D., 2018. Natural fibers for sustainable bio-composites. In: Natural and Artificial Fiber-Reinforced Composites as Renewable Sources. IntechOpen, pp. 107–126, https://doi.org/10.5772/intechopen.71012.

Nordin, N.A., Sulaiman, O., Hashim, R., Salim, N., Nasir, M., Sato, M., Hiziroglu, S., 2015. Effect of adhesive spreading rate on the performance of laminated compressed oil palm trunks. Bioresources 10 (4), 6378–6387.

Nordin, N.A., Sulaiman, O., Hashim, R., Salim, N., Sato, M., Hiziroglu, S., 2013. Properties of laminated panels made from compressed oil palm trunk. Compos. Part B 52, 100–105. https://doi.org/10.1016/j.compositesb.2013.03.016.

Nurjannah, S., 2013. Study on Optimum Manufacturing Condition for Compressed Oil Palm Trunk (Dissertation). Universiti Sains Malaysia, Malaysia.

Nurnabilah, C.A.R., 2016. Properties of Cement Board Made from Various Sizes of Oil Palm Trunk Strands (Dissertation). Universiti Teknologi Mara.

Ofuyatan, O., Olutoge, F., 2013. Flexural characteristics and potentials of oil palm stem as reinforcement in concrete beams. J. Emerg. Trends Eng. Appl. Sci. 4 (4), 642–647. https://hdl.handle.net/10520/EJC142129.

Oksman Niska, K., Sain, M. (Eds.), 2008. Wood-Polymer Composites. Woodhead Publishing, Ltd., Cambridge, UK, p. 384.

Omoniyi, T.E., 2019. Potential of oil palm (*Elaeis guineensis*) empty fruit bunch Fibres cement composites for building applications. AgriEngineering 1 (2), 153–163. https://doi.org/10.3390/agriengineering1020012.

Onoja, E., Chandren, S., Ilyana, F., Razak, A., Mahat, N.A., Abdul Wahab, R., 2019. Oil palm (*Elaeis guineensis*) biomass in Malaysia: the present and future prospects. Waste Biomass Valoriz. 10 (8), 2099–2117.

Othman, N.S., Santiagoo, R., Zainal, M., Mustafa, W.A., Sanapan, V., Zunaidi, I., Razlan, Z.M., Wan, W.K., Shahriman, A.B., 2018. Tensile and morphological studies of polypropylene/empty fruit bunch composite: effect of maleic anhydride-grafted polypropylene. IOP Conf. Ser. Mater. Sci. Eng. 429 (1), 012015.

Peças, P., Carvalho, H., Salman, H., Leite, M., 2018. Natural fibre composites and their applications: a review. J. Compos. Sci. 2 (4), 1–20. https://doi.org/10.3390/jcs2040066.

Peter, P., Nik Soh, N.M.Z., Akasah, Z.A., Mannan, M.A., 2020. Durability evaluation of cement board produced from untreated and pre-treated empty fruit bunch fibre through accelerating ageing. IOP Conf. Ser. Mater. Sci. Eng. 713 (1), 1–9. https://doi.org/10.1088/1757-899X/713/1/012019.

Prabuningrum, D.S., Massijaya, M.Y., Hadi, Y.S., Abdillah, I.B., 2020. Physical-mechanical properties of laminated board made from oil palm trunk (*Elaeis guineensis* Jacq.) waste with various lamina compositions and densifications. J. Korean Wood Sci. Technol. 48 (2), 196–205.

Purwanto, D., 2016. Sifat Fisis dan Mekanis Papan semen dari Serat Tandan Kosong Kelapa Sawit (the physical and mechanical properties of cement board made from oil palm empty fruit bunches fibers). J. For. Prod. Ind. Res. 8 (2), 43–52.

Rasat, M.S.M., Wahab, R., Sulaiman, O., Moktar, J., Mohamed, A., Tabet, T.A., Khalid, I., 2011. Properties of composite boards from oil palm frond agricultural waste. Bioresources 6 (4), 4389–4403.

Ratnasingam, J., Scholz, F., 2009. Rubberwood an Industrial Perspective. World Resource Institute, WRI Publications, Malaysia.

Shahbandeh, M., 2020. Palm oil top global producers 2019/20. https://www.statista.com/statistics/856231/palm-oil-top-global-producers/. (Accessed December 2020).

Sikora, K.S., McPolin, D.O., Harte, A.M., 2016. Effects of the thickness of cross-laminated timber (CLT) panels made from Irish Sitka spruce on mechanical performance in bending and shear. Constr. Build. Mater. 116, 141–150. https://doi.org/10.1016/j.conbuildmat.2016.04.145.

Srivaro, S., Matan, N., Lam, F., 2019. Performance of cross laminated timber made of oil palm trunk waste for building construction: a pilot study. Eur. J. Wood Wood Prod. 77 (3), 353–365.

Stark, N.M., Cai, Z., Carll, C., 2010. Wood-based composite materials: panel products, glued-laminated timber, structural composite lumber, and wood-nonwood composite materials. Chapter 11, In: Centennial (Ed.), Wood handbook: wood as an engineering material. General technical report FPL, GTR-190. US Dept. of Agriculture, Forest Service, Forest Products Laboratory, Madison, WI, pp. 11.1–11.28.

Suhaily, S., Jawaid, M., Abdul Khalil, H.P.S., Rahman Mohamed, A., Ibrahim, F., 2012. A review of oil palm biocomposites for furniture design and applications: potential and challenges. Bioresources 7 (3), 4400–4423.

Sulaiman, O., Salim, N., Nordin, N.A., Hashim, R., Ibrahim, M., Sato, M., 2012. The potential of oil palm trunk as an alternative raw material for compressed wood. Bioresources 7 (2), 2688–2706.

Tazi, M., Sukiman, M.S., Erchiqui, F., Imad, A., Kanit, T., 2016. Effect of wood fillers on the viscoelastic and thermophysical properties of HDPE-wood composite. Int. J. Polym. Sci. 2016, 1–6. https://doi.org/10.1155/2016/9032525.

Tshai, K.Y., Yap, E.H., Wong, T.L., 2016. The effects of weight fraction on mechanical behaviour of thermoset palm EFB composite. Int. J. Mater. Mech. Manuf. 4 (4), 232–236. https://doi.org/10.18178/ijmmm.2016.4.4.262.

Valášek, P., Ruggiero, A., Müller, M., 2017. Experimental description of strength and tribological characteristic of EFB oil palm fibres/epoxy composites with technologically undemanding preparation. Compos. Part B 122, 79–88. https://doi.org/10.1016/j.compositesb.2017.04.014.

Zaaba, N.F., Ismail, H., 2019. Thermoplastic/natural filler composites: a short review. J. Phys. Sci. 30 (1), 81–99. https://doi.org/10.21315/jps2019.30.s1.5.

Zheng, J., 2002. Studies of PF Resole/Isocyanate Hybrid Adhesives (Ph.D. dissertation). Virginia Tech.

PART 4

Current policy, environmental factors, and economic prospects for oil palm biomass and composite panels

19

Policy and environmental aspects of oil palm biomass

R.A. Ilyas[j,k], S.M. Sapuan[a,b], M.S. Ibrahim[c], M.H. Wondi[d],
M.N.F. Norrrahim[e], M.M. Harussani[b], H.A. Aisyah[a],
M.A. Jenol[f], Z. Nahrul Hayawin[l], M.S.N. Atikah[g], R. Ibrahim[h],
S.O.A. SaifulAzry[a], C.S. Hassan[i], and N.I.N. Haris[m]

[a]Laboratory of Biocomposite Technology, Institute of Tropical Forestry and Forest Products (INTROP), Universiti Putra Malaysia, Serdang, Selangor, Malaysia, [b]Advanced Engineering Materials and Composites Research Centre (AEMC), Department of Mechanical and Manufacturing Engineering, Faculty of Engineering, Universiti Putra Malaysia, Serdang, Selangor, Malaysia, [c]Integrated Ganoderma Management, Plant Pathology and Biosecurity Unit, Biology and Sustainability Research (BSR) Division, MPOB, Bandar Baru Bangi, Kajang, Malaysia, [d]Faculty of Plantation and Agrotechnology, Universiti Teknologi MARA, Mukah, Sarawak, Malaysia, [e]Research Centre for Chemical Defence, National Defence University of Malaysia, Kuala Lumpur, Malaysia, [f]Department of Bioprocess Technology, Faculty of Biotechnology and Biomolecular Sciences, Universiti Putra Malaysia, Serdang, Selangor, Malaysia, [g]Department of Chemical and Environmental Engineering, Universiti Putra Malaysia, Serdang, Selangor, Malaysia, [h]Innovation & Commercialization Division, Forest Research Institute Malaysia, Kepong, Selangor, Malaysia, [i]Mechanical Engineering Department, UCSI University, Kuala Lumpur, Malaysia, [j]School of Chemical and Energy Engineering, Faculty of Engineering, Universiti Teknologi Malaysia, Johor Bahru, Malaysia, [k]Center for Advanced Composite Materials (CACM), Universiti Teknologi Malaysia, Johor Bahru, Malaysia, [l]Engineering and Processing Division, Malaysian Palm Oil Board (MPOB), Kajang, Selangor, Malaysia, [m]Institute of Sustainable and Renewable Energy, Universiti Malaysia Sarawak, Kota Samarahan, Sarawak, Malaysia

19.1 Policies regarding oil palm

Palm oil is one of the main crops with the most efficient land use among vegetable oils essential to economic development. High oil demand has caused a massive growth of oil palm plantations. For the record, almost 58.84 million tons of palm oil are produced annually; Indonesia and Malaysia yield about 85% of the total oil palm production.

Oil Palm Biomass for Composite Panels. https://doi.org/10.1016/B978-0-12-823852-3.00001-5
Copyright © 2022 Elsevier Inc. All rights reserved.

As world demand for palm oil is rising, with massive revenues for producers and governments, oil palm companies have focused on increasing palm oil production. Most companies in Indonesia and Malaysia are openly clearing rainforests to expand their plantations at the early stage of the palm oil industry. However, these activities are associated with deforestation, which contributes to biodiversity loss, soil erosion, and water pollution, which subsequently causes climate change and adverse damage to the environment.

For example, this problem is critical in Indonesia as the world's largest palm oil producer. Most of the oil palm plantations in Indonesia are located in swampy and peat soil areas, where dry peat is particularly flame retardant. The cheapest and fastest way to clean land is by burning it. For this reason, oil palm farmers prefer to use the burning method or slash-and-burn practices that cause massive overwhelming fires every year in Southeast Asia, producing heavy toxic haze (Jong, 2019; Varkkey et al., 2018). This haze is not only an air pollution epidemic but also has bad health and economic effects on its neighbors, including Brunei, Malaysia, and Singapore.

19.2 Sustainable policy—NDPE policy

In 2011, a no deforestation, no peat, no exploitation (NDPE) policy with the objective to develop an environmentally and socially responsible palm oil industry was introduced. This policy was first executed when an established Singaporean palm oil company, Golden-Agri Resources (GAR), implemented its forest conservation policy. At first, this policy was only applied to the company to be a high carbon stock (HCS) forest concept, but its scope was restricted. In December 2013, a large agribusiness related to oil palm products, Wilmar International Ltd., introduced their NDPE policy, thus creating standards for its entire supply chain. Wilmar agreed to avoid purchasing from suppliers who burned trees, drained peatland, or abused local people (Scott, 2014). For the following years, other major companies and oil palm mills followed these NDPE policies, and this was recognized as a good turning point for the oil palm industry toward more responsible and sustainable production of palm oil.

To date, most of the large international palm oil companies, traders, and refiners are committed to using NDPE policies in the global supply chain to promote sustainable production. Currently, 99% of palm oil entering Europe is traceable to the oil mill level. Over 84% of all palm oil imports are covered by company sourcing policies that focus on the NDPE role (Anon, 2017). In addition, as of April 2020, it was recorded that almost all large worldwide companies of palm oil supply chains have committed to NDPE policies and cover 83% of Indonesian and Malaysian palm oil companies (Albert, 2020).

In NDPE policy, the principle of "no deforestation" includes protection by not developing on high carbon stock (HCS) forests or high conservation value (HCV) areas, no land burning, and reducing greenhouse gas (GHG) and pollutant emissions. For the "no peat" role, several principles were practiced, such as no new development on peatland regardless of the depth of the peat; adopting best management practices and appropriate agricultural practices on peat; and exploring options for the lasting restoration of peatland and peat forest by working with professional investors and societies. Additionally, palm oil production is also associated with social problems, such as disputes with native groups living on commercial concessions, as well as the exploitation of workers and misuse of labor on oil palm groves. A "no exploitation" policy is related to these problems. Some of the no exploitation policies are to respect and support humans' and workers' rights, as well as their welfare (Pye, 2019).

19.3 Malaysia policy

Palm oil is one of the most extensive and most productive plantations in Malaysia and has been a strong contributor to the Malaysian economy for many decades. In Malaysia, there are giant palm oil companies such as Sime Darby, IOI Corporation, Kuala Lumpur Kepong Berhad, Genting Plantations Berhad, Felda Global Ventures Holdings Berhad, and United Plantations, as well as small plantation companies and private estates. In Malaysia, being one of the largest palm oil exporters globally, it is important to maintain sustainability values and agricultural development of oil palm. This covers new trends and improvements in key sectors of the palm oil industry, namely upstream, midstream, and downstream. The upstream includes oil palm processing sectors such as planting, harvesting, and transporting, while midstream involves crude and palm kernel crushing for oil production. On the other hand, downstream is composed of processing crude palm oil into beneficial products, such as food, fuel, pharmaceutical, lumber, and other value-added products.

Varkkey et al. (2018) highlighted three main policies undertaken by the Malaysian government to provide sustainable stability for the oil palm sector. In 1974, the first step undertaken by the government was established Malaysia's Palm Oil Research Institute (PORIM), which later merged with the Palm Oil Licensing Authority (PORLA) in 1998. This merging formed the Malaysian Palm Oil Board (MPOB) that acts as a center of research and development (R&D) related to oil palm and its products. The emphasis of MPOB has been, primarily, on technological intensification, aiming to close the yield distance between actual and potential oil palm yield (Fairhurst et al., 2010).

MPOB has several units or divisions such as biotechnology, breeding, biology and sustainability research, engineering and processing, etc. All research units are responsible for oil palm research activities.

The economic and industrial development division collects and analyzes data from companies, factories, producers, manufacturers, dealers, and other related parties in the oil palm industry. The technical findings are then shared with all associated parties to represent the status and output of the different industrial sectors. This unit also provides feedback to determine national development policy for palm oil, as well as researching economic growth, downstream processing, marketing, and new technologies in the palm oil industry. The MPOB also offers courses and training officers known as oil palm teaching and advisory in most all growing cities offering smallholder's regulatory, training, and advisory services (Yusof, 2012).

In 1956, the Malaysian government set up collectives of land with complementary political and economic goals, the Federal Land Development Authority (FELDA), eventually aiming to reduce rural poverty. FELDA was reformed in 1960 as a developer of land resettlement programs at the federal level, and the government has given FELDA large areas of agricultural land in Malaysia. Since 1990, FELDA has reinvented itself as a developer of commercial plantations and development projects for settlers, essentially overseeing Malaysia's largest group of smallholder oil palm farms. Such smallholder farmers have benefited from the highly structured FELDA projects, which operate similarly to a commercial oil palm plantation. For example, each plantation has a manager, plantation officer, and an agronomist to ensure the best management and operational practices. Until then, FELDA has established a commercial arm, FELDA Global Ventures (FGV), with investments in China, Indonesia, Pakistan, and Thailand that are actively involved in property fraud charges at home and abroad (O'Donnell et al., 2017).

Strengthening of oil plantations is a priority of the Economic Transformation Program (ETP) initiated by the Malaysian government in 2010 with the primary goal to get the industry closer to the national fresh fruit bunch (FFB) yield target of 26.2 tons per hectare by 2020 by concentrating on plant-level, technology-driven intensification and downstream activities (Anon, 2014). The central concept was to facilitate the replanting of new oil palm, increasing the FFB yield annually, and boosting the efficiency of the workers through technology utilization.

In September 2019, the Malaysian government introduced many policies ensuring sustainable cultivation of oil palms responsibly and sustainably. The sustainable planting practices address the 3Ps, namely profit (economics), people (social), and planet (environment). The policy on sustainable palm oil covers the total cultivated oil palm area up to 6.5 million hectares, avoiding the replantation and cultivation of oil palm in peatland areas, and improving the current regulations on peatland oil palm cultivation. Also, the government also has banned the conversion of reserved forest areas for oil palm cultivation. The policy also committed to making maps of the oil palm plantation

Chapter 19 Policy and environmental aspects of oil palm biomass **343**

available for public access (Anon, 2014). In Indonesia, the government is currently using three policy instruments to influence domestic prices: an export tax, stock buffer operations, and sales driven from public estates (Larson, 1996; Onuigbo et al., 2017).

19.4 A national strategy for the sustainable deployment of this untapped potential

Several initiatives are being implemented to promote the sustainable utilization of Malaysian biomass. In November 2011, the Malaysian Innovation Agency published a National Biomass Strategy that focuses on oil palm biomass as a starting point and might later be extended to include biomass from other sources.

According to the national strategy, from a supply-side perspective, by 2020, the Malaysian palm oil industry is expected to generate about 100 million dry tons of solid biomass. This biomass not only encompasses the empty fruit bunches (EFB), mesocarp fibers (MF), and palm kernel shells (PKS) but also the oil palm fronds (OPF) and trunks (OPT). The palm oil mill effluent (POME) is excluded from this biomass list.

Converting POME into biogas for either powering the mills or selling power to the national grid would potentially allow for an increase of power capacity of 410 MW by 2030. This initiative alone would reduce the nation's carbon dioxide (CO_2) emissions by 12% and free up significant biomass for higher value-added uses.

Assessing the logistic costs related to the mobilization of oil palm biomass, the strategy document concluded that an amount as high as 25 million tons of biomass could be mobilized at globally competitive costs, i.e., at the expense of less than RM 250 (62.5 Euro) per dry-weight ton. Approximately 12 million tons of solid biomass will likely be utilized for nonfertilizer uses by 2020, primarily for wood products and bioenergy, while an additional 20 million tons could be mobilized for pellets, biofuels, and biobased chemical industries. In total, this is approximately 30% of the solid biomass the palm oil industry is expected to generate annually by 2020.

Pellets are a natural entry point; however, the biggest long-term opportunity for Malaysia is in biobased chemicals, with a forecasted global market size of RM 110–175 billion (27.5–43 billion Euro) by 2020.

Mobilization of biomass regarding logistics and competitive costs will be a critical success factor to ensure globally competitive prices. This is why the strategy relies on creating cooperative structures to enable smaller plantations and smallholders to enter the global biomass market. To achieve this, so-called Entry Point Projects (EPP) are foreseen. Two new EPPs have already been defined for pelletization capacity and the launch of an industry consortium to catalyze the development of conversion technologies. In addition, two existing palm oil EPPs have been expanded in scope. Finally, a set of government policies are in the process of being finalized to reduce the risk to the private sector associated with accelerating this opportunity.

19.5 Issues of oil palm/palm oil at national and international levels

The environmental effects and other related issues of mass oil palm plantations are extensive within the national and international levels. The global debate on the sustainability and legitimacy of oil palm production continues to evolve and define the industry. As Malaysia is one of the major palm oil producers, a number of discussions around the issues related to this industry are posed in the face of environmental sustainability, the local and international economy, and human rights. The Malaysian debate on oil palm plantation is a global issue that affects regional, national, and international frameworks. Further research and interventions are necessary to solve the current problem for the future sustainability of the strong oil palm industries. Therefore, in this section, four major issues related to oil palm industries are highlighted.

19.5.1 Deforestation

The activity of deforestation is known for transforming forested areas to other activity purposes, including agricultural, logging, urbanization, etc. Over the past few decades, the oil palm industry has been placed under scrutiny due to its social and environmental effects that overshadow its economic benefits (Tan et al., 2009). Oil palm plantations have been accused of being the major cause of deforestation; in fact, it was responsible for only 3% of global deforestation, which accounted for 17.32 million hectares (Mha) of world palm oil plantation (Khatun et al., 2017). Southeast Asia has unique ecosystems and diversified topography covered by 43% of natural forests. Malaysia, Indonesia, and Thailand have been known as one of the world's most biologically rich ecosystems. Based on Fig. 19.1, deforestation activity reached a peak between 1970 and 1974 due to rubber plantations and was decreasing as palm oil plantations began to increase (Roda, 2019). The oil palm plantation activity was mostly replacing other agricultural plantations, including rubber.

The complex composition of physical, chemical, and biological components is known as the ecosystem that, in return, is responsible for supporting and maintaining these elements. An ecosystem's ability

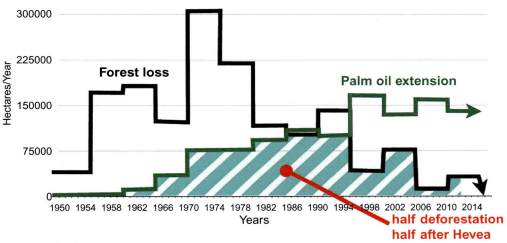

Fig. 19.1 The timeline of Malaysian deforestation from 1950 to 2014. (Source: Reproduced from Roda, J.M. 2019. The Geopolitics of Palm Oil and Deforestation. The Conversation. Available from https://theconversation.com/the-geopolitics-of-palm-oil-and-deforestation-119417.)

to preserve the biological communities, including species assembly, is commensurate to the species found in relatively undisturbed habitats of ecological integrity (the region that is recognized) (Khatun et al., 2017). It includes abiotic (air, soil, water) and biotic (living things) components. Ecological integrity also contains the ecosystem functions within the forest, including nutrient cycling and energy flow. According to Reza and Abdullah (2011), a healthy state of the complex ecosystem is maintained if ecological integrity and resilience are maintained, in turn. Although deforestation activities are insufficient to get alarmed about, the natural habitat of animals is being destroyed, which consequently leads to fauna extinction. For example, in Borneo and Sumatra, deforestation activities in tropical forests were reported to rapidly rise, thus carbon emissions were higher than the emissions from industrial activities in some developed countries. Based on the statistical point of view, palm oil plantations contributed to these scenarios, which harmed ecology stability (Tan et al., 2009). On the other hand, the reason for the planters to clear the tropical forest was to earn income from the sale of harvested timber instead of abandoning the land. Thus as a consequence, this activity endangered wildlife species, including tigers, Sumatran rhinoceros, and Asian elephants. Most of the deforestation activities in Southeast Asia are followed by a land-burning activity, which caused persistent and numerous fires for large-scale clearance. In fact, in 1997, it was claimed that the haze that happened in the Southeast Asia region was due to this activity. It was also reported that the conversion of primary rainforest to oil palm plantation resulted in huge losses of animal species, accounting for more than 80% of species.

19.5.2 International boycott

The palm oil industry has been facing a serious international boycott. In fact, the European Parliament has released a statement to ban the utilization of palm oil in biofuel production for the European Union by 2020. This situation is due to the manifested goal to put an end to the deforestation of rainforest in Southeast Asia, especially in Indonesia and Malaysia. It is interesting to note that these countries (Indonesia and Malaysia) are the largest palm oil plantation and producers, which account for approximately 80% of global production as well as 56% (6.7 million hectares) of the oil palm-cultivated area worldwide (Koh and Wilcove, 2007). Non-Government Organizations (NGOs) claimed that the tremendous growth of the palm oil industry in Southeast Asia contributed solely to colossal damage of tropical forests and further endangered the survivability of native species. In conjunction with this matter, aggressive media campaigns have been launched to ban palm oil-based products. The palm oil producers made an argument in response to the biased accusation of Western NGOs for targeting Southeast Asia's palm oil industry while disregarding other agricultural activities that also caused harm to the biodiversity in other regions. On the other hand, a threat to biodiversity was argued to not be an issue as the conversion of palm oil plantations from existing cropland was done with minimal disturbance to pristine habitats.

Major consumer packaged goods businesses, retailers, as well as NGOs have decided to launch anti-palm oil campaigns, mounting pressure on producers, manufacturers, and retailers to efficiently address these problems. A large number of campaigns have regrettably endorsed the idea of boycotting palm oil products, which might result in significantly larger environmental and social issues. In fact, the International Union of Conservation of Nature discovered that banning palm oil products led to biodiversity loss due to the increase of other oil crop production that required larger land. As it stands at the moment, the palm oil industry has contributed approximately 35% of the world's vegetable oil, concerning 10% of the land allocated to oil crops.

There are two main factors to this matter; lack of awareness associated with the NGOs to the socioeconomic realities in the particular palm oil producer countries and, on the other hand, those who

Chapter 19 Policy and environmental aspects of oil palm biomass **347**

fail to appreciate the uniqueness of Southeast Asia's biodiversity and conservation potential of natural habitats. A new strategy using the revenue from palm oil agricultural sectors to fund land possession to establish private nature reserves is seen to be the solution to the issue.

19.5.3 Waste generation

The palm oil industry leads to massive production of biomass as the oil extraction is only about 10% from palm oil production, with the remaining 90% left as biomass (Harussani and Sapuan, 2021; Kong et al., 2014). Oil palm biomass is the most abundant renewable resource available in Malaysia (Mohd Rafein et al., 2015; Harussani et al., 2021; Norrrahim et al., 2019). The amount is expected to increase each year due to the biotechnological approach in increasing oil palm productivity and less expansion on oil palm cultivation due to rapid industrialization and urbanization. This consequently raised an issue in biomass management, as enormous biomass amounts can create various environmental pollution (Harussani et al., 2022). To overcome this problem, proper biomass management must be applied to minimize or eliminate adverse effects on the environment and human health, support economic development, and improve quality of life.

According to Sulaiman et al. (2011), each kg of palm oil extraction generates approximately 4 kg of dry biomass, which consists of oil palm empty fruit bunch (OPEFB), PKS, and oil palm mesocarp fiber (OPMF) from mills, while OPF and OPT can be obtained from plantations. Umar et al. (2020) reported that there are approximately 80 million tons per year of palm oil biomass generated alone by Malaysia's palm oil industry, and the values are expected to rise to 100 million tons by the year 2020. It is estimated that the total oil palm biomass generated in Malaysia is approximately 44.85 Mt (dry weight basis), which consists of 21.73 Mt of OPF, 7.34 Mt of OPEFB, 4.46 Mt of PKS, 7.72 Mt of OPMF, and 3.60 Mt of OPT, as shown in Fig. 19.2 (Loh, 2016). Given its abundant availability, various industries have started to use oil palm biomass as an alternative raw material for multiple products manufacturing such as biocomposites, biogas, biochar, biocompost, nanocellulose, and many others (Mohd Rafein et al., 2015; Norrrahim et al., 2018, 2019; Yasim-Anuar et al., 2019, 2020).

Besides that, another major waste generated from oil palm plantation is POME, the effluent from the final stages of palm oil production in the mill (Sulaiman et al., 2011). For each ton of crude palm oil produced, an average of 0.9–1.5 m^3 POME is generated. POME has created environmental problems because it makes the largest pollution load into the rivers throughout Malaysia. However, POME contains high concentrations of carbohydrate, protein, nitrogen, lipids, and minerals that could be converted into useful materials via the microbial

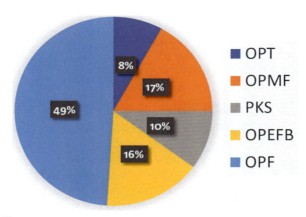

Fig. 19.2 The percentage of oil palm biomass generated in Malaysia.

process (Sulaiman et al., 2011). The biogas and hydrogen produced from POME has drawn a lot of interest (Wu et al., 2017).

19.5.4 Health and safety

The occupational safety and health of palm oil industry workers posed a challenge for the future sustainability of the industry (Myzabella et al., 2019). The palm oil industry, particularly harvesting of the palm fruit, is labor-intensive, and there is little mechanization in the Malaysian, Indonesian, or Nigerian industries, which tend to rely on low-paid workers. The Malaysian industry, in particular, heavily relies on foreign labor, with an estimated 450,000 foreign workers from Indonesia, the Philippines, Nepal, and Bangladesh working on its plantations. Reports from Malaysia highlight labor exploitation, including worker abuse and child labor occurring on oil palm plantations. Based on a survey done by Myzabella et al. (2019), there are four types of common occupational hazards in the palm oil industry: musculoskeletal disorders, infectious diseases, mental disorders, and pesticides/herbicide exposure. They also found that many of the everyday tasks undertaken by workers on oil palm plantations during harvesting are associated with increased risks of musculoskeletal disorders and that the prevalence of musculoskeletal disorders reported in most studies was high. Additionally, oil palm plantation workers are at risk of infectious diseases, stress, and mental disorders because of their working and living environments and exposure to a range of pesticides, particularly paraquat. In general, work on oil palm plantations remains largely unmechanized and relies on low-paid workers. This lack of automation increases the risk of work-related injuries and diseases for these workers.

19.6 Conclusions

The environmental effects and other related issues of mass oil palm plantations are extensive within the national and international levels. As Malaysia is one of the major palm oil producers, a number of discussions around the issues related to this industry are posed in the face of environmental sustainability, the local and international economy, and human rights. Four major issues related to oil palm industries highlighted are wealth and safety, deforestation, international boycott, as well as waste generation. In term of waste generation, oil palm biomass can be used as by-products by transforming these biomasses into valuable products such as biofuels, pellets, food packaging, biochar, filler in composites materials, furniture, pulp and paper, particleboards, polymer composites, medium-density fiber (MDF) and plywood, brake pads, foam, and cement concrete. These have been discussed in previous chapters.

Acknowledgments

This project was funded by Universiti Putra Malaysia through Geran Putra Berimpak (GPB), UPM/800-3/3/1/GPB/2019/9679800.

References

Albert, T.K., 2020. NDPE policies cover 83% of palm oil refineries; implementation at 78%. Chain Reaction Research. Available from https://chainreactionresearch.com/report/ndpe-policies-cover-83-of-palm-oil-refineries-implementation-at-75/.

Anon, 2014. NKEA Palm Oil and Rubber, ETP Annual Report 2014. Available from https://rehdainstitute.com/etp/.

Anon, 2017. ESPO Monitoring Report 2017. Available from https://www.idhsustainabletrade.com/publication/monitoring-report-espo-2017/.

Fairhurst, T.H., Griffiths, W., Donough, C.R., Witt, C., McLaughlin, D., Giller, K.E., 2010. Identification and Elimination of Yield Gaps in Oil Palm Plantations in Indonesia. Available from https://edepot.wur.nl/171498.

Harussani, M.M., Sapuan, S.M., 2021. Development of kenaf biochar in engineering and agricultural applications. Chem. Afr. https://doi.org/10.1007/s42250-021-00293-1.

Harussani, M.M., Sapuan, S.M., Umer, R., Khalina, A., Ilyas, R.A., 2022. Pyrolysis of polypropylene plastic waste into carbonaceous char: priority of plastic waste management amidst COVID-19 pandemic. Sci. Total Environ. 803, 149911. https://doi.org/10.1016/j.scitotenv.2021.149911.

Harussani, M.M., Umer, R., Sapuan, S.M., Khalina, A., 2021. Low-temperature thermal degradation of disinfected COVID-19 non-woven polypropylene-based isolation gown wastes into carbonaceous char. Polymers 13 (22), 3980. https://doi.org/10.3390/polym13223980.

Jong, H.N., 2019. Indonesian Minister Draws Fire for Denial of Transboundary Haze Problem. Mongabay. Available from https://news.mongabay.com/2019/09/indonesian-minister-draws-fire-for-denial-of-transboundary-haze-problem/.

Khatun, R., Reza, M.I.H., Moniruzzaman, M., Yaakob, Z., 2017. Sustainable oil palm industry: the possibilities. Renew. Sust. Energ. Rev. 76, 608–619.

Koh, L.P., Wilcove, D.S., 2007. Cashing in palm oil for conservation. Nature 448 (7157), 993–994.

Kong, S.H., Loh, S.K., Bachmann, R.T., Rahim, S.A., Salimon, J., 2014. Biochar from oil palm biomass: a review of its potential and challenges. Renew. Sust. Energ. Rev. 39, 729–739.

Larson, D., 1996. Indonesia's Palm Oil Subsector. Available from https://elibrary.worldbank.org/doi/abs/10.1596/1813-9450-1654.

Loh, S.K., 2016. The potential of the Malaysian oil palm biomass as a renewable energy source. Energy Convers. Manag. 141, 285–298.

Mohd Rafein, Z., Norrrahim, M.N.F., Hirata, S., Hassan, M.A., 2015. Hydrothermal and wet disk milling pretreatment for high conversion of biosugars from oil palm mesocarp fiber. Bioresour. Technol. 181, 263–269.

Myzabella, N., Fritschi, L., Merdith, N., El-Zaemey, S., Chih, H., Reid, A., 2019. Occupational health and safety in the palm oil industry: a systematic review. Int. J. Occup. Environ. Med. 10 (4), 159–173.

Norrrahim, M.N.F., Ariffin, H., Hassan, M.A., Ibrahim, N.A., Yunus, W.M.Z.W., Nishida, H., 2019. Utilisation of superheated steam in oil palm biomass pretreatment process for reduced chemical use and enhanced cellulose nanofibre production. Int. J. Nanotechnol. 16, 668–679.

Norrrahim, M.N.F., Ariffin, H., Yasim-Anuar, T.A.T., Ghaemi, F., Hassan, M.A., Ibrahim, N.A., Ngee, J.L.H., Yunus, W.M.Z.W., 2018. Superheated steam pretreatment of cellulose affects its electrospinnability for microfibrillated cellulose production. Cellulose 25 (7), 3853–3859.

O'Donnell, M., Mansor, N.B., Yogeesvaran, K., Rashid, A., 2017. Organisational change and success in a government enterprise: Malaysia's Federal Land Development Agency. Econ. Labour Relat. Rev. 28 (2), 234–251.

Onuigbo, D.M., Sinaga, B.M., Harianto, H., 2017. Oil palm policy, land use change and community livelihoods (OLCL) in Indonesia: a sustainability framework. Int. J. Environ. Sci. Dev. 8 (8), 601–605.

Pye, O., 2019. Commodifying sustainability: development, nature and politics in the palm oil industry. World Dev. 121, 218–228.

Reza, M.I.H., Abdullah, S.A., 2011. Regional index of ecological integrity: a need for sustainable management of natural resources. Ecol. Indic. 11 (2), 220–229.

Roda, J.M., 2019. The Geopolitics of Palm Oil and Deforestation. The Conversation. Available from https://theconversation.com/the-geopolitics-of-palm-oil-and-deforestation-119417.

Scott, P., 2014. Wilmar's "no Deforestation" Commitment Could Revolutionise the Way Food Is Grown. The Guardian. Available from https://www.theguardian.com/sustainable-business/wilmar-no-deforestation-commitment-food-production.

Sulaiman, F., Abdullah, N., Gerhauser, H., Shariff, A., 2011. An outlook of Malaysian energy, oil palm industry and its utilization of wastes as useful resources. Biomass Bioenergy 35 (9), 3775–3786.

Tan, K.T., Lee, K.T., Mohamed, A.R., Bhatia, S., 2009. Palm oil: addressing issues and towards sustainable development. Renew. Sust. Energ. Rev. 13 (2), 420–427.

Umar, H.A., Sulaiman, S.A., Ahmad, R.K., Tamili, S.N., 2020. Characterisation of oil palm trunk and frond as fuel for biomass thermochemical. IOP Conf. Ser.: Mater. Sci. Eng. 863, 012011.

Varkkey, H., Tyson, A., Choiruzzad, S.A.B., 2018. Palm oil intensification and expansion in Indonesia and Malaysia: environmental and socio-political factors influencing policy. Forest Policy Econ. 92, 148–159.

Wu, Q., Qiang, T.C., Zeng, G., Zhang, H., Huang, Y., Wang, Y., 2017. Sustainable and renewable energy from biomass wastes in palm oil industry: a case study in Malaysia. Int. J. Hydrog. Energy 42 (37), 23871–23877.

Yasim-Anuar, T.A.T., Ariffin, H., Norrrahim, M.N.F., Hassan, M.A., Andou, Y., Tsukegi, T., Nishida, H., 2020. Well-dispersed cellulose nanofiber in low density polyethylene nanocomposite by liquid-assisted extrusion. Polymers 12, 927.

Yasim-Anuar, T.A.T., Ariffin, H., Norrrahim, M.N.F., Hassan, M.A., Tsukegi, T., 2019. Sustainable one-pot process for the production of cellulose nano fiber and polyethylene/cellulose nano fiber composites. J. Clean. Prod. 207, 590–599.

Yusof, B., 2012. Where lies the future for malaysian palm oil and rubber industry? In: Malaysian Palm Oil Council (MPOC) (Ed.), IKMAS. Universiti Kebangsaan Malaysia, Bangi.

20

Corporate ownership structure of the major oil palm plantation companies in Malaysia and biomass agenda

K. Norfaryanti[a], A. Adhe Rizky[a], H.M. Omar Shaiffudin[a], A.M. Amira Nabilah[a], C.L. Ong[a], and J.M. Roda[b]

[a]*Institute of Tropical Forestry and Forest Products (INTROP), Universiti Putra Malaysia, Serdang, Selangor, Malaysia,* [b]*UR 105 Forests & Societies, The French Agricultural Research Centre (CIRAD), Montpellier, France*

20.1 Introduction

A corporate ownership structure of a business involves a multitude of shareholding relations among stakeholders. The stakeholders vary from individuals, government, and financial institutions. Among these stakeholders, one will have significant ownership of the business. As the structure of a company grows more complex, the power and control flow from the ultimate shareholder may be concealed, and identifying the actual controller of decisions becomes more difficult. Cross-shareholdings and a pyramidal structure play important roles in transferring information when making decisions. An assessment of cross-shareholdings can reveal the amount of control held by a company/individual in a structure, where they own each other but have different amounts of control. For example, company A owns company B by 51%, while company B owns company A by 10%, but these interlocking shareholdings may hide the identity of the actual controller of decisions. The pyramidal structure reflects the hierarchical level of information flow in decision-making. The more pyramidal the company, the bigger the flow of information, a factor that can result in inflexibility in the decision-making process.

Given that the ownership structure of companies involved in diversified business activities vary, it is crucial to analyze them in greater detail to gain a better understanding of their decision-making behavior. Generally, when the behavior is better understood, the policy intervention is better designed.

Oil Palm Biomass for Composite Panels. https://doi.org/10.1016/B978-0-12-823852-3.00011-8
Copyright © 2022 Elsevier Inc. All rights reserved.

Agribusiness and plantations are the leading agents in Malaysia's agriculture industry. The oil palm industry constitutes most of the agribusiness and plantations sector and is the major contributor to the gross domestic product (GDP) in the agriculture sector. In 2021, the oil palm industry contributed 37.1% to the sector, compared to forestry logging and rubber that only contributed 5.2% and 2.5%, respectively (Department of Statistics Malaysia DOSM, 2021).

One of the major sources of biomass in Malaysia is from oil palm crops. Oil palm mills generate several oil palm wastes. The oil palm wastes contribute about RM 6379 million of energy, annually (Sulaiman et al., 2011). Several strategies need to be mobilized by the government to achieve the optimum utilization of oil palm waste for energy generation as renewable energy and the fifth fuel. This includes increasing its development by providing good incentives to promote the utilization of biomass wastes and encouraging collaborative efforts between government and private institutions to implement the technical and commercial aspects of its commission, which should include further research development of biomass wastes (Sulaiman et al., 2011).

This chapter reviews the development process of green energy policies and programs. We analyzed and highlighted the topology of the major oil palm actors' corporate ownership to fundamentally understand their decision-making behavior in adapting to policies and government programs.

20.2 Ownership structure

The ownership structure of a company has a crucial bearing on its corporate strategies (Chandler, 1962; Miles and Snow, 1978). The ownership structures of large corporations are complex and diverse, which informs their decision-making behavior. Decision-making control by a corporate entity is complex because of the convoluted shareholding structures within the company. The ownership structures that can be employed include cross-shareholdings and pyramidal structures that can contribute to the complexity of the decision-making process. Such structures could also lead to devious decision-making.

The board of directors and the management team are key actors in the decision-making structure. They belong to the hierarchical managerial decision-making structure, also regarded as the power structure (Martz and Semple, 1985). Each hierarchy level has a range of control in the decision-making, and it carries a decision's load. For example, a business development department has its hierarchy levels that are responsible for making decisions on future investment of the company. These decisions have a load in the hierarchy. The decision's load in the hierarchy somewhat reflects the decision-making behavior of the company.

Connections in the structure convey the flow and concentration of corporate control as well as the flow of information. High concentration of share ownership reflects high influence over decision-making.

In the agribusiness and plantation sectors, the decisions of a company are often directly related to environmental sustainability. These important sectors also recorded high volumes of trade and investment in the country and region. Most of the time, it is difficult to decipher a decision made by a company because many aspects could drive decision-making, from political intervention, oil prices, and environmental issues, to international trade and global financial market trends. Understanding the reason for decision-making by a company, be it to reinforce control patterns or other factors, is crucial as this provides insights into issues such as the sustainability of the agricultural sector worldwide. This research is designed to understand and analyze the decision-making patterns by major agribusiness and plantation companies in Malaysia.

Most of the ownership structures of companies are becoming complex networks. With the advancement of network studies, network analysis is emerging as an important tool to understand the interactions between actors in complex networks. Complex networks are present in a wide range of systems in nature and society, such as the Internet, movie actor collaborations, cellular networks, ecological networks, citation networks, linguistics networks, power and neural networks, financial networks, and many others (Albert and Barabasi, 2002).

A company's structure determines its corporate behavior, and it can emerge as a complex structure, which is why network analysis serves as a tool to link corporate strategies, ownership, and control. This logical framework is the basis for the theoretical framework that will be discussed in this chapter.

Due to the complex corporate structure that informs decision-making influence, this research is designed to quantify and analyze the decision-making behavior of each firm based on its corporate structure using network analysis tools. As mentioned, network analysis is emerging as a tool to analyze a complex network such as the shareholding structures of large companies.

20.3 Biomass utilization

It is believed that there are a few potentials of palm biomass utilization in Malaysia: (i) the support and policies from the government, (ii) national and international business opportunities, and (iii) research and development activities from Malaysian agencies and higher institutions. Malaysia is determined to promote itself as a major biomass hub in the Southeast Asia region. Supportive policies such as the Small Renewable Energy Power (SREP) and the National Renewable Energy

Policy of 2010 were introduced to enhance implementation and investment from the private sector (Ng et al., 2012).

The economic feasibility and sustainability of converting oil palm biomass to biobased commercial products, synthetic biofuels, and for power generation have been reported. The findings show that Malaysia has the potential to be one of the major contributors of renewable energy in the world via oil palm biomass. Subsequently, Malaysia can then become a model to other countries that has huge biomass feedstock (Shuit et al., 2009) (Table 20.1).

On the business side, Lahad Datu Biomass JV Cluster Berhad was established in July 2014, following a joint venture agreement signed

Table 20.1 Industrial applications of oil palm biomass.

No	Type of biomass	Utilization	References
1	Oil palm trunks	Paper making	Abdul Khalil et al. (2010a,b)
		Production of plywood	Nordin et al. (2004)
		Production of laminated veneer lumber	Hashim et al. (2012)
		Production of particle board panels	Abnisa et al. (2013)
		Production of bio-oil and biochar	
2	Oil palm fronds	Paper making	Wan-Daud et al. (2011)
		Production of bindless boards	Hashim et al. (2012)
		Production of syngas	Nipattummakul et al. (2012)
		Roughage for ruminant animals	Abnisa et al. (2013)
		Production of bio-oil and biochar	
3	Empty fruit bunch	Paper making	Ghazali et al. (2006)
		Production of syngas/hydrogen	Uemura et al. (2011)
		Production of fertilizer	Seenivasagam et al. (2015)
		Production of medium density fiberboard	Oviasogie et al. (2010)
			Abdul Khalil et al. (2010a,b)
		Production of briquettes	Sethupathi et al. (2008)
		Production of biocomposite	Abu-Bakar et al. (2008)
		Production of biochar	Abdul Khalil et al. (2012)
		Production of activated carbon	Abdul Khalil et al. (2009)
		Electricity generation	Abdul Khalil et al. (2011)
		Production of hydrogen	Bernama (2010)
			Firoozian et al. (2011)
			Kelly-Yong et al. (2007)
			Awaludin et al. (2005)

Table 20.1 Industrial applications of oil palm biomass—cont'd

No	Type of biomass	Utilization	References
4	Mesocarp fiber	Briquettes	Hussain et al. (2002)
		For erosion control	Geng (2014)
		For making mattress cushion	Kormin et al. (2016)
		For soil stabilization	Geng (2014)
		For horticulture and landscaping	Geng (2014)
		For ceramic and brick production	Shuit et al. (2009)
		For paper production	Haron et al. (2001)
		For acoustic control	Mayulu (2014)
		As fertilizer	Abdul Khalil et al. (2012)
		For animal feed	Abdul Khalil et al. (2011)
		As filler in thermoplastic and thermo-setting composite	Salema and Ani (2011)
		Production of bio-oil	
5	Palm kernel shell	Syngas production	Hussain et al. (2002)
		Briquettes production	Nomanbhay and Palanisamy (2005)
		For producing PKS/chitosan composite	Tan and Nasir (2004)
		for removal of heavy metal from waste	Ahmad et al. (2007)
		water	Wan-Daud et al. (2007)
		Production of carbon molecular sieve	Ahmad et al. (2008)
		Production of bio-oil	Nomanbhay and Palanisamy (2005)
			Tan and Nasir (2004)
			Ahmad et al. (2007)
			Wan-Daud et al. (2007)
			Ahmad et al. (2008)
			Kim et al. (2014)
6	POME	Biogas synthesis	Esther (1997)
		For producing fertilizer	Ahmadun et al. (2005)
		Syngas synthesis	
7	Oil palm boiler ash	For silica gel synthesis	Kow et al. (2015)
		As biofertilizer	Haron et al. (2008)
		For producing cement bricks	Tangchirapat et al. (2009)
		As absorbent for sulfur dioxide, for removal of disperse dyes and Cr(VII) from aqueous solutions	Mohamed et al. (2005)
			Isa et al. (2007)
			Isa et al. (2008)
		As filler in polypropylene composite	Bhat and Abdul Khalil (2011)
		As filler in natural rubber composite	Ooi et al. (2013)

Source: Onoja, E., Chandren, S., Razak, F.I.A., Mahat, N.A., Wahab, R.A. 2019. Oil palm (*Elaeis guineensis*) biomass in Malaysia: the present and future prospects. Waste Biomass Valoriz. 10 (8), 2099–2117. https://doi.org/10.1007/s12649-018-0258-1.

by five biomass leading companies in Malaysia. It is to boost conversion of oil palm biomass or any other forms of biomass in Malaysia into value-added products. The five biomass companies that formed a consortium include Genting Sdn Bhd, Kelas Wira Sdn Bhd, Bell Corp Sdn Bhd, Teck Guan Industries Sdn Bhd, and Golden Elate Sdn Bhd.

Commercial-scale production of bioethanol from oil palm biomass in Malaysia was first achieved in 2014, and commercial biofuel from cellulosic ethanol first became available in Malaysia. Production of cellulosic ethanol from oil palm biomass started as a pilot-scale project through a partnership between Sime Darby Plantation and Mitsui Engineering Shipbuilding of Japan (Admin, 2011). The facility producing cellulosic ethanol formally co-located within Sime Darby's facilities in Selangor was later taken over and operated by Tech Guan Group in Malaysia (Biofuel Digest). Correspondingly, Beta Renewable state-of-the-art technology is another foreign technology that specializes in converting oil palm biomass into biofuels and biochemical in Malaysia (Onoja et al., 2019).

On the government side, the establishment of a cluster (led by the Agensi Inovasi Malaysia (AIM)) by the Malaysian Government in 2014 is relevant to highlight the bottleneck of the long existing gap in the market between owners of biomass (oil palm farms, mills, processing industries) and the downstream users. In addition to the present existing cluster, the Malaysian Prime Minister in February 2016 launched the Sabah and Sarawak Biomass Industries Development Plan (Onoja et al., 2019).

Despite all, implementation of these policies in the industry is still at its nascent stage as the rate of oil palm biomass conversion to value-added products remains rather low. The obstacles responsible for low conversion of oil palm biomass to desired end products include (i) low participation of small- and medium-sized enterprises (SMEs) in the utilization of oil palm biomass to produce value-added product and (ii) low quality of the products (fibers, compost, fuel, and fuel products) from oil palm biomass that renders them inferior and can easily be displaced by other substitute products, hence affecting its market value and patronage.

To achieve this, it has been suggested that, in addition to the existing technologies and government policies, newer technologies and policies could be formulated and enacted to encourage private sector participation (Onoja et al., 2019).

20.4 Case study on the ownership structure of oil palm corporations

Shareholding data was collected from eight publicly listed palm oil corporations for the year 2013. They are government-linked companies (GLCs) and family-owned companies. The data was obtained from verified sources: Osiris and Oriana databases developed by Bureau Van Dijk, the Malaysian Registrar of Company (RoC), and the companies' audited published annual reports. Data were collected for 10 levels of shareholdings for each company. Data on biomass utilization was collected based on their annual reports and web pages.

To understand the effect of forest-related trade and investment in Southeast Asia, this chapter decided to focus on agribusiness and oil palm plantation companies and analyze their involvement in biomass-related activity. This chapter analyzed eight companies among the top 12 companies listed in Table 20.2. These eight companies are global players in the industry. As of 2019, there were 48 agribusiness and plantation corporations listed on the Bursa Kuala Lumpur (Table 20.2). Their market capitalization was then RM 130 billion, 7% of the total market capital value of the Bursa Kuala Lumpur, which was RM 1.7 trillion (Gomez et al., 2017).

Table 20.2 Agribusiness and plantation corporations listed in Bursa Kuala Lumpur in 2019.

No.	Plantation corporations	Market capital (RM billion)	Ownership type	Estimated planted area (ha)
1.	Sime Darby Plantation Berhad[a]	33.597	Federal GLC	20,264
2.	IOI Corporation[a]	27.528	Family	177,279
3.	Kuala Lumpur Kepong Berhad (KLK)[a]	23.739	Family	224,454
4.	Batu Kawan Berhad	6.073	Family	109,911
5.	Genting Plantations[a]	8.740	Family	159,183
6.	United Plantations	5.953	Family	51,284
7.	Kulim (M) Berhad[a]	Not listed	State GLC	
8.	TSH Resources	1.333	Family	42,109
9.	IJM Plantations Berhad[a]	1.435	Federal GLC	60,633
10.	Sarawak Oil Palms Berhad	2.033	Family	87,571
11.	Boustead Plantations Berhad	1.098	Federal GLC	79,406
12.	Jaya Tiasa Holdings[a]	0.696	Family	69,589
13.	Hap Seng Plantations Holdings	1.336	Private	36,103
14.	TH Plantations	0.389	Federal GLC	69,891
15.	MKH Plantations	Not listed	Private	
16.	Ta Ann Holdings Berhad	1.192	Private	49,028
17.	United Malacca Berhad	0.999	Family	33,569
18.	TDM Berhad	0.396	State GLC	42,022
19.	Far East Holdings Berhad	1.514	State GLC	16,160

Continued

360 Chapter 20 Biomass policies and strategies of Malaysian oil palm plantation corporations

Table 20.2 Agribusiness and plantation corporations listed in Bursa Kuala Lumpur in 2019—cont'd

No.	Plantation corporations	Market capital (RM billion)	Ownership type	Estimated planted area (ha)
20.	Kretam Holdings	1.234	Private	19,517
21.	Rimbunan Sawit Berhad	0.541	Family	48,765
22.	Chin Tek Plantations Bhd	0.548	Family	9380
23.	Kim Loong Resources Bhd	1.291	Family	14,946
24.	Tanah Makmur Berhad	Not listed	Private	
25.	BLD Plantation	0.683	Private	38,800
26.	Sarawak Plantation Berhad	0.546	Private	35,076
27.	Kwantas Corporation Bhd	0.433	Family	19,665
28.	WTK Holdings Berhad	Not listed	Private	
29.	Dutaland Berhad	0.279	Family	Na
30.	PLS Plantation	0.341	Private	10,516
31.	Negri Sembilan Oil Palms Bhd	0.198	Family	5618
32.	Inch Kenneth Kajang Rubber PLC	0.210	Private	Na
33.	NPC Resources Bhd	0.216	Private	5,900,000
34.	Cepatwawasan Group Berhad	0.191	Family	8440
35.	Golden Land Berhad	0.086	Family	Na
36.	Riverview Rubber Estates Bhd	0.176	Private	Na
37.	Sungei Bagan Rubber Co (M) Bhd	0.175	Family	4928
38.	Kluang Rubber Co (M) Bhd	0.192	Family	6502
39.	Harn Len Corporation Bhd	0.093	Family	16,837
40.	MHC Plantations Berhad	0.117	Family	11,403
41.	Gopeng Berhad	0.188	Private	1434
42.	Astral Asia Berhad	0.083	Private	4019
43.	Malpac Holdings Bhd	0.045	Private	Na
44.	Pinehill Pacific Berhad	0.065	Private	3438
45.	FGV Holdings Berhad	3.794	Federal GLC	349,638
46.	Innoprise Plantations Berhad	0.512	Private	12,208
47.	Matang Berhad	0.145	Private	1080
48.	Sin Heng Chan (Malaya) Berhad	0.049	Private	Na
Total		**130.482**		

[a] Companies involved in the detail shareholding structure analysis.
Source: 2019 Companies' Annual Report, 2019 Stock Performance Guide, and www.malaysiastock.biz.

After shareholdings data were gathered, we analyzed their structure using network analysis software, Cytoscape. This chapter visualized the topology of the ownership structure of the major oil palm companies in Malaysia based on their shareholding's ownership. The software is able to generate various types of layouts, and we chose the hierarchical layout to visualize the corporate ownership topology. Their topology provides insights on the decision-making control pattern by each company.

20.5 Findings

We observed that ownership structure of a corporation explains more of their decision-making control than the ownership type. Fig. 20.1 illustrates four GLC's ownership topology, three corporations are federal government-owned (Sime Darby, IJM Plantations, and Boustead Holdings), and one corporation is a state government-owned corporation (Kulim).

Sime Darby Berhad illustrates that most of the subsidiaries are concentrated at the flagship company, Sime Darby Berhad. It shows that the decision-making process and control are very much centralized at the main holding company, as Sime Darby is a highly diversified business group. From the topology structure described earlier, it is clear that the decision-making control mainly comes from the main holding company at the top. This topology implies that the main holding group is powerful and dictates the decision-making control and process. It is a highly centralized firm as the hierarchical index is 92%. The higher hierarchical index means the companies have many layers of hierarchy in its structure. It makes the holding company highly centralized and heavy in maneuvering the whole group in synergy. Sime Darby, which was founded in 1910 and is the largest plantation business group in Malaysia in terms of market capitalization, is a global player in the agribusiness sector. It operates in 26 countries through more than 500 subsidiaries through five sectors: plantation, motors, property development, industrial, and energy and utilities. This explains the highly centralized and hierarchical structure of the company.

Boustead Holdings can be seen that most of the companies are spread throughout the network, forming many clusters. Boustead Holdings as the holding company owns majority shares in other public-listed companies. Boustead Holdings is 94% hierarchic, the highest hierarchical structure among GLCs. It also has the largest number of companies visualized in the network. Based on these features, Boustead has a heavy decision-making load as a diversified main holding company in various economic sectors.

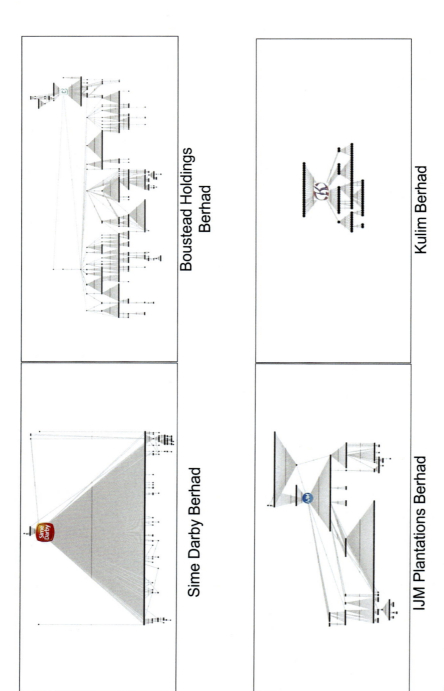

Fig. 20.1 Government linked corporations' ownership topology.

IJM Plantations is majority owned by its holding company, IJM Corporations. IJM has an incredibly complex interlocking stock ownership pattern as this group has interests in three public-listed companies: IJM Corp, IJM Land, and IJM Plantations. The major shareholder in IJM Corporations is PNB, whose stake is through the ASNB's various trust fund schemes. IJM Plantations is 84% hierarchic, which is not so high compared to the other GLCs of a similar size. It has 421 companies visualized in the network. Based on the topology metrics, IJM Plantations is reasonably proportionate compared to Sime Darby and Boustead. This could be due to its feature as the subsidiary of IJM Corporation, whilst both Sime Darby and Boustead are main holding companies.

Kulim has a smaller ownership structure compared to three other plantations company, Sime Darby, Boustead, and IJM Plantations. There are only two big business groups, which are majority owned by Kulim. It wholly owns EPA Management, an investment holding providing management services and consultancy, and a mechanical equipment assembly company. Kulim is 81% hierarchic, also the least among all GLCs. It has 101 companies visualized in the network. Based on the topology metrics, Kulim is smaller than the other three GLCs. It shows the features of a lean structure, which could have an efficient flow of decision-making. This could be due to its feature as a state-owned GLC compared to the other three companies, which are owned by the federal government-linked investment companies (GLICs) that have a higher amount of investments and higher social obligations.

It is clear that, within GLCs, there are variations of share ownership structure. Kulim has the lowest value for hierarchical index at 81% compared to Sime Darby, Boustead, and IJM at 92%, 94%, and 84%, respectively. This reflects its size of business operation. This is due to its ownership through the state-owned investment company, compared to Sime Darby, Boustead, and IJM Plantations, which are owned by the federal GLICs. The amount of investment for the federal GLICs is more than the state GLIC. In terms of decision-making process, the state-owned investment company has a shorter process, as indicated by Kulim's pyramidal tier and hierarchical index.

Fig. 20.2 illustrates the family-owned corporations' ownership topology which consists of four major corporations; Genting Plantations Berhad, IOI Corporation Berhad, Kuala Lumpur Kepong Berhad (KLK), and Jaya Tiasa Berhad. Genting Plantations is owned by Genting Berhad, which is owned by Kien Huat Realty Sdn Bhd, which is in turn owned by Parkview Management Sdn Bhd. Lim Goh Tong's family owns Parkview Management. Employees Provident Fund (EPF), a statutory body that acts as a government investment company, also has a 16% share in Genting Plantations. Given its smaller size of subsidiaries, Genting Plantations is 90% hierarchic, higher than KLK, the largest pyramid, in terms of subsidiaries and shareholdings

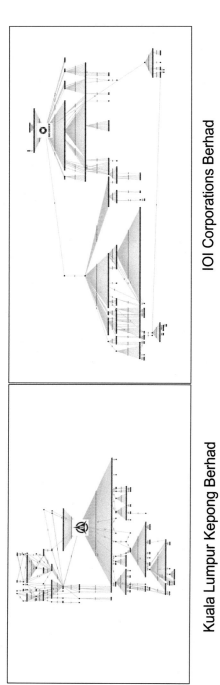
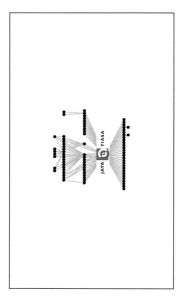

Fig. 20.2 Family-owned corporations' ownership topology.

degree. It reflects that the decision-making control is centralized at the Genting Plantations group. It has 483 companies visualized in the network. While the oil palm plantation business remains the core activity of Genting Plantations, the company has diversified into property development to unlock the value of its strategically-located land bank through its wholly-owned subsidiary Genting Property Sdn Bhd.

The company is majority-owned by Tiong Toh Siong Holdings with 22%. The other two major shareholders are Genine Chain Limited with 18% and Double Universal Limited with 15% through Amanas Sdn Bhd, Nustinas Sdn Bhd, and Insan Anggun Sdn Bhd. Genine Chain Limited and Double Universal Limited were incorporated in 1992 as a private company limited by shares registered in Hong Kong. Jaya Tiasa is 75% hierarchic, the lowest among family-owned businesses (FOBs). It suggests that the decision-making control is decentralized, where the subsidiaries are autonomous. It has 81 companies visualized in the network.

Vertical Capacity Sdn Bhd is the majority shareholder in IOI Corporation. Lee Shin Cheng is the founder of IOI Corporation and owns Progressive Holdings Sdn Bhd, which owns Vertical Capacity Sdn Bhd. Lee Shin Cheng's ownership of Progressive Holdings Sdn Bhd gives him high control in IOI Corporations' business decision-making. IOI Corp is 93% hierarchic, the highest among the FOBs. It is proposed that IOI Corp's decision-making control is more centralized than KLK. Both IOI Corp and KLK have a similar number of companies, but the respective hierarchy levels vary. IOI Corp has 639 companies visualized in the network, making it the largest FOB in this research.

KLK Berhad's ownership structure has an indirect link to the ultimate owner, Batu Kawan Bhd and Arusha Enterprise. Batu Kawan is an investment holding company with the majority of its investments in chemical manufacturing, transportation, property development and plantations. Arusha Enterprise is also an investment company. KLK is 81% hierarchic, the second least hierarchical among all the FOBs. It suggests that the size of the shareholdings degree does not reflect the decision-making control in KLK. It has 520 companies visualized in the network, the second largest among the FOBs.

The shareholdings topologies are similar for both GLCs and family-owned corporations. They are all financialized by the financial investment institutions. However, the degree of the financialization differs from each other. GLCs' major shareholders are linked to most of the other companies, including the family-owned firms. The GLCs are linked via GLICs, such as Permodalan Nasional Berhad (PNB), Employees Provident Fund (EPF), Lembaga Tabung Angkatan Tentera (LTAT), and Kumpulan Wang Persaraan (KWAP). The GLICs have a stake in the four plantation companies. The small percentage of shareholdings suggests that its motive is to focus on the return on investment instead of controlling the business's decision-making. PNB is the major shareholder for Sime Darby and IJM Plantation, which are the big players in oil palm plantation. The links via GLICs show the agribusiness corporations are heavily financialized by the financial institutions. This could shift their strategies in the decision-making for sustainable development and future adoptions of policies in biomass and green energy.

In contrast to that, the family-owned firms' major shareholders are not linked to any other companies. GLCs' shareholders are mainly the investment houses that focus on building the nation's wealth. They have the ability and the capacity to invest widely in other companies to maximize their returns. On the other hand, family-owned firms' shareholders mainly depend on the independent investing institutions, which have limitations in investing at a bigger scale. They tend to focus on creating and circulating their wealth internally within their holdings group.

Fig. 20.3 shows the hierarchical layout of the major oil palm companies. There are five companies, Sime Darby, Boustead, IOI Corp, KLK, and Jaya Tiasa, situated at the fourth level of the hierarchy. Three other companies, IJM Plantations, Kulim, and Genting Plantations, are at the mid-level of the hierarchy. This could be due to their main holdings company as the major shareholders. There are many cross-linkages among the top level in the hierarchy involving family-owned companies. It is evident that there are various proportions of the hierarchical structure for all companies. These variations lead to their decision-making behaviors. The family-owned structure topology is relatively smaller than the government-owned topology. This is due to the size of companies and the fact that family-owned firms are smaller than government-owned firms. Although the business activities are similar for both family-owned and government-owned, family-owned companies are less hierarchical.

The explanation given by the ownership structure of a corporation demonstrates their decision-making pattern, whether it is very hierarchical or less hierarchical, and very centralized or decentralized. Our result shows the ownership structure of a corporation explains more of their decision-making control than the ownership type, compared to vast literature supporting otherwise. Their adaptation to the new policies and government programs may depend on their corporate ownership topology.

20.6 Biomass development progress and challenges

The key challenge to achieving the renewable energy target is the inadequate grid infrastructure that inhibits palm oil developers from benefiting from the Feed-in-Tariff (FiT) payment scheme. One way forward, a strategic partnership between government and industrial players, offers a promising outcome, depending on an economic feasibility study

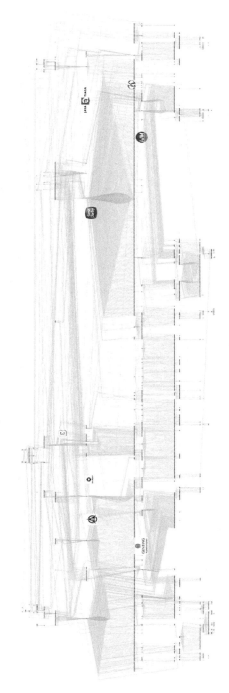

Fig. 20.3 Ownership hierarchy structure of the major oil palm plantations companies.

(Aghamohammadi et al., 2016). FiT system, introduced in 2011, is a tool that will help to support the growth of the renewable energy industry.

The high investment cost is the main challenge restricting their participation in the renewable energy business. The cost of setting up a biogas plant can amount to RM 10 million per mill, which will impose a heavy burden on the mill developers due to the long payback period and uncertainty of recouping their investment costs. The unattractive tariff means they are not interested in investing in the biomass renewable energy business, and 18% of respondents agreed that the inconsistency of the biomass supply chain was one of the challenges (Aghamohammadi et al., 2016).

The conducive and limiting factors for development and proliferation of a palm oil biomass waste-to-energy niche in Malaysia were examined during the period 1990–2011. Rising oil prices, strong pressure on the palm oil industry from environmental groups, and a persisting palm oil biomass waste disposal problem in Malaysia appear to have been conducive to niche proliferation. On top of this, national renewable energy policies and large-scale donor programs have specifically supported the utilization of palm oil biomass waste for energy. However, in spite of this, the niche development process has only made slow progress. The paper identifies reluctant implementation of energy policy, rise in biomass resource prices, limited network formation, and negative results at the niche level as the main factors hindering niche development (Hansen and Nygaard, 2014).

Although more were planned during this period, 39 plants had reached some level of progression from the initial planning stage toward full operation. The plants analyzed are located in Peninsular Malaysia and Sabah, respectively, in west and east Malaysia (no plants were identified in Sarawak) and include combined heat and power (CHP) plants, stand-alone grid-connected power plants, and off-grid steam-generating plants for various industrial users (Hansen and Nygaard, 2014).

There are three phases involved in analyzing the main conducive and limiting factors in the development of palm oil to biomass waste-to-energy niche in Malaysia from 1990 to 2011. The phases and the period are (i) Slow introduction (1990–2001), (ii) Growth (2002–2006), and (iii) Fragmentation (2007–2011). Three domains were used to analyze the each of these phases: (i) Landscape conditions, (ii) Regime conditions, and (iii) Niche processes. In the regime conditions, energy, palm oil, and cross-cutting elements were included. In the niche processes, expectations, actor network, and learning elements were analyzed (Hansen and Nygaard, 2014).

Based on the analysis, four enabling factors were found to enable the niche areas' development: rising oil prices, donor interventions, environmental pressures on the oil palm industry, and waste disposal problems in the palm oil industry. Increasing oil prices from the late

1990s to 2008 is the most important incentive to fuel-switching. Donor intervention programs contributed significantly to creating and sustaining the expectations for investors and technology suppliers by engaging in advocacy, knowledge dissemination, and financial support to specific projects. In 2004, Roundtable of Sustainable Palm Oil (RSPO) was established to reduce the environmental effect of palm oil industry and directly influence the dynamics of the biomass niche development. The biomass waste disposal problem is a significant incentive throughout the three periods, which forced palm oil companies to adopt alternative waste management practices.

Four factors limited the niche development, such as reluctance to implement the energy policy, rise in biomass resource prices, limited network formation, and negative results at the niche level. There are difficulties in introducing a private-sector model for electricity production in Malaysia and to conflicting political interests in providing the necessary incentives limiting the biomass niche development. The high cost of biomass waste, such as EFB and PKS, reduced the economic viability of producing green energy. Throughout the periods, the network linking actors across the value chain was weak and led to poor intercompany coordination of activities, cooperation, and interaction in terms of knowledge sharing. To conclude the slow progress in the niche development by 2011, limited strategic interest from investors and producers of equipment in engaging the sufficient investment was the negative result at the niche level (Hansen and Nygaard, 2014).

However, the development of palm biomass utilization and an evolution in the palm biomass industry in the country are in progress but at a slow rate. It was estimated that the advancement or further breakthroughs in palm biomass utilization can be achieved by 2020. This advancement can be observed from the progressive switch of palm biomass product value from low-value palm biomass products to higher-value palm biomass products (Ng et al., 2012). Other factors that led to the underdevelopment of the palm biomass industry are the lack of available proven processing technologies and "closed-door" attitudes exhibited by industrial players (Ng et al., 2012) The main challenge for the development of the palm biomass industry relates to the growth of oil palm industry, as the development of the palm biomass industry is directly affected by the existence of the oil palm industry. In Malaysia, the projects that utilize palm oil as feedstock have been criticized by some environmentalists (Ng et al., 2012).

20.7 Conclusions

Corporate ownership structure analysis gives a new perspective in understanding the decision-making behavior by the business corporations. Even though the ownership structure differs in terms of government-owned or family-owned, the analysis shows their topologies are similar. Both groups have a high to low hierarchical index, which explained the high to low level of bureaucracy for strategic decision-making within the corporation. It indicates a government-owned corporation could have a low bureaucracy level and a family-owned corporation could have a high bureaucracy level. The corporate ownership topology depends on many factors, such as the history of the establishment, the investment return of the corporations, and others.

Oil palm biomass valorization and its development is promising and has high potentials due to its abundance in supply. Government provides many programs and policies to stimulate its growth. Various agencies involved in the development of the agenda drive progress, and some are showing positive results. However, the rate of the biomass agenda development is rather slow as its duration has passed a decade.

The chapter linked the findings of corporate ownership topology with the biomass development agenda adoption by the major agribusiness corporation in Malaysia. Our results highlight the ability of explaining the decision-making control of a corporation by their ownership structure. Most of the corporations, despite being either government-owned or family-owned, have a high to medium hierarchical index. The bigger the corporation, the higher the hierarchical index, which is linked with the top level strategic decision-making control. This suggests that it takes a longer time for a corporation to implement progress and drive a new development agenda.

References

Abdul Khalil, H.P.S., Jawaid, M., Hassan, A., Tahir, P.M., Ashaari, Z., 2012. Oil Palm Biomass Fibres and Recent Advancement in Oil Palm Biomass Fibres Based Hybrid Biocomposites. Intech, Rijeka. https://doi.org/10.5772/48235.

Abdul Khalil, H.P.S., Kang, C.W., Awang, K., Ramli, R., Omar, A.T., 2009. The effect of different laminations on mechanical and physical properties of hybrid composites. J. Reinf. Plast. Compos. 28 (9), 1123–1137. https://doi.org/10.1177/0731684407087755.

Abdul Khalil, H.P.S., Poh, B.T., Issam, A.M., Jawaid, M., Ramli, R., 2010a. Recycled polypropylene-oil palm biomass: the effect on mechanical and physical properties. J. Reinf. Plast. Compos. 29 (8), 1117–1130. https://doi.org/10.1177/0731684409103058.

Abdul Khalil, H.P.S., Rawi, N.-F.M., Bhat, A.H., Jawaid, M., Nik-Abdillah, N.-F., 2010b. Development and material properties of new hybrid plywood from oil palm biomass. Mater. Des. 31 (1), 417–424. https://doi.org/10.1016/j.matdes.2009.05.040.

Abdul Khalil, H.P.S., Rawi, N.-F.M., Jawaid, M., Bhat, A.H., Abdullah, C.K., 2011. Empty fruit bunches as a reinforcement in laminated bio-composites. J. Compos. Mater. 45 (2), 219–236. https://doi.org/10.1177/0021998310373520.

Abnisa, F., Niya, A.A., Daud, W.-M.A.W., Sahu, J.N., Noor, I.M., 2013. Utilization of oil palm tree residues to produce bio-oil and bio-char via pyrolysis. Energy Convers. Manag. 76, 1073–1082. https://doi.org/10.1016/j.enconman.2013.08.038.

Abu-Bakar, N., Ma, A.N., May, C.Y., Mohamad, S., Halim, R.M., Awaludin, A., Zainal, Z., 2008. Oil palm biomass as potential substitution raw materials for commercial biomass briquettes production. Am. J. Appl. Sci. 5 (3), 179–183.

Admin, 2011. Sime Darby, Mitsui Launch Malaysian Palm Waste Biofuel Project. The Digest: The World's Most Widely-Read Bioeconomy Daily (Jan 3). https://www.biofuelsdigest.com/bdigest/2011/01/03/sime-darby-mitsui-launch-malaysian-palm-waste-biofuel-project/. (Accessed 25 February 2021).

Aghamohammadi, N., Reginald, S.S., Shamiri, A., Zinatizadeh, A.A., Wong, L.P., Sulaiman, N.M.N., 2016. An investigation of sustainable power generation from oil palm biomass: a case study in Sarawak. Sustainability (Switzerland) 8 (5), 416. https://doi.org/10.3390/su8050416.

Ahmad, M.A., Wan-Daud, W.-M.A., Aroua, M.K., 2007. Synthesis of carbon molecular sieves from palm shell by carbon vapor deposition. J. Porous. Mater. 14, 393–399.

Ahmad, M.A., Wan-Daud, W.-M.A., Aroua, M.K., 2008. Adsorption kinetics of various gases in carbon molecular sieves (CMS) produced from palm shell. Colloids Surf. A Physicochem. Eng. Asp. 312 (2–3), 131–135. https://doi.org/10.1016/j.colsurfa.2007.06.040.

Ahmadun, F.'l.-R., Yassin, A.A.A., Lyuke, S.E., Ngan, M.A., Masayoshi, M., 2005. Bio-hydrogen synthesis from wastewater by anaerobic fermentation using microflora. Int. J. Green Energy 2, 387–396. https://doi.org/10.1080/01971520500288014.

Albert, R., Barabasi, A.L., 2002. Statistical mechanics of complex networks. Rev. Mod. Phys. 74 (1), 47–97. https://doi.org/10.1103/RevModPhys.74.47.

Awaludin, A., Bakar, N.A., May, C.Y., Adam, N.M., Salit, M.S., 2005. Development of gasification system fuelled with oil palm fibres and shells. Am. J. Appl. Sci. 9, 72–75.

Bernama, 2010. Biochar Malaysia: UPM-NASMECH Effect of Producing EFB Biochar: World's First. 7 January 2010 http://biocharmalaysia.blogspot.com/2010/01/upm-nasmech-effort-of-producing-efb.html. (Accessed 4 April 2011).

Bhat, A.H., Abdul Khalil, H.P.S., 2011. Exploring, nano filler based on oil palm ash in polypropylene composites. Bioresources 6 (2), 1288–1297.

Chandler, A., 1962. Strategy and Structure. MIT Press, Cambridge, MA.

Department of Statistics Malaysia (DOSM), 2021. Indikator Pertanian Terpilih 2021. Available at: https://newss.statistics.gov.my/newss-portalx/ep/epFreeDownloadContentSearch.seam?cid=60824 (Accessed 14 December 2021).

Esther, O.U., 1997. Anaerobic digestion of palm oil mill effluent and its utilization as fertilizer for environmental protection. Renew. Energy 10 (2–3), 291–294. https://doi.org/10.1016/0960-1481(96)00080-8.

Firoozian, P., Bhat, I.H., Abdul Khalil, H.P.S., Noor, A.M., Akil, H.M., Bhat, A.H., 2011. High surface area activated carbon prepared from agricultural biomass: empty fruit bunch (EFB), bamboo stem and coconut shells by chemical activation with H3PO4. Mater. Technol. 26 (5), 222–228.

Geng, A., 2014. Upgrading of oil palm biomass to value- added products. In: Hakeem, K.R., Jawaid, M., Rashid, U. (Eds.), Biomass and Bioenergy: Applications. Springer, pp. 187–209.

Ghazali, A., Wan-Daud, W.-R., Law, K.-N.K., 2006. Alkaline peroxide mechanical pulping (APMP) of oil palm lignocellulosics: part 2 empty fruit bunch (EFB) responses to pretreatments. Appita J. 59 (1), 65–70.

Gomez, E.T., Padmanabhan, T., Kamaruddin, N., Bhalla, S., Fisal, F., 2017. Minister of Finance Incorporated: Ownership and Control of Corporate Malaysia. Institute for Democracy and Economic Affairs (IDEAS) and Strategic Information and Research Development Centre (SIRD), Petaling Jaya, Kuala Lumpur and Petaling Jaya.

Hansen, U.E., Nygaard, I., 2014. Sustainable energy transitions in emerging economies: the formation of a palm oil biomass waste-to-energy Niche in Malaysia 1990–2011. Energy Policy 66, 666–676. https://doi.org/10.1016/j.enpol.2013.11.028.

Haron, K., Darus, A., Zakaria, Z.Z., Tarmizi, A.M., 2001. An innovative technique on management of biomass during oil palm plantation. MOPB Inf. Ser. 101.

Haron, K., Mohammed, A.T., Halim, R.M., Din, A.K., 2008. Palm-based bio-fertilizer from decanter cake and boiler ash of palm oil mill. Inf. Ser. 412, 1–4.

Hashim, R., Nadhari, W.N.A.W., Sulaiman, O., Sato, M., Hiziroglu, S., Kawamura, F., Sugimoto, T., Seng, T.G., Tanaka, R., 2012. Properties of binderless particleboard panels manufactured from oil palm biomass. Bioresources 7 (1), 1352–1365.

Hussain, Z., Zainac, Z., Abdullah, M.Z., 2002. Briquetting of palm fibre and shell from the processing of palm nuts to palm oil. Biomass Bioenergy 22 (6), 505–509.

Isa, M.H., Lang, L.S., Asaari, F.A.H., Aziz, H.A., Ramli, N.A., Dhas, J.P., 2007. Low cost removal of disperse dyes from aqueous solution using palm ash. Dyes Pigments 74 (2), 446–453. https://doi.org/10.1016/j.dyepig.2006.02.025.

Isa, M.H., Ibrahim, N., Aziz, H.A., Adlan, M.N., Sabiani, N.H.M., Zinatizadeh, A.A.L., Kutty, S.R.M., 2008. Removal of chromium (VI) from aqueous solution using treated oil palm fibre. J. Hazard. Mater. 152 (2), 662–668. https://doi.org/10.1016/j.jhazmat.2007.07.033.

Kelly-Yong, T.L., Lee, K.T., Mohamed, A.R., Bhatia, S., 2007. Potential of hydrogen from oil palm biomass as a source of renewable energy worldwide. Energy Policy 35 (11), 5692–5701. https://doi.org/10.1016/j.enpol.2007.06.017.

Kim, S.W., Koo, B.S., Lee, D.H., 2014. Catalytic pyrolysis of palm kernel shell waste in a fluidized bed. Bioresour. Technol. 167, 425–432. https://doi.org/10.1016/j.biortech.2014.06.050.

Kormin, S., Rus, A.Z.M., Azahari, M.S.M., 2016. Preparation of polyurethane foams using liquefied oil palm mesocarp fibre (OPMF) and renewable monomer from waste cooking oil. In: 4th International Conference on the Advancement of Materials and Nanotechnology (ICAMN IV 2016), AIP Conference Proceedings, 1877, 060006-1-060006-7., https://doi.org/10.1063/1.4999885.

Kow, K.-W., Mun, L.Y., Yusoff, R., 2015. Silica gel synthesized from oil palm boiler ash. J. Miner. Metal Mater. Eng. 1, 14–18.

Martz, D.J., Semple, R.K., 1985. Hierarchical corporate decision-making structure within the Canadian urban system: the case of banking. Urban Geogr. 6 (4), 316–330.

Mayulu, H., 2014. The nutrient potency of palm oil plantation and mill's by-product processed with Amofer technology as ruminant feed. Int. J. Sci. Eng. 6 (2), 112–116. https://doi.org/10.12777/ijse.6.2.112-116.

Miles, R.E., Snow, C.C., 1978. Organizational Strategy, Structure, and Process. McGraw-Hill, New York.

Mohamed, A.R., Lee, K.T., Noor, N.M., Zainudin, N.F., 2005. Oil palm ash/Ca (OH)2/CaSO4 absorbent for flue gas desulfurization. Chem. Eng. Technol. 28 (8), 939–945. https://doi.org/10.1002/ceat.200407106.

Ng, W.P.Q., Lam, H.L., Ng, F.Y., Kamal, M., Lim, J.H.E., 2012. Waste-to-wealth: green potential from palm biomass in Malaysia. J. Clean. Prod. 34, 57–65. https://doi.org/10.1016/j.jclepro.2012.04.004.

Nipattummakul, N., Ahmed, I.I., Kerdsuwan, S., Gupta, A.K., 2012. Steam gasification of oil palm trunk waste for clean syngas production. Appl. Energy 92, 778–782. https://doi.org/10.1016/j.apenergy.2011.08.026.

Nomanbhay, S.M., Palanisamy, K., 2005. Removal of heavy metal from industrial wastewater using chitosan coated oil palm shell charcoal. Electron. J. Biotechnol. 8 (1), 44–53.

Nordin, K., Jamaludin, M.A., Ahmad, M., Samsi, H.W., Salleh, A.H., Jallaludin, Z., 2004. Minimizing the environmental burden of oil palm trunk residues through the development of laminated veneer lumber products. Manag. Environ. Qual. 15 (5), 484–490. https://doi.org/10.1108/14777830410553924.

Onoja, E., Chandren, S., Razak, F.I.A., Mahat, N.A., Wahab, R.A., 2019. Oil palm (*Elaeis guineensis*) biomass in Malaysia: the present and future prospects. Waste Biomass Valoriz. 10 (8), 2099–2117. https://doi.org/10.1007/s12649-018-0258-1.

Ooi, Z.X., Ismail, H., Abu-Bakar, A., 2013. Optimisation of oil palm ash as reinforcement in natural rubber vulcanisation: a comparison between silica and carbon black fillers. Polym. Test. 32 (4), 625–630. https://doi.org/10.1016/j.polymertesting.2013.02.007.

Oviasogie, P.O., Aisueni, N.O., Brown, G.E., 2010. Oil palm composted biomass a review of the preparation, utilization, handling and storage. Afr. J. Agric. Res. 5 (13), 1553–1571.

Salema, A.A., Ani, F.N., 2011. Microwave induced pyrolysis of oil palm biomass. Bioresour. Technol. 102 (3), 3388–3395. https://doi.org/10.1016/j.biortech.2010.09.115.

Seenivasagam, S., Zainal, Z., Ali, S., Hin, T.-Y.Y., 2015. Supercritical water gasification of empty fruit bunches from oil palm for hydrogen production. Fuel 143, 563–569. https://doi.org/10.1016/j.fuel.2014.11.073.

Sethupathi, S., Piao, C.S., Mohamed, A.R., 2008. Utilization of oil palm as a source of renewable energy in Malaysia. Renew. Sust. Energ. Rev. 12 (9), 2404–2421. https://doi.org/10.1016/j.rser.2007.06.006.

Shuit, S.H., Tan, K.T., Lee, K.T., Kamaruddin, A.H., 2009. Oil palm biomass as a sustainable energy source: a Malaysian case study. Energy 34 (9), 1225–1235. https://doi.org/10.1016/j.energy.2009.05.008.

Sulaiman, F., Nurhayati, A., Gerhauser, H., Shariff, A.R.M., 2011. An outlook of Malaysian energy, oil palm industry and its utilization of wastes as useful resources. Biomass Bioenergy 35 (9), 3775–3786. https://doi.org/10.1016/j.biombioe.2011.06.018.

Tan, J.S., Nasir, A.F., 2004. Carbon molecular sieves produced from oil palm shell for air separation. Sep. Purif. Technol. 35 (1), 47–54. https://doi.org/10.1016/S1383-5866(03)00115-1.

Tangchirapat, W., Jaturapitakkul, C., Chindaprasirt, P., 2009. Use of palm oil fuel ash as a supplementary cementitious material for producing high-strength concrete. Constr. Build. Mater. 23 (7), 2641–2646. https://doi.org/10.1016/j.conbuildmat.2009.01.008.

Uemura, Y., Omar, W.N., Tsutsui, T., Yusup, S., 2011. Torrefaction of oil palm wastes. Fuel 90 (8), 2585–2591. https://doi.org/10.1016/j.fuel.2011.03.021.

Wan-Daud, W.-M.A., Ahmad, M.A., Aroua, M.K., 2007. Carbon molecular sieves from palm shell: effect of the benzene deposition times on gas separation properties. Sep. Purif. Technol. 57 (2), 289–293. https://doi.org/10.1016/j.seppur.2007.04.006.

Wan-Daud, W.-R., Mazlan, I., Law, K.N., Razali, N., 2011. Influences of the operating variables of acetosolv pulping on pulp properties of oil palm frond fibers. Maderas Cienc. Tecnol. 13 (2), 193–202. https://doi.org/10.4067/S0718-221X2011000200007.

Index

Note: Page numbers followed by *f* indicate figures and *t* indicate tables.

A

Accelerated air drying, 119
Activated carbon, 207
Adhesion theory, 102–103
Aerobic granules
  formation in POME, 200
  in wastewater treatment, action
    mechanism of, 199–200
Agensi Inovasi Malaysia (AIM),
    358
Agribusiness and plantation
    corporations, 354,
    359–360*t*
Air drying, 118
Air velocity, 119
Alkali pretreatment,
    lignocellulosic fiber,
    161–162
Anaerobic baffled filter (ABF)
    reactor, 197
Anaerobic downflow filter (ADF)
    reactor, 197
Anaerobic hybrid reactor (AHR),
    197
Angiosperms, 57
ANOVA analysis on strength and
    physical properties of
    COPF composite, 185, 185*t*
Auto infeed and outfeed for hot-
    press, 264
Axial swelling, 136

B

Binderless particleboard from
    OPT, 106–107, 285
Bioethanol, commercial-scale
    production of, 358
Biohydrogen, 203
Biological pretreatment,
    lignocellulosic fiber,
    163–164
Biomass development progress
    and challenges, 366–369

Biomass resource prices, 369
Biomass types, in Malaysia, 5*f*
Biomass utilization, 355–358,
    356–357*t*
Bleaching pretreatments,
    lignocellulosic fiber, 163
Bonding of oil palm biomass
    (OPB), 103–108
Boustead Holdings, 361
Brown rot, 58
  characteristics of, 59*t*
By-products
  of oil palm kernel, 206–207
  from oil palm mills, 13–24
  from oil palm plantations, 12–13

C

Carbohydrates, 101, 105
Cellulose structure unit with
    hydroxyl groups, 50*f*
Cement composites, 328–332, 329*t*
  empty fruit bunches, properties
    of, 331–332
  oil palm frond, properties of,
    330–331
  oil palm trunk, properties of, 330
Chemical degradation of
    lignocellulosic materials,
    300
Chemical mechanical pulp (CMP),
    300
Chemical pretreatment,
    lignocellulosic fiber, 161
Chemical thermomechanical pulp
    (CTMP), 300
Clipper, 145
Cocoa, 113
Coconut, 113
Commercial-scale production of
    bioethanol, 358
Composite panel products,
    classification and application
    of, 215–240, 219–221*t*

Compressed oil palm fronds
    (COPF) composite board
  ANOVA analysis, strength and
    physical properties, 185,
    185*t*
  basic density investigation,
    179–180, 179*t*
  board strength properties, 180–185
  compressive strength, 183–184,
    184*t*
  density measurement, 178,
    178*t*
  microstructure, 186–187, 186*f*
  physical properties, 177–180
  processing, 176, 177*f*
  resin penetration on, 187, 188*f*
  static bending strength,
    180–183, 181*t*
Continuous sterilizer, 16*f*
Conveyor, 145
Corporate ownership structure of
    business, 354–355
  case study on, 358–361
  shareholding relations, 353
  topology, 362*f*
Coupling agents, lignocellulosic
    fiber, 162
Cross-cutting (bucking), OPT, 143,
    143*f*
Cross-laminated timber (CLT), 324
Cross-shareholdings and a
    pyramidal structure, 353
Crude palm oil (CPO), 10–12

D

Debarking of OPT using rotary
    lathe machine, 224*f*
Deforestation, 344–346, 345*f*
Dryer, 148
Drying
  of oil palm trunk and lumber,
    117–121
  of OPT veneers, 225*f*

375

376 Index

E

Ecological integrity, 344–346
Economic Transformation
　　Program (ETP), 3–4, 342
Electron radiation, lignocellulosic
　　fiber, 161
Employees Provident Fund (EPF),
　　365–366
Empty fruit bunch (EFB) fiber,
　　13–19, 13f, 41–42f, 63–64,
　　290–292, 291–292f, 291t
　cement composites, properties
　　of, 331–332
　chemical characteristics,
　　155–156
　complex nature of, 154–156
　in composite panel industry,
　　156–158
　constraints in producing
　　composite panels, 156–158
　fiberboard, 313–315
　hydrophilic, 157–158
　internal bond strength of, 294f
　lignocellulosic fiber
　　pretreatment method,
　　types of, 158–164
　mechanical characteristics, 156
　morphological characteristics,
　　154–155
　polymer-based composites, 328
　pretreatment of, 153–174
　with silica bodies, 43f
　vascular bundles, composition
　　of, 157
Energy Commission of Malaysia,
　　9–10
Epoxy with short oil palm fiber,
　　104–105

F

Family-owned corporations'
　　ownership topology, 364f
Feed-in-Tariff (FiT) payment
　　scheme, 366–368
FELDA projects, 342
Felling and bucking, OPT veneer
　　processing, 140–141, 141f
Felling and cross-cutting,
　　plantation, 223–224, 223f

Fiber length, 46–47
Fiber processing, 22
Fiber separation and development
　　in refining process, 302,
　　302f
Fiber-reinforced composites,
　　pretreated EFB fibers,
　　164–168
Fiberboard, 72–73, 84–85, 84f
　from oil palm biomass, 305
　　government policies,
　　　297–298
　　manufacturing process,
　　　298–304
　from oil palm empty fruit
　　bunches, 313–315
　from oil palm frond, 305–307t,
　　307–311
　from oil palm trunk, 311–313
Five-step processing method
　　(sawing, drying,
　　impregnation, re-drying,
　　and hot pressing), 124, 127t
Flowering plants, 57
Forced-air drying, 118–119
Formaldehyde-based adhesives
　fabrication of oil palm-based
　　panels, 88
　hybrid composites, 93–94
　melamine-urea-formaldehyde,
　　88
　for oil palm-based panels,
　　87–97, 93t
　phenol-formaldehyde, 88
　resin properties, 87–88
　synthetic polymeric material,
　　87–88
　urea-formaldehyde, 88
Formaldehyde-based resin,
　　100–102, 100f
Fresh fruit bunch (FFB), 41f
　harvesting for palm oil
　　production, 10–12
　solid waste materials and by-
　　products, 42f
Fruit digestion, 17–19, 18f
Fruit sterilization process, 14,
　　14–15f
Fruit stripping process, 14–17

G

Genting plantations, 363–365
Gluing of OPT veneers, 225f
Golden-Agri resources (GAR), 340
Green veneer production, 144–146
Green veneer segregator, 262, 263f

H

Hardboard, 298
Health and safety, 348
Heating/semicuring, 126
High carbon stock (HCS) forests,
　　341
High conservation value (HCV)
　　areas, 341
High-density fiberboard (HDF), 298
Horizontal sterilizer, 14–15f
Hot-pressing/densification, 126

I

IJM Plantations, 363
Infeed racks, 264
Inorganic-bonded composites,
　　328–332, 329t
Integrated system, 197
Internal bonding (IB), 272, 275t,
　　277f
International boycott, 346–347
Isocyanate, 101, 105

J

Japanese Agricultural Standard
　　(JAS), 322–324
Japanese Industry Standard (JIS),
　　285

K

KLK Berhad's ownership structure,
　　365
Kumpulan Wang Persaraan
　　(KWAP), 365–366

L

Laminated-based composites,
　　322–326, 323t
　oil palm frond, 325–326
　oil palm trunk, 322–325
Laminated veneer lumber (LVL)
　adhesives, 247
　dimensional stability and
　　durability, 247–248

manufacturing process, 242–243
from oil palm trunk, 243–245,
244*t*
physical and mechanical
properties, 246–247, 246*t*
properties of, 245–248
raw materials, 242
thickness swelling, 248
Lathe machine, 144–145, 144*f*
Lembaga Tabung Angkatan
Tentera (LTAT), 365–366
Lignin degradation, 57–58
Lignocellulosic biomass, structure
of, 102*f*
Lignocellulosic fiber pretreatment
method, 153
alkali pretreatment, 161–162
biological pretreatment,
163–164
bleaching pretreatments, 163
chemical pretreatment, 161
coupling agents, 162
electron radiation, 161
mechanical process, 158–159
physical pretreatment, 158–161
solvent extraction, 159
steam explosion, 160
thermal process, 160
types of, 158–164
ultrasonication pretreatment,
160–161
Lignocellulosic materials, 48
Log storage, OPT, 141–143
Log-end processing, OPT veneer
processing, 140–143
Low molecular weight phenol
formaldehyde (LMwPF),
64
Low-density fiberboards (LDF),
298

M

Malaysia policy, 341–343
Malaysian Palm Oil Board
(MPOB), 4*f*, 27–28
using EFB fibers, 314–315*f*, 315
Malaysian timber industry, 253
Malaysian Timber Industry Board
(MTIB), 254

Maleated polypropylene (MAPP),
107
Mechanical dryer, 224–225
Mechanical process,
lignocellulosic fiber,
158–159
Mechanized infeed rack for platen
dryer, 264
Mechanized stacking rack, 264
Medium-density fiberboard
(MDF), 139, 298, 298–299*f*,
312*f*
dimensional stability behavior,
313–314
EFB fibers, 316*f*
oil palm biomass, 226, 227*f*
performance properties, 314
physical and mechanical
properties, 310*t*
process flow, 309*f*
Melamine-urea-formaldehyde
(MUF), 88–90
resin levels, 285
Mesocarp fiber, 20–21
Methane gas, 198
Methanogenesis, 195–196
Microorganism treatment system,
195
Microstructure of COPF composite
board, 186–187, 186*f*
Mobilization of biomass, 343–344
Modified super-fast drying
method for OPL, 120–121
Modulus of elasticity (MOE), 272,
275*t*
empty fruit bunch, 292, 293*f*
Modulus of rupture (MOR), 272,
275*t*
empty fruit bunch, 292, 293*f*
Monocotyledon *versus*
dicotyledon, 57, 58*t*

N

Nanocellulose from plant fiber,
24–26
National Green Technology Policy
in 2009, 3–4
National Key Economic Areas
(NKEAs), 9–10

National strategy for sustainable
deployment of untapped
potential, 343–344
National Timber Industry Census
Study, 78
Natural plant fiber
advantages, 218
utilization, 218
No deforestation, no peat, no
exploitation (NDPE) policy,
340–341
Nonbiological method, 196–197
Nonformaldehyde-based
adhesives, for bonding oil
palm biomass, 99–110

O

Oblique sterilizer, 16*f*
Occupational safety and health of
palm oil industry, 348
Oil extraction, 19
Oil palm (OP) waste, 8, 113
Oil palm biomass (OPB)
biological durability and
deterioration, 57–67
biological durability, 61–65
bonding, 103–108
cell wall structure, 47–48
chemical properties, 48–51,
48–49*f*, 50*t*
fiber morphology properties, 47*t*
lignocellulose content of, 60*t*
mechanical characteristics,
51–52, 51*t*
for medium-density fiberboard
(MDF), 226, 227*f*
for molded particleboard, 222,
222*f*
morphological properties of, 47
nonformaldehyde-based
adhesives, 99–110
from oil palm tree, 40*f*
panels, application of, 332–333
physical and anatomical
properties, 41–48
physicochemical properties, 59–61
trunk enhancement against
termites and fungal decay,
65*t*

Oil palm fronds (OPF), 10–12, 12*f*, 26, 46*f*, 63, 113, 175
  blending ratios, 309
  cement composites, properties of, 330–331
  fiberboard, 305–307*t*, 307–311
  laminated-based composites, 325–326
  petiole of, 46*f*
  polymer-based composites, 327
Oil palm fruit, 21*f*
Oil palm kernel shells, 21*f*
Oil palm lignocellulosic fiber biomass, 10–12, 11*f*
Oil palm lumbers (OPL), 114–116
  drying, 117–121
  five-step processing method, 122
  imperfections, 115–116, 117*f*
  with inherently inferior properties, 122
  modified super-fast drying method, 120–121
  quality enhancement, 121–126, 122–123*t*
  sawing methods, 116–117, 117*f*
  substitution for wood timber, 226, 228*f*
  super-fast drying, 119–120
Oil palm mesocarp fiber (OPMF) biomass, 20–21, 22*f*
Oil palm mills, 354
Oil palm parenchyma tissues, 276–277
Oil palm plywood, 223–225
Oil palm productions, 27–28, 27*t*
Oil palm residues after oil extraction, 8*f*
Oil palm tree, 43
Oil palm trunk (OPT), 10–13, 13*f*, 61–62, 113, 115–116
  anatomical structure, 133–134, 134*f*
  average density, 136, 137*f*
  cement composites, properties of, 330
  characteristics, 132–139
  chemical properties, 138
  fiberboard, 311–313
    for sound absorption, 310*f*

illustration, 44*f*
laminated veneer lumber (*see* Laminated veneer lumber (LVL))
laminated-based composites, 322–325
moisture content, 45–46
mechanical properties, 138–139
moisture content, 45–46, 134–136, 135*f*
morphological features, 132, 133*f*
OPT-based panel products, 131–132
at plantation, 256
physical properties, 134–139
polymer-based composites, 326
rotary veneer processing, 131–152
swelling, 136, 138*f*
vascular bundles, 45*f*
volumetric shrinkage, 136, 138*f*
woody parts of, 134
Oil palm wood (OPW)
  drying, 117–121
  imperfections, 115–116, 117*f*
  moisture content, 114
  processing method, five-step *versus* six-step, 127*t*
  sawing methods, 116–117, 117*f*
  utilization, 114
Oil palm-based composites, 91–94
Oil palm-based panels, formaldehyde-based adhesives, 87–97, 93*t*
Olefins, 107
Open pond concept, 195
Organic acids, 64–65
Oriented strand board (OSB)
  panels, 70, 73–74
  mechanical properties of wood panels, 268, 274–276
  moisture content, 272–273, 272*f*
  morphological analysis, 276–278
  from oil palm biomass, 267–281
  physical properties, 272–278
  plywood elements and properties, 267–268
  preparation of, 270–272, 270–271*f*

substitute materials for structural applications, 269
thickness swelling, 273–274, 273*t*
value-added goods, 269
water absorption, 273–274, 273*t*
wood-engineered product, 268–269
Ownership structure of oil palm corporations, 354–355
  case study on, 358–361
  of major oil palm plantations companies, 367*f*
  shareholding relations, 353
  topology, 362*f*
Ozone pretreatment method, anaerobic treatment system, 197

P

Palm kernel shell (PKS), 21–22, 26–27
Palm oil mill effluent (POME), 22–24, 23*f*, 23*t*
  with aerobic system, 196
  biochemical oxygen demand, 194
  biohydrogen generation, 204–205, 205*t*
  biological treatment, 195–196
  chemical oxygen demand, 194
  composition, 203
  features, 194*t*
  Malaysian crude oil palm production and estimation, 193*t*
  nonbiological method, 196–197
  nontoxic material, 194
  oily waste in, 192, 192*f*
  products from, 198–199, 198*t*
  source and characteristics, 193–194, 193*t*
  treatment methods, 195–197
  treatment systems, 24*f*
Palm oil prices, 28–30, 28*f*, 29*t*
Palm oil production by country in 1000 MT, 7*f*
Palm oil solid residues, properties of, 24–27

Index **379**

Particleboard, 72, 139
  deforestation, 293–294
  from oil palm biomass, 283–296
  oil palm fronds, 286–290, 287f, 288t
  oil palm trunk, 284–286, 286–287t
  thickness swelling, 285–286, 285t
  water absorption, 285–286, 285t
Peeling, 145–146
Peeling equipment, 144–145
Permodalan Nasional Berhad (PNB), 365–366
PHA-accumulating microorganisms, 203
Phenol-formaldehyde (PF), 88, 90–91
Phenolic compregnation technique, 64
Phenolic resin treatment, 64–65
Physical pretreatment, lignocellulosic fiber, 158–161
Plywood from oil palm trunk, 70–72, 83, 84f, 131–132, 255, 256t
  cold-pressing process, 257
  enhanced process flow, 262–264, 263f
  enhancement of, manufacturing process, 259–264, 261–262t
  glue spreader, 257
  grading and packing, 259, 260f
  hot-pressing, 258, 259f
  log peeling process, 257, 258f
  log preparation, 256
  panel sanding, 259
  panel sizing, 259
  prepressing, 257
  production process of, 256–259
  veneer drying, 257, 258f
Plywood manufacturing process, 140, 142t
Policies regarding oil palm, 339–340
Policy and environmental aspects, 337–351
Polyhydroxyalkanoate (PHA), 201–204

accretion using POME, 203–204
biohydrogen from, 206
and biohydrogen processing issues, 205–206
mixed culture, 202–203
from POME, 205–206
production method, 201
quantity of, 204, 204–205t
volatile fatty acids, 201
Polymer-based composites, 326–328, 327t
  empty fruit bunches, 328
  oil palm frond, 327
  oil palm trunk, 326
Polyol and diisocyanate, reaction of, 103f
Polyurethane (PU), 104
Predrying, 149
Preheating chamber, 264
Pretreated EFB fibers for fiber-reinforced composites, 164–168
Protein utilization, 107
Pulping, 299–300, 301t

R

Radial swelling, 136
Reductive bleaching, 163
Refiner disc gap, zones in, 301f
Refiner mechanical pulp (RMP), 300
Refining process, 299
Renewable resources, 5–6
Resin for wood based industry, 100–102, 100f
Resin penetration on COPF composite board from compressed oil palm fronds, 187, 188f
Reversed cant sawing method, 117f
Rice, 113
Rotary OPT veneer processing, 139–150
  challenges in, 150
  definition, 139–140
  green veneer production, 144–146

jointing/composing, 149–150
log-end processing, 140–143
manufacturing OPT plywood, 140
of oil palm trunk (*see* Rotary OPT veneer processing)
sorting and repairing, 146
thickness, 146
veneer drying, 146–149
Rubber, 113
Rubberwood (RW), 283–284, 289f, 297–298

S

Sawing, 125
  of oil palm trunk and lumber, 116–117
Sawn timber, 83, 83f
Sequencing batch reactor (SBR) system, 196
Six-step processing method (Sawing → Steaming and Compression → Drying → Impregnation → Semicuring → Densification), 124–126, 125f, 126–127t
Sludge, 198–199, 199t
Soaking, 126
Soft rot, 58, 59t
Solid oil palm biomass, 9f
Solid production, 198–199, 199t
Solvent extraction, lignocellulosic fiber, 159
Sound barrier fiberboard from OPF, 308, 308f
Static bending strength of COPF composite board, 180–183, 181t
Steam explosion, lignocellulosic fiber, 160
Steaming, compression, and drying, 125
Sterilizer cage loaded with FFB, 14f
Stone groundwood (SGW), 300
Stripped palm fruitlets, 17–18
Sugar and starch found in oil palm p, 60–61

Sugarcane bagasse (SCB) fiber, 93–94
Super-fast drying of OPL, 119–120
Sustainable deployment of untapped potential, national strategy, 343–344
Sustainable policy, 340–341

T

Tangential swelling, 136
Tannin, 108
10th Malaysia Plan (10MP), 3–4
Thermal process, lignocellulosic fiber, 160
Thermo-mechanical pulping (TMP), 300–301
  fiber properties, 302
  fracture zones, 304*f*
Threshing station, 17*f*
Timber mills, 79*t*
Trade value of Malaysian major wood products, 80–85, 81–82*f*

U

Ultrasonication pretreatment, lignocellulosic fiber, 160–161

Upflow anaerobic filtration (UAF) system, 195
Upflow anaerobic sludge blanket reactor (UASB), 195
Upflow fix system (UFF), 196
Urea formaldehyde (UF) value, 88–89, 100–101

V

Veneer drying, 146–149, 149*f*
  dryer and requirements, 148–149
  drying rate, 148
  natural characteristics of palm wood, 147
  principle, 146–147
Veneer jointing/composing, 149–150
Vertical Capacity Sdn Bhd, 365
Volumetric shrinkage, oil palm trunk, 136, 138*f*

W

Waste aerobic granules (WAG), 199–200
  application of, 200
  generation, 200
Waste generation, 347–348
Water transfer to air, 119

White-rot fungi (*Pycnoporous sanguineus*), 58–59, 59*t*
Wood-based composites
  classifications, 321–322
  composite, 321–322
  laminates, 321–322
  preparation, 99–100
  veneer-based, 321–322
  wood-non-wood composite, 321–322
Wood-based industry in Malaysia, 78–85
Wood-based panel export, 76*f*
Wood-based panel import, 76*f*
Wood-based panel industries, 69–70
  nonstructural panel, 69–70
  structural panel, 69–70
  types, 70–74
Wood-based panel production by category, 77*f*
Wood-based panels industry market analysis, 74–78, 74–75*f*
Wood fibers, 47*t*
Wood-processing mills, 78–80, 80–81*t*
Wooden furniture, 82–83, 82*f*

Printed in the United States
by Baker & Taylor Publisher Services